*Proceedings*, 2019, MaxEnt 2019

*Proceedings*, 2019, MaxEnt 2019

Special Issue Editors

**Udo von Toussaint**
**Roland Preuss**

MDPI • Basel • Beijing • Wuhan • Barcelona • Belgrade • Manchester • Tokyo • Cluj • Tianjin

*Special Issue Editors*
Udo von Toussaint
Max-Planck-Institute for Plasma Physics
Germany

Roland Preuss
Max-Planck-Institute for Plasma Physics
Germany

*Editorial Office*
MDPI
St. Alban-Anlage 66
4052 Basel, Switzerland

This is a reprint of articles from the Special Issue published online in the open access journal *Proceedings* (ISSN 2504-3900) (available at: https://www.mdpi.com/2504-3900/33/1).

For citation purposes, cite each article independently as indicated on the article page online and as indicated below:

LastName, A.A.; LastName, B.B.; LastName, C.C. Article Title. *Journal Name* **Year**, *Article Number*, Page Range.

**ISBN 978-3-03928-476-4 (Pbk)**
**ISBN 978-3-03928-477-1 (PDF)**

Cover image courtesy of Roland Preuss.

© 2020 by the authors. Articles in this book are Open Access and distributed under the Creative Commons Attribution (CC BY) license, which allows users to download, copy and build upon published articles, as long as the author and publisher are properly credited, which ensures maximum dissemination and a wider impact of our publications.
The book as a whole is distributed by MDPI under the terms and conditions of the Creative Commons license CC BY-NC-ND.

# Contents

**About the Special Issue Editors** . . . . . . . . . . . . . . . . . . . . . . . . . . . . . . . . . . . . . . ix

**Udo von Toussaint and Roland Preuss**
Bayesian Inference and Maximum Entropy Methods in Science and Engineering—MaxEnt 2019
Reprinted from: *Proceedings* **2019**, *33*, 8, doi:10.3390/proceedings2019033008 . . . . . . . . . . . . 1

**Christopher G. Albert**
Gaussian Processes for Data Fulfilling Linear Differential Equations
Reprinted from: *Proceedings* **2019**, *33*, 5, doi:10.3390/proceedings2019033005 . . . . . . . . . . . . 3

**Arthur Baraov**
Electromagnetic Induction and Relativistic Double Layer: Mechanism for Ball Lightning Formation
Reprinted from: *Proceedings* **2019**, *33*, 3, doi:10.3390/proceedings2019033003 . . . . . . . . . . . . 13

**Scott A. Cameron, Hans C. Eggers and Steve Kroon**
A Sequential Marginal Likelihood Approximation Using Stochastic Gradients
Reprinted from: *Proceedings* **2019**, *33*, 18, doi:10.3390/proceedings2019033018 . . . . . . . . . . . . 23

**Nicholas Carrara**
Quantum Trajectories in Entropic Dynamics
Reprinted from: *Proceedings* **2019**, *33*, 25, doi:10.3390/proceedings2019033025 . . . . . . . . . . . . 33

**Nicholas Carrara and Jesse Ernst**
On the Estimation of Mutual Information
Reprinted from: *Proceedings* **2019**, *33*, 31, doi:10.3390/proceedings2019033031 . . . . . . . . . . . . 43

**Ariel Caticha**
The Information Geometry of Space-Time
Reprinted from: *Proceedings* **2019**, *33*, 15, doi:10.3390/proceedings2019033015 . . . . . . . . . . . . 53

**Nestor Caticha**
Entropic Dynamics for Learning in Neural Networks and the Renormalization Group
Reprinted from: *Proceedings* **2019**, *33*, 10, doi:10.3390/proceedings2019033010 . . . . . . . . . . . . 61

**Camille Chapdelaine, Ali Mohammad-Djafari, Nicolas Gac and Estelle Parra**
Variational Bayesian Approach in Model-Based Iterative Reconstruction for 3D X-Ray Computed Tomography with Gauss-Markov-Potts Prior
Reprinted from: *Proceedings* **2019**, *33*, 4, doi:10.3390/proceedings2019033004 . . . . . . . . . . . . 69

**Natalya Denisova**
Bayesian Approach with Entropy Prior for Open Systems
Reprinted from: *Proceedings* **2019**, *33*, 1, doi:10.3390/proceedings2019033001 . . . . . . . . . . . . 79

**Deniz Gençağa and Sevgi Şengül Ayan**
Effects of Neuronal Noise on Neural Communication
Reprinted from: *Proceedings* **2019**, *33*, 2, doi:10.3390/proceedings2019033002 . . . . . . . . . . . . 87

**Fabrizia Guglielmetti, Eric Villard and Ed Fomalont**
Bayesian Reconstruction through Adaptive Image Notion
Reprinted from: *Proceedings* **2019**, *33*, 21, doi:10.3390/proceedings2019033021 . . . . . . . . . . . . 95

**R. Wesley Henderson and Paul M. Goggans**
TI-Stan: Adaptively Annealed Thermodynamic Integration with HMC
Reprinted from: *Proceedings* 2019, *33*, 9, doi:10.3390/proceedings2019033009 . . . . . . . . . . . . 103

**Sayedeh Marjaneh Hosseini, Ali Mohhamad-Djafari, Adel Mohammadpour, Sobhan Mohammadpour and Mohammad Nadi**
Carpets Color and Pattern Detection Based on Their Images
Reprinted from: *Proceedings* 2019, *33*, 28, doi:10.3390/proceedings2019033028 . . . . . . . . . . 113

**Selman Ipek and Ariel Caticha**
An Entropic Dynamics Approach to Geometrodynamics
Reprinted from: *Proceedings* 2019, *33*, 13, doi:10.3390/proceedings2019033013 . . . . . . . . . . 121

**David N. John, Michael Schick and Vincent Heuveline**
Learning Model Discrepancy of an Electric Motor with Bayesian Inference
Reprinted from: *Proceedings* 2019, *33*, 11, doi:10.3390/proceedings2019033011 . . . . . . . . . . 131

**Yannis Kalaidzidis, Hernán Morales-Navarrete, Inna Kalaidzidis and Marino Zerial**
Intracellular Background Estimation for Quantitative Fluorescence Microscopy
Reprinted from: *Proceedings* 2019, *33*, 22, doi:10.3390/proceedings2019033022 . . . . . . . . . . 141

**Robert K. Niven, Ali Mohammad-Djafari, Laurent Cordier, Markus Abel and Markus Quade**
Bayesian Identification of Dynamical Systems
Reprinted from: *Proceedings* 2019, *33*, 33, doi:10.3390/proceedings2019033033 . . . . . . . . . . 149

**Kevin H. Knuth, Robert M. Powell and Peter A. Reali**
Estimating Flight Characteristics of Anomalous Unidentified Aerial Vehicles in the 2004 Nimitz Encounter
Reprinted from: *Proceedings* 2019, *33*, 26, doi:10.3390/proceedings2019033026 . . . . . . . . . . 157

**Marcelo S. Lauretto, Rafael Stern, Celma Ribeiro and Julio Stern**
Haphazard Intentional Sampling Techniques in Network Design of Monitoring Stations
Reprinted from: *Proceedings* 2019, *33*, 12, doi:10.3390/proceedings2019033012 . . . . . . . . . . 169

**Masrour Makaremi, Camille Lacaule and Ali Mohammad-Djafari**
Determination of the Cervical Vertebra Maturation Degree from Lateral Radiography
Reprinted from: *Proceedings* 2019, *33*, 30, doi:10.3390/proceedings2019033030 . . . . . . . . . . 179

**Ali Mohammad-Djafari**
Interaction between Model Based Signal and Image Processing, Machine Learning and Artificial Intelligence
Reprinted from: *Proceedings* 2019, *33*, 16, doi:10.3390/proceedings2019033016 . . . . . . . . . . 189

**Donald W. Nelson and Udo von Toussaint**
Radiometric Scale Transfer Using Bayesian Model Selection
Reprinted from: *Proceedings* 2019, *33*, 32, doi:10.3390/proceedings2019033032 . . . . . . . . . . 199

**Dirk Nille and Udo von Toussaint**
2D Deconvolution Using Adaptive Kernel
Reprinted from: *Proceedings* 2019, *33*, 6, doi:10.3390/proceedings2019033006 . . . . . . . . . . . 211

**Roland Preuss, Rodrigo Arredondo and Udo von Toussaint**
Bayesian Determination of Parameters for Plasma-Wall Interactions
Reprinted from: *Proceedings* 2019, *33*, 27, doi:10.3390/proceedings2019033027 . . . . . . . . . . 221

**Sascha Ranftl, Gian Marco Melito, Vahid Badeli, Alice Reinbacher-Köstinger, Katrin Ellermann and Wolfgang von der Linden**
On the Diagnosis of Aortic Dissection with Impedance Cardiography: A Bayesian Feasibility Study Framework with Multi-Fidelity Simulation Data
Reprinted from: *Proceedings* **2019**, *33*, 24, doi:10.3390/proceedings2019033024 . . . . . . . . . . . 227

**Guillaume Revillon and Ali Mohammad-Djafari**
A Complete Classification and Clustering Model to Account for Continuous and Categorical Data in Presence of Missing Values and Outliers
Reprinted from: *Proceedings* **2019**, *33*, 23, doi:10.3390/proceedings2019033023 . . . . . . . . . . . 237

**Olivia Saa and Julio Michael Stern**
Auditable Blockchain Randomization Tool
Reprinted from: *Proceedings* **2019**, *33*, 17, doi:10.3390/proceedings2019033017 . . . . . . . . . . . 247

**John Skilling**
Galilean and Hamiltonian Monte Carlo
Reprinted from: *Proceedings* **2019**, *33*, 19, doi:10.3390/proceedings2019033019 . . . . . . . . . . . 253

**John Skilling**
Information Geometry Conflicts With Independence
Reprinted from: *Proceedings* **2019**, *33*, 20, doi:10.3390/proceedings2019033020 . . . . . . . . . . . 261

**Hellinton H. Takada, Sylvio X. Azevedo, Julio M. Stern and Celma O. Ribeiro**
Using Entropy to Forecast Bitcoin's Daily Conditional Value at Risk
Reprinted from: *Proceedings* **2019**, *33*, 7, doi:10.3390/proceedings2019033007 . . . . . . . . . . . . 265

**Martino Trassinelli**
The Nested_fit Data Analysis Program
Reprinted from: *Proceedings* **2019**, *33*, 14, doi:10.3390/proceedings2019033014 . . . . . . . . . . . 273

**Marnix Van Soom and Bart de Boer**
A New Approach to the Formant Measuring Problem
Reprinted from: *Proceedings* **2019**, *33*, 29, doi:10.3390/proceedings2019033029 . . . . . . . . . . . 281

**Keith Earle and Oleks Kazakov**
The Spin Echo, Entropy, and Experimental Design
Reprinted from: *Proceedings* **2019**, *33*, 34, doi:10.3390/proceedings2019033034 . . . . . . . . . . . 291

# About the Special Issue Editors

**Udo von Toussaint**, PD, is group leader in the Numerical Methods in Plasma Physics (NMPP) division in the Max Planck Institute for Plasma Physics (IPP). His scientific career led him to the University of Bayreuth and the Research Institute for Advanced Computer Science (RIACS) at NASA Ames Research Center (Mountain View, CA, USA). Since 2003, he has worked on Bayesian experimental design, data fusion, uncertainty quantification, artifical neural networks, and inverse problems in general. Recent research has focused on the development of reduced-complexity models and global optimization.

**Roland Preuss**, Dr., M.S. (SUNY Albany), is scientist in the Numerical Methods in Plasma Physics (NMPP) division at the Max Planck Institute for Plasma Physics (IPP). Graduating from the University of Würzburg in the field of strongly correlated fermionic systems, he had first contacts with Bayesian probability theory to transform ill-conditioned matrices. Thereafter, Bayesian inference guided his path through various scientific fields and numerical methods, among them inverse problems, model comparison, Gaussian processes, and uncertainty quantification.

*Editorial*

# Bayesian Inference and Maximum Entropy Methods in Science and Engineering—MaxEnt 2019 [†]

**Udo von Toussaint** [*,‡] **and Roland Preuss**

Max-Planck-Institut for Plasmaphysics, D-85748 Garching, Germany; preuss@ipp.mpg.de
* Correspondence: udo.v.toussaint@ipp.mpg.de; Tel.: +49-89-3299-1817
† Presented at the 39th International Workshop on Bayesian Inference and Maximum Entropy Methods in Science and Engineering, Garching, Germany, 30 June–5 July 2019.
‡ Chair of MaxEnt 2019.

Published: 22 November 2019

---

As key building blocks for modern data processing and analysis methods—ranging from AI, ML and UQ to model comparison, density estimation and parameter estimation—Bayesian inference and entropic concepts are in the center of this rapidly growing research area. Beyond the general interest in the underlying foundations of inference, the relevance of this subject is due to the demonstrated success of these concepts in areas like pattern recognition, optimization, experimental design or event prediction. Devoted to the development and application of new and innovative concepts the 39th International Workshop on Bayesian Inference and Maximum Entropy Methods in Science and Engineering took place in Garching near Munich, Germany, from June 30th to July 5th 2019. The Max-Planck-Institute for Plasma Physics was hosting this conference for the fourth time after 1998, 2004 and 2012. About 60 participants from Europe and the US, but also from Brasil, South Africa, Australia, Russia and China, attended the conference and had intense discussions about the ongoing development. In this volume, 33 contributed papers are presented.

The workshop invited contributions on all aspects of probabilistic inference, including novel techniques and applications, and work that sheds new light on the foundations of inference. The scientific topics of the conference have been:

1. Inverse problems
2. Uncertainty quantification (UQ)
3. Gaussian process (GP) regression
4. Optimal experimental design
5. Data assimilation and Causal Inference
6. Data mining, ML algorithms
7. Numerical integration
8. Information geometry
9. Real world applications in various fields of science and engineering (e.g., earth science, astrophysics, material and plasma science, imaging in geophysics and medicine, nondestructive testing, density estimation, remote sensing)

The conference started on Sunday, June 30th with a tutorial by Romke Bontekoe titled "Bayes' Theorem, a toolbox for data analysis", followed on Monday, July 1st by a tutorial from Ariel Caticha about "Where do Hamiltonians come from". After that a total of 39 talks were presented by 35 participants till Friday, July 5th. In the poster session on Tuesday, July 2nd two poster prices were awarded to

1st place: Martino Trassinelli, CNRS, Institute of NanoSciences, Sorbonne Univ., Paris, France, "Nested sampling for atomic physics data: the nested_fit program".

2nd place: Scott Cameron, Stellenbosch University, South Africa, "A Sequential Marginal Likelihood Approximation Using Stochastic Gradients".

The follow-up conference MaxEnt2020 will take place in Graz in July 2020. It will be organized by Prof. Dr. Wolfgang von der Linden, Institute for Theoretical Physics, Technical University Graz, Austria, vonderlinden@tugraz.at.

**Acknowledgments:** We thank all persons who contributed to the organization of the conference. Special thanks also to the E.T. Jaynes Foundation and the local institute IPP for financial and logistic support.

© 2019 by the authors. Licensee MDPI, Basel, Switzerland. This article is an open access article distributed under the terms and conditions of the Creative Commons Attribution (CC BY) license (http://creativecommons.org/licenses/by/4.0/).

*Proceedings*

# Gaussian Processes for Data Fulfilling Linear Differential Equations †

### Christopher G. Albert

Max-Planck-Institut für Plasmaphysik, Boltzmannstr. 2, 85748 Garching, Germany; albert@alumni.tugraz.at
† Presented at the 39th International Workshop on Bayesian Inference and Maximum Entropy Methods in Science and Engineering, Garching, Germany, 30 June–5 July 2019.

Published: 21 November 2019

**Abstract:** A method to reconstruct fields, source strengths and physical parameters based on Gaussian process regression is presented for the case where data are known to fulfill a given linear differential equation with localized sources. The approach is applicable to a wide range of data from physical measurements and numerical simulations. It is based on the well-known invariance of the Gaussian under linear operators, in particular differentiation. Instead of using a generic covariance function to represent data from an unknown field, the space of possible covariance functions is restricted to allow only Gaussian random fields that fulfill the homogeneous differential equation. The resulting tailored kernel functions lead to more reliable regression compared to using a generic kernel and makes some hyperparameters directly interpretable. For differential equations representing laws of physics such a choice limits realizations of random fields to physically possible solutions. Source terms are added by superposition and their strength estimated in a probabilistic fashion, together with possibly unknown hyperparameters with physical meaning in the differential operator.

**Keywords:** gaussian process regression; field reconstruction; partial differential equations; meshless methods

## 1. Introduction

The larger context of the present work is the goal to construct reduced complexity models as emulators or surrogates that retain mathematical and physical properties of the underlying system. Similar to usual numerical models, such methods aim to represent infinite systems by exploiting finite information in some optimal sense. In the spirit of structure preserving numerics the aim here is to move errors to the "right place", in order to retain laws such as conservation of mass, energy or momentum.

This article deals with Gaussian process (GP) regression on data with additional information known in the form of linear, generally partial differential equations (PDEs). An illustrative application is the reconstruction of an acoustic sound pressure field and its sources from discrete microphone measurements. GPs, a special class of random fields, are used in a probabilistic rather than a stochastic sense: approximate a fixed but unknown field from possibly noisy local measurements. Uncertainties in this reconstruction are modeled by a normal distribution. For the limit of zero measured data a prior has to be chosen whose realizations take values in the expected order of magnitude. An appropriate choice of a covariance function or kernel guarantees that all fields drawn from the GP at any stage fulfill the underlying PDE. This may require to give up stationarity of the process.

Techniques to fit GPs to data from PDEs has been known for some time, especially in the field of geostatistics [1]. A general analysis including a number of important properties is given by [2]. In these earlier works GPs are usually referred to as Kriging and stationary covariance functions/kernels as covariograms. A number of more recent works from various fields [3–5] use the linear operator of the problem to obtain a new kernel function for the source field by applying it twice to a generic, usually squared exponential, kernel. In contrast to the present approach, that method is suited best for source fields that are non-vanishing across the whole domain. In terms of deterministic numerical methods one could say that the approach correspond to meshless variants of the finite element method (FEM). The approach in the present work instead represents a probabilistic variant of a procedure related to the boundary element method (BEM), also known as the *method of fundamental solutions* (MFS) or regularized BEM [6–8]. As in the BEM, the MFS also builds on fundamental solutions, but allows to place sources outside the boundary rather than localizing them on a layer. Thus the MFS avoids singularities in boundary integrals of the BEM while retaining a similar ratio of numerical effort and accuracy for smooth solutions. To the author's knowledge the probabilistic variant of the MFS via GPs has first been introduced by [9] to solve the boundary value problem of the Laplace equation and dubbed *Bayesian boundary elements estimation method ((BE)²M)*. This work also provides a detailed treatment of kernels for the 2D Laplace equation. A more extensive and general treatment of the Bayesian context as well as kernels and their connection to fundamental solutions is available in [10] under the term *probabilistic meshless methods (PMM)*.

While Mendes et al. [9] is focused on boundary data of a single homogeneous equation, and Cockayne et al. [10] provides a detailed mathematical foundation, the present work aims to explore the topic further for application and extend the recent work in [11]. Starting from general notions some regression techniques are introduced with emphasis on the role of localized sources. For this purpose Poisson, Helmholtz and heat equation are considered and several kernels are derived and tested. To fit a GP to a homogeneous (source-free) PDE, kernels are built via according fundamental solutions. Possible singularities (sources) are moved outside the domain of interest. In particular, boundary conditions on a finite domain can be either supplied or reconstructed in this fashion. In addition contributions by internal sources are superimposed, using again fundamental solutions in the free field. For that part boundary conditions of the actual problem are irrelevant. The specific approach taken here is most efficient for source-free regions with possibly few localized sources that are represented by monopoles or dipoles.

## 2. GP Regression for Data from Linear PDEs

Gaussian processes (GPs) are a useful tool to represent and update incomplete information on scalar fields (The more general case of complex valued fields and vector fields is left open for future investigations in this context.) $u(\mathbf{x})$, i.e., a real number $u$ depending on a (multi-dimensional) independent variable $\mathbf{x}$. A GP with mean $m(\mathbf{x})$ and covariance function of kernel $k(\mathbf{x}, \mathbf{x}')$ is denoted as

$$u(\mathbf{x}) \sim \mathcal{G}(m(\mathbf{x}), k(\mathbf{x}, \mathbf{x}')). \tag{1}$$

The choice of an appropriate kernel $k(\mathbf{x}, \mathbf{x}')$ restricts realizations of (1) to respect regularity properties of $u(\mathbf{x})$ such as continuity or characteristic length scales. Often regularity of $u$ does not appear by chance, but rather reflects an underlying law. We are going to exploit such laws in the construction and application of Gaussian processes describing $u$ for the case described by linear (partial) differential equations

$$\hat{L}u(\mathbf{x}) = q(\mathbf{x}). \tag{2}$$

Here $\hat{L}$ is a linear differential operator, and $q(\mathbf{x})$ is an inhomogeneous source term. In physical laws dimensions of $\mathbf{x}$ usually consist of space and/or time. Physical scalar fields $u$ include e.g. pressure $p$, temperature $T$ or the electrostatic potential $\phi_e$. Corresponding laws under certain conditions include Gauss' law of electrostatics for $\phi_e$ with Laplacian $\hat{L} = \varepsilon \Delta$, frequency-domain acoustics for $p$ with Helmholtz operator $\hat{L} = \Delta - k_0^2$ or thermodynamics for $T$ with heat/diffusion operator $\hat{L} = \frac{\partial}{\partial t} - D\Delta$. These operators contain free parameters, namely permeability $\varepsilon$, wavenumber $k_0$, and diffusivity $D$, respectively. While $\varepsilon$ may be absorbed inside $q$ in a uniform material model of electrostatics, estimation of parameters $k_0$ or $D$ is useful for material characterization.

For the representation of PDE solutions the weight-space view of Gaussian process regression is useful. There the kernel $k$ is represented via a tuple $\boldsymbol{\phi}(\mathbf{x}) = (\phi_1(\mathbf{x}), \phi_2(\mathbf{x}), \dots)$ of basis functions $\phi_i(\mathbf{x})$ that underlie a linear regression model

$$u(\mathbf{x}) = \boldsymbol{\phi}(\mathbf{x})^T \mathbf{w} = \sum_i \phi_i(\mathbf{x}) w_i. \tag{3}$$

Bayesian inference starting from a Gaussian prior with covariance matrix $\Sigma_p$ for weights $\mathbf{w}$ yields a Mercer kernel

$$k(\mathbf{x}, \mathbf{x}') \equiv \boldsymbol{\phi}^T(\mathbf{x}) \Sigma_p \boldsymbol{\phi}(\mathbf{x}') = \sum_{i,j} \phi_i(\mathbf{x}) \Sigma_p^{ij} \phi_j(\mathbf{x}'). \tag{4}$$

The existence of such a representation is guaranteed by Mercer's theorem in the context of reproducing kernel Hilbert spaces (RKHS) [8]. More generally one can also define kernels on an uncountably infinite number of basis functions in analogy to (3) via

$$f(\mathbf{x}) = \hat{\phi}[w(\boldsymbol{\zeta})] = \langle \phi(\mathbf{x}, \boldsymbol{\zeta}), w(\boldsymbol{\zeta}) \rangle = \int \phi(\mathbf{x}, \boldsymbol{\zeta}) w(\boldsymbol{\zeta}) \, d\boldsymbol{\zeta}, \tag{5}$$

where $\hat{\phi}$ is a linear operator acting on elements $w(\boldsymbol{\zeta})$ of an infinite-dimensional weight space parametrized by an auxiliary index variable $\boldsymbol{\zeta}$, that may be multi-dimensional. We represent $\hat{\phi}$ via an inner product $\langle \phi(\mathbf{x}, \boldsymbol{\zeta}), w(\boldsymbol{\zeta}) \rangle$ in the respective function space given by an integral over $\boldsymbol{\zeta}$. The infinite-dimensional analogue to the prior covariance matrix is a prior covariance operator $\hat{\Sigma}_p$ that defines the kernel as a bilinear form

$$k(\mathbf{x}, \mathbf{x}') \equiv \langle \phi(\mathbf{x}, \boldsymbol{\zeta}), \hat{\Sigma}_p \phi(\mathbf{x}', \boldsymbol{\zeta}') \rangle \equiv \int \phi(\mathbf{x}, \boldsymbol{\zeta}) \Sigma_p(\boldsymbol{\zeta}, \boldsymbol{\zeta}') \phi(\mathbf{x}', \boldsymbol{\zeta}') \, d\boldsymbol{\zeta} \, d\boldsymbol{\zeta}'. \tag{6}$$

Kernels of the form (6) are known as convolution kernels. Such a kernel is at least positive semidefinite, and positive definiteness follows in the case of linearly independent basis functions $\phi(\mathbf{x}, \boldsymbol{\zeta})$ [8].

*2.1. Construction of Kernels for PDEs*

For treatment of PDEs possible choices of index variables in (4) or (6) include separation constants of analytical solutions, or the frequency variable of an integral transform. In accordance with [10], using basis functions that satisfy the underlying PDE, a probabilistic meshless method (PMM) is constructed. In particular, if $\boldsymbol{\zeta}$ parameterizes positions of sources, and $\phi(\mathbf{x}, \boldsymbol{\zeta}) = G(\mathbf{x}, \boldsymbol{\zeta})$ in (6) is chosen to be a fundamental solution/Green's function $G(\mathbf{x}, \boldsymbol{\zeta})$ of the PDE, one may call the resulting scheme a *probabilistic method of fundamental solutions (pMFS)*. In [10] sources are placed across the whole computational domain, and the resulting kernel is called *natural*. Here we will instead place sources in the exterior to fulfill the homogeneous interior problem, as in the classical MFS [6–8]. Technically, this is also achieved by setting $\Sigma_p(\boldsymbol{\zeta}, \boldsymbol{\zeta}') = 0$ for either $\boldsymbol{\zeta}$ or $\boldsymbol{\zeta}'$ in the interior. For discrete sources localized $\boldsymbol{\zeta} = \boldsymbol{\zeta}_i$ one obtains again discrete basis functions $\phi_i(\mathbf{x}) = G(\mathbf{x}, \boldsymbol{\zeta}_i)$ for (4).

More generally, according to theorem 2 of [2], for linear PDE operators $\hat{L}$ in (2) with $q \neq 0$ we require a Gaussian process of non-zero mean $m(\mathbf{x})$ with

$$\hat{L} m(\mathbf{x}) = q(\mathbf{x}), \tag{7}$$
$$\hat{L} k(\mathbf{x}, \mathbf{x}') = 0. \tag{8}$$

Here $\hat{L}$ acts on the first argument of $k(\mathbf{x}, \mathbf{x}')$. Sources affect only the mean $m(\mathbf{x})$ of the Gaussian process, whereas the kernel $k(\mathbf{x}, \mathbf{x}')$ should be based on the homogeneous equation. This hints to the technique of [12] discussed in [13] chapter 2.7 to treat $m(\mathbf{x})$ via a linear model added on top of a zero-mean process for the homogeneous equation. In that case we consider is the superposition

$$u(\mathbf{x}) = u_h(\mathbf{x}) + u_p(\mathbf{x}), \tag{9}$$
$$u_h(\mathbf{x}) \sim \mathcal{G}(0, k(\mathbf{x}, \mathbf{x}')), \tag{10}$$
$$u_p(\mathbf{x}) = \mathbf{h}^T(\mathbf{x})\mathbf{b}, \tag{11}$$
$$\mathbf{b} \sim \mathcal{N}(\mathbf{b}_0, B). \tag{12}$$

where $\mathbf{h}^T(\mathbf{x})\mathbf{b}$ is a linear model for $m(\mathbf{x})$ with Gaussian prior mean $\mathbf{b}_0$ and covariance $B$ for the model coefficients. The homogeneous part (10) corresponds to a random process $u_h(\mathbf{x})$ where a source-free $k$ is constructed according to (8). The inhomogeneous part (11) may be given by any particular solution $u_p(\mathbf{x})$ for arbitrary boundary conditions. Using the limit of a vague prior with $\mathbf{b}_0 = 0$ and $|B^{-1}| \to 0$, i.e., minimum information/infinite prior covariance [12,13], posteriors for mean $\bar{u}$ and covariance matrix $\mathrm{cov}(u, u)$ based on given training data $\mathbf{y} = u(X) + \sigma_\mathrm{n}$ with measurement noise variance $\sigma_\mathrm{n}^2$ are

$$\bar{u}(X_\star) = K_\star^T K_y^{-1}(\mathbf{y} - H^T \bar{\mathbf{b}}) + H_\star^T \bar{\mathbf{b}} = K_\star^T K_y^{-1} \mathbf{y} + R^T \bar{\mathbf{b}}, \tag{13}$$
$$\mathrm{cov}(u(X_\star), u(X_\star)) = K_{\star\star} - K_\star^T K_y^{-1} K_\star + R^T (H K_y^{-1} H^T)^{-1} R. \tag{14}$$

Here $X = (\mathbf{x}_1, \mathbf{x}_2, \ldots \mathbf{x}_N)$ contains the training points, $X_\star = (\mathbf{x}_{\star 1}, \mathbf{x}_{\star 2}, \ldots, \mathbf{x}_{\star N_\star})$ the evaluation or test points. Functions of $X$ and $X^\star$ are to be understood as vectors or matrices resulting from evaluation at different positions, i.e., $\bar{u}(X_\star) \equiv (\bar{u}(\mathbf{x}_{\star 1}), \bar{u}(\mathbf{x}_{\star 2}), \ldots, \bar{u}(\mathbf{x}_{\star N_\star}))$ is a tuple of predicted expectation values. The matrix $K \equiv k(X, X)$ is the kernel covariance of the training data with entries $K_{ij} \equiv k(\mathbf{x}_i, \mathbf{x}_j)$ and $\mathrm{cov}(u(X_\star), u(X_\star))_{ij} \equiv \mathrm{cov}(u(\mathbf{x}_{\star i}), u(\mathbf{x}_{\star j}))$ are entries of the predicted covariance matrix for $u$ evaluated in the test points $\mathbf{x}_{\star i}$. Furthermore $K_y \equiv k(X, X) + \sigma_\mathrm{n}^2 I$, $K_\star \equiv k(X, X_\star)$, $K_{\star\star} \equiv k(X_\star, X_\star)$, $R \equiv H_\star - H K_y^{-1} K_{\star\star}$, and entries of $H$ are $H_{ij} \equiv h_i(\mathbf{x}_j)$, $H_{\star ij} \equiv h_i(\mathbf{x}_{\star j})$, and $\equiv (H K_y^{-1} H^T)^{-1} H K_y^{-1} \mathbf{y}$.

## 2.2. Linear Modeling of Sources

A linear model for $m(\mathbf{x})$ fulfilling a PDE according to (8) follows directly from the source representation. Consider sources to be modeled as a linear superposition over basis functions

$$q(\mathbf{x}) = \sum_i \varphi_i(\mathbf{x}) q_i \tag{15}$$

with unknown source strength coefficients $\mathbf{q} = (q_i)$. To model the mean instead of the source functions themselves, one uses an according superposition

$$m(\mathbf{x}) = \sum_i u_{pi}(\mathbf{x}) q_i \tag{16}$$

of particular solutions $u_{pi}(\mathbf{x})$ from inhomogeneous equations

$$\hat{L} u_{pi}(\mathbf{x}) = \varphi_i(\mathbf{x}). \tag{17}$$

For the linear Model (9) this means that $\mathbf{b} = \mathbf{q}$ and $h_i(\mathbf{x}) = u_{pi}(\mathbf{x})$. Posterior mean of source strengths and their uncertainty are

$$\bar{\mathbf{q}} = (H K_y^{-1} H^T)^{-1} H K_y^{-1} \mathbf{y}, \tag{18}$$
$$\mathrm{cov}(\mathbf{q}, \mathbf{q}) = (H K_y^{-1} H^T)^{-1}. \tag{19}$$

One can easily check that the predicted mean $\bar{u}(\mathbf{x}_\star) = \bar{u}_h(\mathbf{x}_\star) + \bar{u}_p(\mathbf{x}_\star)$ at a specific point $\mathbf{x}_\star$ in (13) fulfills the linear differential Equation (2). In the homogeneous part $\bar{u}_h(\mathbf{x}_\star) = k(\mathbf{x}_\star, X) K_y^{-1}(\mathbf{y} - H^T \bar{\mathbf{q}})$ sources are absent with $\hat{L} \bar{u}_h(\mathbf{x}_\star) = 0$, with $\hat{L}$ acting on $\mathbf{x}^\star$ here. The particular solution $\bar{u}_p(\mathbf{x}_\star) =$

$\mathbf{h}^T(\mathbf{x}_\star)\bar{\mathbf{q}} = \sum_i u_{p\,i}(\mathbf{x}_\star)\bar{q}_i$ adds source contributions $q_i\,\varphi_i(\mathbf{x}^\star)$ due to (17). For point monopole sources $\varphi_i(\mathbf{x}) = \delta(\mathbf{x} - \mathbf{x}_{q\,i})$ placed at at positions $\mathbf{x}_{q\,i}$, the particular solution $u_{p,\,i}(\mathbf{x})$ equals the fundamental solution $G(\mathbf{x}, \mathbf{x}_{q\,i})$ evaluated for the respective source. In the absence of sources the part described in this subsection isn't modeled and (13) and (14) reduce to posteriors of a GP with prior mean $m(\mathbf{x}) = 0$ where matrix $R$ vanishes.

## 3. Application Cases

Here the general results described in the previous section are applied to specific equations. Regression is performed based on values measured at a set of sampling points $\mathbf{x}_i$ and may also include optimization of hyperparameters $\beta$ appearing as auxiliary variables inside the kernel $k(\mathbf{x}, \mathbf{x}'; \beta)$. The optimization step is usually performed in a maximum a posteriori (MAP) sense, choosing $\beta_{\mathrm{MAP}}$ as fixed rather than providing a joint probability distribution function including $\beta$ as random variables. We note that depending on the setting this choice may lead to underestimation of uncertainties in the reconstruction of $u$, in particular for sparse, low-quality measurements.

### 3.1. Laplace's Equation in Two Dimensions

First we explore construction of kernels in (10) for a purely homogeneous problem in a finite and infinite dimensional index space, depending on the mode of separation. Consider Laplace's equation

$$\Delta u(\mathbf{x}) = 0. \tag{20}$$

In contrast to the Helmholtz equation, Laplace's equation has no scale, i.e., permits all length scales in the solution. In the 2D case using polar coordinates the Laplacian becomes

$$\frac{1}{r}\frac{\partial}{\partial r}\left(r\frac{\partial u}{\partial r}\right) + \frac{1}{r^2}\frac{\partial^2 u}{\partial \theta^2} = 0. \tag{21}$$

A well-known family of solutions for this problem based on the separation of variables is

$$u = r^{\pm m} e^{\pm im\theta}, \tag{22}$$

leading to a family of solutions

$$r^m \cos(m\theta),\ r^m \sin(m\theta),\ r^{-m} \cos(m\theta),\ r^{-m} \sin(m\theta). \tag{23}$$

Since our aim is to work in bounded regions we discard the solutions with negative exponent that diverge at $r = 0$. Choosing a diagonal prior that weights sine and cosine terms equivalently [9] and introducing a length scale $s$ as a free parameter we obtain a kernel according to (4) with

$$k(\mathbf{x}, \mathbf{x}'; s) = \sum_{m=0}^{\infty} \left(\frac{rr'}{s^2}\right)^m \sigma_m^2 \left(\cos(m\theta)\cos(m\theta') + \sin(m\theta)\sin(m\theta')\right) = \sum_{m=0}^{\infty} \left(\frac{rr'}{s^2}\right)^m \sigma_m^2 \cos\left(m(\theta - \theta')\right). \tag{24}$$

A flat prior $\sigma_m^2 = 1$ for all polar harmonics and a characteristic length scale $s$ as a hyperparameter, yields

$$k(\mathbf{x}, \mathbf{x}'; s) = \frac{1 - \frac{rr'}{s^2}\cos(\theta - \theta')}{1 - 2\frac{rr'}{s^2}\cos(\theta - \theta') + \frac{(rr')^2}{s^4}} = \frac{1 - \frac{\mathbf{x}\cdot\mathbf{x}'}{s^2}}{1 - 2\frac{\mathbf{x}\cdot\mathbf{x}'}{s^2} + \frac{|\mathbf{x}|^2|\mathbf{x}'|^2}{s^4}}. \tag{25}$$

This kernel is not stationary, but isotropic around a fixed coordinate origin. Introducing a mirror point $\bar{\mathbf{x}}'$ with polar angle $\bar{\theta}' = \theta'$ and radius $\bar{r}' = s^2/r'$ we notice that (25) can be written as

$$k(\mathbf{x}, \mathbf{x}'; s) = \frac{|\bar{\mathbf{x}}'|^2 - \mathbf{x}\cdot\bar{\mathbf{x}}'}{(\mathbf{x} - \bar{\mathbf{x}}')^2}, \tag{26}$$

making a dipole singularity apparent at $\mathbf{x} = \bar{\mathbf{x}}'$. In addition $k$ is normalized to 1 at $\mathbf{x} = 0$. Choosing $s > R_0$ larger than the radius $R_0$ of a circle centered in the origin and enclosing the computational domain, we have $\bar{r}' > s^2/s = s > R_0$. Thus all mirror points and the according singularities are moved outside the domain.

Choosing a slowly decaying $\sigma_m^2 = 1/m$, excluding $m = 1$ and adding a constant term yields a logarithmic kernel instead [9] with

$$k(\mathbf{x}, \mathbf{x}'; s) = 1 - \frac{1}{2} \ln\left(1 - 2\frac{\mathbf{x} \cdot \mathbf{x}'}{s^2} + \frac{|\mathbf{x}|^2 |\mathbf{x}'|^2}{s^4}\right) = 1 - \ln\left(\frac{|\mathbf{x} - \bar{\mathbf{x}}'|}{|\bar{\mathbf{x}}'|}\right). \tag{27}$$

Instead of a dipole singularity that expression features a monopole singularity at $\mathbf{x} - \bar{\mathbf{x}}'$ that is avoided as mentioned above.

Using instead Cartesian coordinates $x, y$ to separate the Laplacian provides harmonic functions like

$$u = e^{\pm \kappa x} e^{\pm i \kappa y}. \tag{28}$$

Here all solutions yield finite values at $x = 0$, so we don't have to exclude any of them *a priori*. Introducing again a diagonal covariance operator in (6) and taking the real part yields

$$k(\mathbf{x}, \mathbf{x}') = \int \varphi(\mathbf{x}, \kappa) \sigma^2(\kappa) \varphi(\mathbf{x}', \kappa) \, d\kappa = \operatorname{Re} \int_{-\infty}^{\infty} \sigma^2(\kappa) e^{\kappa(x \pm x')} e^{i\kappa(y \pm y')} \, d\kappa. \tag{29}$$

Setting $\sigma^2(\kappa) \equiv e^{-2\kappa^2}$ and choosing a characteristic length scale $s$ together with a possible rotation angle $\theta_0$ of the coordinate frame yields the kernel

$$k(\mathbf{x}, \mathbf{x}'; s, \theta_0) = \frac{1}{2} \operatorname{Re} \exp\left(\frac{((x + x') \pm i(y - y'))^2 e^{i 2 \theta_0}}{s^2}\right). \tag{30}$$

Other sign combinations do not yield a positive definite kernel – similar to the polar kernel (26) before we couldn't obtain an fully stationary expression that depends only on differences between coordinates of $\mathbf{x}$ and $\mathbf{x}'$.

For demonstration purposes we consider an analytical solution to a boundary value problem of Laplace's equation on a square domain $\Omega$ with corners at $(x, y) = (\pm 1, \pm 1)$. The reference solution is

$$u_{\mathrm{ref}}(x, y) = \frac{1}{2} e^y \cos x + 2x \cos(2y) \tag{31}$$

and depicted in the upper left of Figure 1 together with the extension outside the boundaries. This figure also shows results from a GP fitted based on data with artificial noise of $\sigma_n = 0.1$ measured at 8 points using kernel (26) with $s = 2$. Inside $\Omega$ the solution is represented with errors below 5%. This is also reflected in the error predicted by the posterior variance of the GP that remains small in the region enclosed by measurement points. The analogy in classical analysis is the theorem that the solution of a homogeneous elliptic equation is fully determined by boundary values.

In comparison, a reconstruction using a generic squared exponential kernel $k \propto \exp((\mathbf{x} - \mathbf{x}')^2 / (2s^2))$ yields a result of similar approximation quality in Figure 2. The posterior covariance of that reconstruction is however not able to capture the vanishing error inside the enclosed domain due to given boundary data. More severely, in contrast to the previous case, the posterior mean $\bar{u}$ doesn't satisfy Laplace's equation $\Delta \bar{u} = 0$ exactly. This leads to a violation of the classical result that (differences of) solutions of Laplace's equation may not have extrema inside $\Omega$, showing up in the difference to the reconstruction in Figure 2. This kind of error is quantified by computation of the reconstructed charge density $\bar{q} = \Delta \bar{u}$. This is fine if data from Poisson's equation $\Delta u = q$ with distributed charges should be fitted instead. However, to keep $\Delta u = 0$ exact in $\Omega$, one requires more specialized kernels such as (26).

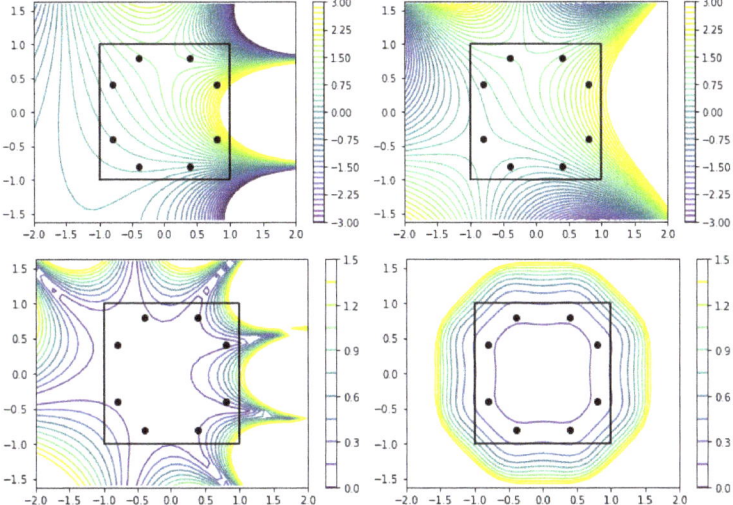

**Figure 1.** Analytical solution of Laplace equation (**top left**) and GP reconstruction with source-free Mercer kernel (26) (**top right**) with absolute error (**bottom left**) and predicted 95% confidence interval (**bottom right**). Sources lie outside the black square region and measurement positions are marked by black dots.

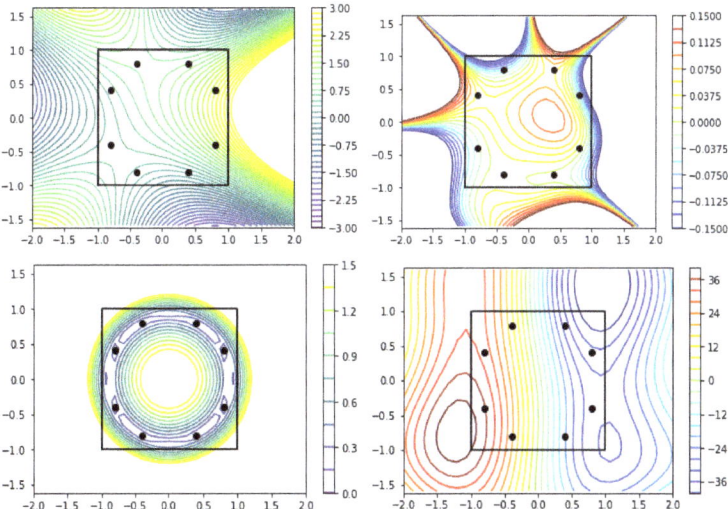

**Figure 2.** GP reconstruction of case in Figure 1 with generic squared exponential kernel (**top left**) with predicted 95% confidence interval (**bottom left**). Difference to reconstruction with source-free kernel (26) (**top right**) and source density $\bar{q} = \Delta \bar{u}$ of prediction (**bottom right**).

## 3.2. Helmholtz Equation: Source and Wavenumber Reconstruction

To demonstrate the proposed method in full we now consider the Helmholtz equation with sources

$$\Delta u(\mathbf{x}) + k_0^2 u(\mathbf{x}) = q(\mathbf{x}). \tag{32}$$

Stationary kernels based on Bessel functions for the homogeneous equation have been presented in [11]. These functions provide smoothing regularization on the order of the wavelength $\lambda_0 = 2\pi/k_0$ and have been demonstrated to produce excellent field reconstruction from point measurements. Here we consider the two-dimensional case. The method of source strength reconstruction is improved compared to [11], as it constitutes a linear problem according to (18) and (19). Non-linear optimization is instead applied to wavenumber $k_0$ as a free hyperparameter to be estimated during the GP regression.

The setup is the same as in [11]: a 2D cavity with various boundary conditions and two sound sources of strengths 0.5 and 1, respectively. Results for sound pressure fulfilling (32) are normalized to have a maximum of $p/p_0 = 1$. Figure 3 shows reconstruction error in field reconstruction depending on the number of measurement positions. Here noise of $\sigma_n = 0.01$ has been added to the samples. The obtained negative log-likelihood depending on $k_0$ permits an accurate reconstruction of this quantity that has the physical meaning of a wavenumber. A generic squared exponential kernel $k \propto \exp((\mathbf{x} - \mathbf{x}')^2/(2(\pi/k_0)^2))$ leads to results of similar quality and a slightly less peaked spatial length scale hyperparameter without a direct physical interpretation.

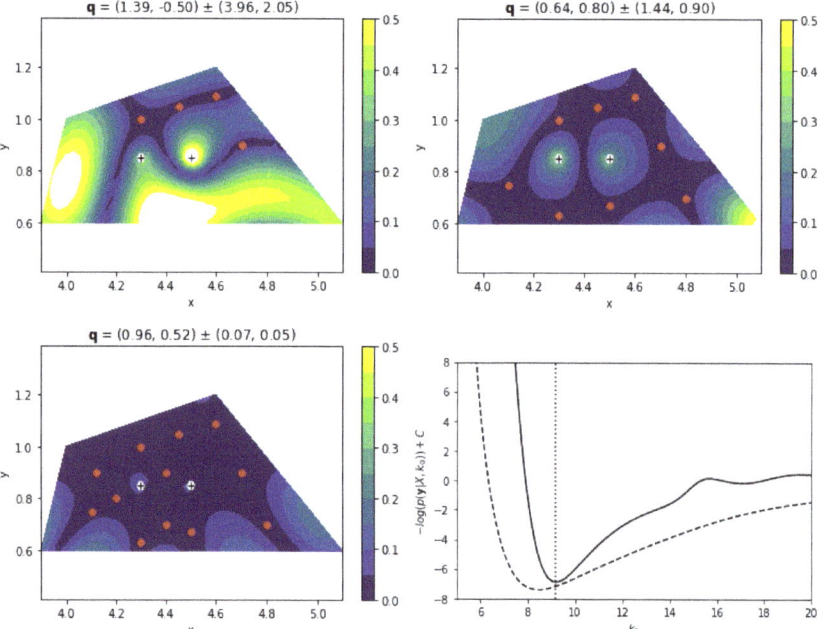

**Figure 3.** Reconstruction error for Helmholtz equation with different sensor count (top, bottom left) and reconstructed source strengths $\mathbf{q}$ with 95% confidence interval according to posterior (18) and (19). Negative log likelihood (bottom right) with optimum at $k_0^{\mathrm{ML}} = 9.19$ for Bessel kernel [11] (solid line), whereas the actual value (dotted line) is $k_0 = 9.16$. The length scale of a squared exponential kernel (dashed line) is less peaked.

### 3.3. Heat Equation

Consider the homogeneous heat/diffusion equation

$$\frac{\partial u}{\partial t} - D\Delta u = 0. \tag{33}$$

for $(x, t) \in \mathbb{R} \times \mathbb{R}^+$. Integrating the fundamental solution $G = 1/\sqrt{4\pi(t-\tau)}\exp((x-\xi)^2/(4(t-\tau)))$ from $\xi = -\infty$ to $\infty$ at $\tau = 0$, i.e., placing sources everywhere at a single point in time, leads to the kernel

$$k_\mathrm{n}(x - x', t + t'; D) = \frac{1}{\sqrt{4\pi D(t + t')}} e^{-\frac{(x-x')^2}{4D(t+t')}}. \tag{34}$$

In terms of $x$ this is a stationary squared exponential kernel and the natural kernel over the domain $x \in \mathbb{R}$. The kernel broadens with increasing $t$ and $t'$. Non-stationarity in time can also be considered natural to the heat equation, since its solutions show a preferred time direction on each side of the singularity $t = 0$. The only difference of (34) to the singular heat kernel is the positive sign between $t$ and $t'$. If both of them are positive, $k$ is guaranteed to takes finite values.

As for the Laplace equation it is also convenient to define a spatially non-stationary kernel by cutting out a finite source-free domain. Evaluating the integral over the fundamental solution in $\mathbb{R}\setminus(a,b)$ without our domain interval $(a,b)$ we obtain

$$k_n(x,t,x',t') = k_n(x-x',t+t';D)\left[1 - \frac{g(x,t,x',t';D,b) - g(x,t,x',t';D,a)}{2}\right]. \tag{35}$$

where

$$g(x,t,x',t';D,s) \equiv \operatorname{erf}\left(\frac{(s-x)/t + (s-x')/t'}{2\sqrt{D}\sqrt{1/t + 1/t'}}\right). \tag{36}$$

Incorporating the prior knowledge that there are no domain sources could potentially improve the reconstruction. Initial investigations on the initial-boundary value problem of the heat equation based on those kernels produce stable results showing natural regularization within the limits of the strongly ill-posed setting. Reconstruction of diffusivity $D$ has proven to be a difficult task and requires further investigations.

## 4. Summary and Outlook

A framework for application of Gaussian process regression to data from an underlying partial differential has been presented. The method is based on Mercer kernels constructed from fundamental solutions and produces realizations that match the homogeneous problem exactly. Contributions from sources are superimposed via an additional linear model. Several examples for suitable kernels have been given for Laplace's equation, Helmholtz equation and heat equation. Regression performance has been shown to yield results of similar or higher quality to a squared exponential kernel in the considered application cases. Advantages of the specialized kernel approach are the possibility to represent exact absence of sources as well as physical interpretability of hyperparameters.

In a next step reconstruction of vector fields via GPs could be formulated, taking laws such as Maxwell's equations or Hamilton's equations of motion into account. A starting point could be squared exponential kernels for divergence- and curl-free vector fields [14]. Such kernels have been used in [15] to perform statistical reconstruction, and Cobb et al. [16] apply them to GPs for source identification in the Laplace/Poisson equation. In order to model Hamiltonian dynamics in phase-space, vector-valued GPs could possibly be extended to represent not only volume-preserving (divergence-free) maps but retain full symplectic properties, thereby conserving all integrals of motion such as energy or momentum.

**Acknowledgments:** I would like to thank Dirk Nille, Roland Preuss and Udo von Toussaint for insightful discussions. This study is a contribution to the *Reduced Complexity Models* grant number ZT-I-0010 funded by the Helmholtz Association of German Research Centers.

## References

1. Dong, A. Kriging Variables that Satisfy the Partial Differential Equation $\Delta Z$ = Y. *Geostatistics* **1989**, 237–248. doi:10.1007/978-94-015-6844-9_17.
2. van den Boogaart, K.G. Kriging for processes solving partial differential equations. In Proceedings of the IAMG2001, Cancun, Mexico, 6–12 September 2001; pp. 1–21.
3. Graepel, T. Solving noisy linear operator equations by Gaussian processes: Application to ordinary and partial differential equations. In Proceedings of the International Conference on Machine Learning, Washington, DC, USA, 21–24 August 2003; Fawcett, T., Mishra, N., Eds.; 2003; pp. 234–241.
4. Särkkä, S. Linear Operators and Stochastic Partial Differential Equations in Gaussian Process Regression. In *Artificial Neural Networks and Machine Learning*; Lecture Notes in Computer Science; Springer: Berlin/Heidelberg, Germany, 2011; Volume 6792, pp. 151–158. doi:10.1007/978-3-642-21738-8_20.
5. Raissi, M.; Perdikaris, P.; Karniadakis, G.E. Inferring solutions of differential equations using noisy multi-fidelity data. *J. Comput. Phys.* **2017**, *335*, 736–746. doi:10.1016/j.jcp.2017.01.060.
6. Lackner, K. Computation of ideal MHD equilibria. *Comput. Phys. Commun.* **1976**, *12*, 33–44. doi:10.1016/0010-4655(76)90008-4.

7. Golberg, M.A. The method of fundamental solutions for Poisson's equation. *Eng. Anal. Bound. Elem.* **1995**, *16*, 205–213. doi:10.1016/0955-7997(95)00062-3.
8. Schaback, R.; Wendland, H. Kernel techniques: From machine learning to meshless methods. *Acta Numer.* **2006**, *15*, 543–639. doi:10.1017/S0962492906270016.
9. Mendes, F.M.; da Costa Junior, E.A. Bayesian inference in the numerical solution of Laplace's equation. *AIP Conf. Proc.* **2012**, *1443*, 72–79. doi:10.1063/1.3703622.
10. Cockayne, J.; Oates, C.; Sullivan, T.; Girolami, M. Probabilistic Numerical Methods for Partial Differential Equations and Bayesian Inverse Problems. *arXiv* **2016**, arXiv:1605.07811.
11. Albert, C. Physics-informed transfer path analysis with parameter estimation using Gaussian processes. In Proceedings of the 23rd International Congress on Acoustics, Aachen, Germany, 9–13 September 2019.
12. O'Hagan, A. Curve Fitting and Optimal Design for Prediction. *J. R. Stat. Soc. Ser. B* **1978**, *40*, 1–24. doi:10.1111/j.2517-6161.1978.tb01643.x.
13. Rasmussen, C.E.; Williams, C.K.I. *Gaussian Processes for Machine Learning*; MIT Press: Cambridge, MA, USA, 2006; doi:10.1142/S0129065704001899.
14. Narcowich, F.J.; Ward, J.D. Generalized Hermite Interpolation via Matrix-Valued Conditionally Positive Definite Functions. *Math. Comput.* **1994**, *63*, 661. doi:10.2307/2153288.
15. Macêdo, I.; Castro, R. Learning divergence-free and curl-free vector fields with matrix-valued kernels. Available online: http://preprint.impa.br/FullText/Macedo__Thu_Oct_21_16_38_10_BRDT_2010/macedo-MVRBFs.pdf (accessed on 30 June 2019).
16. Cobb, A.D.; Everett, R.; Markham, A.; Roberts, S.J. Identifying Sources and Sinks in the Presence of Multiple Agents with Gaussian Process Vector Calculus. In Proceedings of the 24th ACM SIGKDD International Conference on Knowledge Discovery & Data Mining, London, UK, 19–23 August 2018; pp. 1254–1262, doi:10.1145/3219819.3220065.

 © 2019 by the author. Licensee MDPI, Basel, Switzerland. This article is an open access article distributed under the terms and conditions of the Creative Commons Attribution (CC BY) license (http://creativecommons.org/licenses/by/4.0/).

*Proceedings*

# Electromagnetic Induction and Relativistic Double Layer: Mechanism for Ball Lightning Formation [†]

## Arthur Baraov

Hewlett-Packard Inc., Houston, TX 77389, USA; arthur.baraov@gmail.com
† Presented at the 39th International Workshop on Bayesian Inference and Maximum Entropy Methods in Science and Engineering, Garching, Germany, 30 June–5 July 2019.

Published: 21 November 2019

**Abstract:** What is the probability that ball lightning (BL) is a real phenomenon of nature? The answer depends on your *prior information*. If you are one of those lucky men who had a close encounter with a BL and escaped unscathed, your probability that it is real equals, of course, unity. On the other hand, if you are a theoretical physicist deeply involved in the problem of controlled thermonuclear fusion, your probability is likely to be zero. In this study, an attempt is being made to raise the likelihood of reality of BL phenomenon for everyone, plasma physicists included. BL is conceived here as highly structured formation of air, at roughly atmospheric pressure, with a set of nested sheaths, each of which is a double electrical layer with voltage drop in the order of 100 kV.

**Keywords:** prior information; ball lightning; fireball; bead lightning; double electrical layer; dynamic capacitor; controlled nuclear fusion

---

## 1. Introduction

*Ball lightning*, or *fireball*, is an atmospheric phenomenon in the form of a long-lived luminous sphere. Floating slowly in the air, or hovering over the ground, BL is observed most frequently in close proximity of a lightning strike during intense thunderstorm activity. There are numerous eyewitness accounts from around the world that are quite consistent with each other. This fact alone is a strong evidence for the reality of this phenomenon. Nevertheless, the reported characteristics and features of BL appear not only contradictory, but seem to be at odds with the well-established laws of nature, which makes it hard for some down-to-earth physicists to take the reality of this phenomenon seriously. The most puzzling feature of BL is its longevity—it can last for seconds to minutes. At the same time, it is the most obvious and undeniable attribute of BL. It was the longevity of BL that made the Nobel Prize winning physicist Pyotr Kapitsa to draw the following conclusion [1]:

> Since the energy stored in the cloud [of nuclear detonation] is proportional to the volume $d^3$, and the emission of the surface is $\sim d^2$, energy radiation from the ball will last for time interval proportional to $d$, its linear size. The mushroom cloud of a nuclear explosion with a diameter $d$ of 150 m lasts for less than 10 s, so the energy of a ball with a diameter of 10 cm shall be exhausted in less than 0.01 s. But in fact, as indicated in the literature, ball lightning of this size most often lives for a few seconds, sometimes even a minute. Thus, if there are no sources of energy in nature not yet known to us, due to the law of conservation of energy we have to accept that energy is continuously supplied to ball lightning as long as it glows, and we are forced to look for this source of energy outside the body of ball lightning.

Kapitsa suggested that the external source of energy for BL is a short-wave radio emission in the 35 to 70 cm range, and that the presence of ionized air facilitates the creation of radio waves, while the causative agent of these oscillations is the strike of a thunderbolt. However, despite numerous attempts, no one ever succeeded in detecting the indicated radio emission.

If we are to assert that the energy of BL is self-contained, we are confronted with a baffling feature of BL in the form of incredibly high density of energy content. Based on eyewitness reports, the energy content of most fireballs must be in the order of 10 or 100 kJ. However, a few sightings were reported in the literature which suggests that the energy of a fireball can be as high as 10 MJ or even higher. The most widely known evidence for the possibility of extraordinary high density of energy stockpiled in a BL is the publication in *The Daily Mail* on Nov 5, 1936 of a letter to the editor from one Mr. W. Morris with a title *A thunderstorm mystery* [2]:

> Sir, during a thunderstorm I saw a large, red hot ball come down from the sky. It struck our house, cut the telephone wires, burnt the window frame, and then buried itself in a tub of water which was underneath. The water boiled for some minutes afterwards, but when it was cool enough for me to search I could find nothing in it...

Specific heat of water is about 4 200 J/kg·K, so, if water in the amount of 18 liters (four British gallons, as indicated in the letter) was heated from, say, 20° C to the boiling point of 100° C, it follows that the energy of the BL was over 6 MJ. To visualize the enormity of this energy for a small globular object of 10 cm in diameter, which is capable of hovering freely over the ground, suffice to note that it matches the kinetic energy of a 5-ton truck dashing at 176 km/h! Just think of the destructive power of this "bullet". Is it possible to fit somehow that much energy in a luminous ball of air weighing less than one gram? For instance, what temperature is required in order to dissociate all the $N_2$ and $O_2$ molecules in one gram of air, then singly ionize all nitrogen and oxygen atoms, and, finally, bring the thermal energy of the resulting plasma to 6 MJ? Simple calculations point to a temperature of nearly 4,000,000° K. Yet, according to eyewitness reports, BL does not produce a marked sensation of heat at arm's length or even closer!

Kapitsa's assertion about the source of BL energy being outside its body is based on the premise that fireball is nothing but a fully ionized air, i.e., a regular plasma. So, it overlooks the idea that the energy in question could be, quite simply, ordinary kinetic energy of ultrarelativistic electrons, the motion of which is *coordinated* in some intricate way at the inception of a fireball. In other words, perhaps BL is not your ordinary plasma, but rather a *highly structured* object comprised of both neutral and charged particles. Nurbey Gulia, the inventor of the so-called *superflywheel*, disagrees with the main thesis of Kapitsa that the source of energy in BL is to be sought outside its body. He gives a number of compelling objections to the model in general, and then proceeds to suggest that BL is a kind of plasma flywheel created by nature itself [3]. And that is precisely what we hope to demonstrate on the following pages, concluding with a fairly detailed experimental schema for creating fireballs in a lab to test the idea. But our "fire flywheel" will differ profoundly from its mechanical cousin.

## 2. In Search for a Mechanism of High Energy Accumulation in BL

Consider an annular glass tube of small cross-section. The tube is evacuated to a high vacuum, and there is a free electron inside it that can slide along the tube with no friction. Suppose there is a magnetic field, which is orthogonal to the plane of the ring and distributed uniformly through space. Let the magnetic induction rise from zero to $B_a = 1.5$ T (magnetic field in the vicinity of lightning discharge channel can reach this level). How much energy will our free electron acquire? Electron's motion is determined by the second law of motion and the law of electromagnetic induction:

$$eE_{vrtx} = \frac{dp}{dt}, \quad 2\pi R E_{vrtx} = \frac{d(\pi R^2 B)}{dt},$$

where $e$ and $p$ are electron's charge and momentum, respectively, $E_{vrtx}$ is a vortex electric field, and $R$ is the radius of the ring. With the assumption that the electron was initially at rest, this yields

$$p = \frac{eRB}{2}. \tag{1}$$

Now, recalling the relationship between the momentum and the energy of relativistic particles, $(m_0c^2)^2 + (pc)^2 = E^2$, for the kinetic energy of the electron at the height of its acceleration we have:

$$K = E - m_0c^2 = (\sqrt{1+\alpha^2} - 1)m_0c^2, \text{ where } \alpha \equiv \frac{eRB_a}{2m_0c}. \tag{2}$$

Substituting $R = 0.05$ m and $B_a = 1.5$ T into this equation, we get $K = 10.7$ MeV. That is, our electron accumulates kinetic energy in the amount comparable to that of deuteron-triton fusion event, 17.6 MeV! This estimate is based on idealized and simplified schema, of course. Nevertheless, the fact that the vortex electric field, which is generated by a lightning discharge, is capable of accelerating a nearby free electron to ultrarelativistic velocities gives us a real hope of understanding how a BL accumulates enormous amount of energy at its inception. Indeed, all is left to do is to demonstrate that, starting with a few random seed electrons, electromagnetic induction can cause an avalanche ionization in the air and accelerate not one, but many electrons, forming in the process a stable configuration of charge particles. This is easier said than done, but we shall give it a try.

The idea of vortex field as the mechanism behind charging BL with energy at its birth is not entirely new [4]. The betatron model has a serious difficulty though, which boils down to the following. While the magnetic flux through a hypothetical vacuum ring is rising, the electron gains in speed and energy, but when this flux inevitably subsides and vanishes, the electron's speed shall fall back to its original value. That is precisely why electrons must be moved out of the betatron's vacuum chamber as soon as the magnetic flux is peaked. But how does nature prevent electrons from losing all the acquired energy when acceleration turns inevitably to deceleration? We won't find an answer to this question in [4] for it is not even raised there. We'll postpone this question for now and answer it later.

Let's try to imagine in outline how fireballs are created in nature. Consider lightning discharge channel as a long straight tube—horizontal when lightning bolt strikes between the clouds, or vertical when lightning hits the ground. It takes only a few microseconds for the discharge current to reach its peak value $\sim 100$ kA. Rapidly rising electric current generates a rapidly rising magnetic flux around the streamline. Let the diameter of the discharge channel be $d = 5$ cm, while the amplitude of the current $I_a = 200$ kA. Magnetic induction around the discharge channel has its peak value at the surface of the channel,

$$B_a = \frac{\mu_0 I_a}{\pi d} = \frac{4\pi 10^{-7} \cdot 2 \cdot 10^5}{0.05\pi} = 1.6 \text{ T}.$$

Consider now a thick toroid, which is tightly embracing the discharge channel. Magnetic field line at any point inside the toroid is perpendicular then to the toroid's cross-section at that point, so we have an intense magnetic flux along the toroid's central line. Suppose a free electron has appeared on the surface of our toroid, and let its frictionless motion be restricted to that surface. Assuming that it was initially at rest, the electron will be spun by a vortex electric field along the perimeter of toroid's cross-section. Assume now that we have not just one, but many such electrons. Then, neglecting the interaction of electrons with each other, each electron will be spun along a poloidal circle, and we'll get a BL in the shape of a... doughnut. We've been looking for a *fireball*, but have found a *firebagel* instead—what a bad luck!

But have we? Even if we get a bagel-shaped glowing object, being highly unstable, it will quickly disintegrate into a few pieces. But why wouldn't it be stable? After all, we know that skillful smokers can easily launch stable rings of smoke into the air, and the rings created by dolphins in the water are so stable that one can play with them as with elastic balls. Why then the ring of electrons rotating rapidly along the poloidal tracks on the bagel's surface is lacking topological stability? Well, because the forces that act between these charge particles are quite different—in both nature and scale—from the forces acting between electrically neutral molecules that make up rings created in the water or in the air. Indeed, parallel currents attract each other while anti-parallel currents repel. So, each individual poloidal turn of current tends to expand in the radial direction while the adjacent turns of current attract each other. Such a configuration is, obviously, highly unstable and the slightest violation of symmetry will lead to local constrictions in the bagel, splitting it into several pieces. Perhaps this is how the *bead lightning* is formed—a phenomenon of nature which is even more rare than BL.

At this stage of idealization, where we have extended the freedom of motion of charge particles to the surface of a torus, we are moving away from the betatron model. Electron flux in the betatron is highly rarefied, so ignoring the interaction between the individual charges while deriving the condition of keeping each electron in a *fixed circular* orbit—the so called Widerøe's condition—is quite justified. But the density of charges that is required to form BL is so high that the interaction between the charges cannot be ignored: electric and magnetic forces, as well as direct collisions of charge particles, will alter the trajectory of each particle in unpredictable way. Consequently, each electron is losing its circular orbit and engaging in a complicated pattern of motion, in both poloidal and toroidal directions. That's why the energy, which the electrons gain at the stage of acceleration by the vortex field, cannot be taken away entirely from them when the field changes its direction and deceleration sets in. Indeed, part of the kinetic energy of accelerated electron has already been passed to the *toroidal* component of its velocity, and this part cannot be taken away from it while decelerating in the *poloidal* direction.

## 3. Field Emission and Balance of Forces

Imagine a smooth ball made of highly conductive material—copper, for example—in the air. How many excess electrons can hold this ball? It can be charged until the field strength on the surface of the ball has reached the threshold value of 30 kV/cm, at which point air breakdown is triggered. If the ball is in a vacuum, there is no medium for the electrical breakdown to take place in. Nevertheless, the force of electrostatic repulsion of excess electrons, crowded in a thin surface layer on the ball, will become unbearable at some point, and they'll start leaving the ball in a hurry. This phenomenon is called *field emission*. In close to ideal vacuum conditions, field emission won't commence until the field strength on the smooth surface of the ball rises to $\sim 10^9$ V/m.

How thin is the thin surface layer, where excess electrons keep crowding until they decide that this injustice can no longer be tolerated? And why do they show such longanimity in the first place—why don't they simply run away as soon as we start charging our ball? After all, it seems there are no forces opposing the electrostatic repulsion of excess electrons cramped on the surface of the ball. And, yes, they have gathered there precisely because there are no forces inside the ball to counter their mutual dislike for each other. But as soon as the electrons reach the surface, they—for some mysterious reason—wilt as if an invisible, but very powerful barrier had sprung in front of them. Perhaps these questions may sound childish to some. But a child may ask such a simple question that no adult can answer sensibly. And how do adults answer these childish questions? They say that field emission is a quantum-mechanical effect, refer the inquiring child to the Fowler-Nordheim equation, solving of which—they add—is a highly complicated task. This "explanation" is not particularly illuminating and, frankly, hardly it can convince an electrical engineer or your ordinary physicist.

The retention of free charges on the surface of a conducting body cannot be explained if one assumes that charge carriers are really at rest on the surface of electrified body. In fact, these carriers—electrons in our case—are surely not at rest; they zip around at an average speed of $\sqrt{3kT/m_e}$, which is a whopping 100 km/s at room temperature! And when one of the electrons tries to escape from the crowded company of his brothers, it provokes an instantaneous rearrangement of its neighbors, which is equivalent to the appearance of a virtual mirror charge, resulting in immediate suppression of the attempt to escape. That is, the origin of forces, which hold excess electrons on the surface of a conductor, is of purely *dynamic* nature. The mirror-charge mechanism is, in my opinion, the clearest and most convincing explanation for the obstacle that prevents electrons from leaving a negatively charged conducting body. Anyhow, the fact remains whether we can explain it or not: an enormous amount of excess charge can stay on the surface of a conductor in *electrically* equilibrium state, i.e., the carriers of these charges can neither penetrate into the conductor nor escape from it. As the result of this, we get an *electrostatic* charge distribution on the surface of the conductor, and this is the most important thing for us in the whole story.

Suppose we charge our copper ball positively now, i.e., we take electrons from it instead of adding. Then we won't get electron emission, of course, but when a certain level of field strength is reached, the emission of positive ions commence. This level is of the same order as for the electron field emission, i.e., $10^9$ V/m. Imagine now that we launch an electron into a circular orbit around the copper ball charged positively to the limit. How much energy this "sputnik" must have in order to stay in the orbit? Let $E_s$ stand for the electric field strength on the surface of the ball. Then electron's motion in the orbit is described by the following equation

$$eE_s = \frac{K}{R}\left(1 + \sqrt{1 - \frac{v^2}{c^2}}\right), \text{ i.e. } K = \frac{eE_s R}{\mu}, \tag{3}$$

where $K$ and $v$ are electron's kinetic energy and its velocity, respectively; $\mu \equiv 1 + \sqrt{1 - \frac{v^2}{c^2}}$ is a factor that varies in narrow limits ($1 < \mu < 2$).

Let $E_s = 10^9$ V/m, then the energy required for the electron to stay in a circular orbit of radius 5 cm is about 50 MeV. The orbiting electron is losing energy to synchrotron radiation. How long will it take for our electron to lose 90% of its initial energy, i.e., for its energy to fall from 50 MeV to 5 MeV? The intensity of synchrotron radiation for ultrarelativistic electrons ($\gamma \equiv E/m_0c^2 \gg 1$) is given by [5]:

$$W = \frac{ce^2\gamma^4}{6\pi\epsilon_0 R^2}. \tag{4}$$

So, the time interval, which takes $\gamma$ to fall from $\gamma_1$ to $\gamma_2$, is

$$\tau = -\int_{\gamma_1}^{\gamma_2} \frac{dE}{W} = \frac{2\pi\epsilon_0 R^2 m_0 c}{e^2}\left(\frac{1}{\gamma_2^3} - \frac{1}{\gamma_1^3}\right). \tag{5}$$

Recalling that electron's rest energy is 0.51 MeV, and substituting $\gamma_1 = 100, \gamma_2 = 10, R = 0.05$ m into this equation, we get $\tau = 50$ s, which is of the right order for the lifespan of fireballs.

To accelerate an electron to 50 MeV via the mechanism of electromagnetic induction in one shot, so to speak, the required change of magnetic flux is so large that it is apparently beyond the ability of the most powerful lightning bolt. However, it should be remembered that the lightning discharge takes place in several steps of rapid succession. There are several direct and return strikes along the path laid by the leader, and each of these surges of current can charge BL with energy. We have estimated earlier that a one-time surge of magnetic induction from zero to a peak value of 1.5 T can spin up electron to 10.7 MeV. So, due to cumulative effect of direct and reverse strikes of lightning, the vortex field can accelerate electrons up to 50 MeV. From the moment of fireball's inception to the point when it catches casual observer's attention, at least a few seconds shall pass. So, it is highly unlikely that the energy of the fastest electrons in any BL to exceed a few dozen MeV. Indeed, even if we assume that an electron with a next to impossible value of 300 MeV (betatron limit) is present in a BL at its inception, it would take only a fraction of a second for its energy to fall to 50 MeV due to synchrotron radiation.

Imagine now that we have launched not one, but a whole lot of electrons along various geodesic lines, which are distributed uniformly all over our copper ball. Assume that this cloud of electrons is confined to a spherical shell of thickness $h$. Due to shielding of positively charged "core" by negative electron cloud, the energy of each electron is determined then by its position in the cloud—the lower the electron orbit the higher its energy. As you have probably guessed by now, huge amounts of energy can be accumulated in systems like this. Moreover, in terms of force balance, this system forms a completely legitimate and stable configuration. Note also that, in addition to the net kinetic energy of electrons, a certain amount of electrostatic energy is accumulated in this *dynamic capacitor*. Let's see which component—the net kinetic energy of electrons or the electrostatic energy of the capacitor—makes greater contribution to the stockpile of energy accumulated in this system.

Suppose a ball, with a uniform distribution of positive charges, $\sigma$, on its surface, is surrounded by a cloud of electrons orbiting the ball. Let the orbits of electrons lie within a thin shell of thickness $h \ll R$, directly above the surface of the ball, and let, finally, the density of electron distribution be a function of the height of the orbit only: $n = n(x)$. The system as a whole might be electrically neutral, or it may have an excess *positive* charge. Consider the neutral case,

$$\int_0^h en(x)\,dx = \sigma. \tag{6}$$

Due to the assumed symmetry, magnetic forces cancel out. Then, neglecting at this point the possible collisions of electrons, the motion of each electron is determined by play of two opposing forces—the centripetal force of electrostatic attraction to the positive ball and the centrifugal force of inertia:

$$\frac{e}{\epsilon_0}\left(\sigma - \int_0^x en(x)\,dx\right) = \frac{\mu K(x)}{R}, \tag{7}$$

where $K(x)$ is electron's kinetic energy in the orbit of hight $x$; $\mu$ is a function of electron's speed that varies, as noted above, in narrow limits. Therefore, when calculating the net kinetic energy $K_\Sigma$ of electrons, $\mu$ can be treated as a constant with some effective value, $1 < \mu_{eff} < 2$:

$$K_\Sigma = 4\pi R^2 \int_0^h n(x)K(x)\,dx = \frac{4\pi R^3}{\mu_{eff}\epsilon_0}\left[\int_0^h \left(\sigma - \int_0^x en(x)\,dx\right)en(x)\,dx\right] = \frac{2\pi R^3 \sigma^2}{\mu_{eff}\epsilon_0}.$$

We have obtained an interesting result: with a fixed $R$, the total kinetic energy of electrons depends only on the integral of the density function $n(x)$, i.e. the details of this function, which are not known, do not matter. Since the field strength on the surface of the ball $E_s = \sigma/\epsilon_0$, we may right this result as:

$$K_\Sigma = \frac{2\pi\epsilon_0 R^3 E_s^2}{\mu_{eff}}. \tag{8}$$

Let $E_s = 5 \times 10^9$ V/m, then for a ball of radius $R = 0.05$ m we get $K_\Sigma \approx 0.6$ MJ. The exact value of electrostatic energy, $E_c$, of this system depends, of course, on the details of the density function $n(x)$; regardless, it makes only a small fraction of the energy stored in the system:

$$E_c < 4\pi R^2 h(\epsilon_0 E_s^2/2) = \mu_{eff} K_\Sigma (h/R) \ll K_\Sigma. \tag{9}$$

Imagine now that our positively charged copper ball, with a cloud of electrons rapidly revolving around it, is placed inside a copper spherical shell of thickness 1 mm or less. This copper shell won't "feel" the presence of electrically neutral copper ball inside it. So, we can repeat our trick and build another double layer on the outer surface of this shell. Building up this "matreshka doll" by adding more and more copper shells to it, we can get a compact and highly dense energy storage system with a capacity in the order of 10 MJ or even higher.

But what is the relevance of this all to the phenomenon of BL? We have demonstrated the feasibility of large-orange-size material object with seemingly fantastic properties that some fireballs have according to eyewitness accounts: long life, incredibly high energy content, and low temperature. True, our "object" is heavier than air, thus it cannot hover over the ground like a fireball. To turn our copper-electronic battery into a real fireball, we need to show that a similar structure can be created from... a thin air, literally. In the design of our energy capsule above, a double electrical layer was the key element, which we have imagined rather than built in practice, while double layer sheath is a real attribute of plasma. Moreover, this attribute is created naturally by plasma itself as a way of protection from the environment. So, the transition from an imaginary double layer on the surface of a copper ball to a natural double layer in the air shall not come as a miracle to us.

With the transition from a copper ball to a ball of air, we no longer have free electrons and lose the conductivity associated with them, so it seems that the entire construct is collapsing. However, in the air—especially, in thunderstorm conditions—a small count of free electrons is always present. We have seen earlier that, with each strike of a thunderbolt, a rapidly rising flux of magnetic induction takes place around the discharge channel, thereby creating an intense vortex field. Seed electrons, which happen to be in the area of this strong field, get accelerated to energies that might be high enough to cause an avalanche ionization. At the same time, electrons—in virtue of their greater mobility—gain quickly in tangential velocity, with a tendency to scatter in radial directions due to the combined action of centrifugal force of inertia and the Lorentz force, thus leaving behind the heavy and sluggish positive ions and creating a gap in the air in the form of a thin sheath. This peculiar *rupture* in the air *is* a relativistic double layer, with an almost perfect vacuum and high intensity electrostatic field between the oppositely charged layers. It is precisely this electrostatic field that prevents electrons from scattering and confines them to the double layer sheath. This mechanism will, most likely, lead to stratification, i.e. to formation not just one, but a set of nested sheaths.

The occasional collisions of electrons in the sheaths will result in continuous ejection of highly energetic electrons from the fireball at various angles. These eruptions may manifest themselves as tiny crackling flashes of electric discharge all over the surface of BL, which can account for the "boiling" ("hairy") appearance of fireball in eyewitness reports. Besides, the electron ejections result in fireball becoming periodically net positive then neutral by attracting electrons back from the surrounding air, which is quite consistent with the erratic and highly unpredictable motion of the fireball due to electromagnetic interaction with the earth at large and metal objects in particular.

## 4. Schema for Creating BL in a Modestly Equipped Laboratory

The above ideas of the mechanism of BL formation suggest a schema for generating fireballs in a lab. Basically, it is as follows. In a small volume of air, create a powerful flux of magnetic induction, which peaks in a matter of a few dozen microseconds. To implement it successfully, the schema must be fleshed out, of course, with a multitude of important details.

First and foremost, experiments shall be conducted in a vessel with an air rarefied to about 1 mbar. An avalanche ionization, which—according to our model—is a requisite for creating a fireball, is much easier to trigger in a rarefied air. If we succeed in creating relativistic double layers in a rarefied air, pressure in the vessel can be raised back to atmospheric by leaking air slowly into it.

The next detail is concerned with the shape of the vessel. A hollow torus with external and internal diameters of 25 and 5 cm, respectively, might be the right choice. By passing a powerful pulse of current through a toroidal coil with a few turns of a thick rod, tightly embracing the torus, one can hope to get a few fireballs at once. To keep the coil inductance low, the number of turns of the coil shall be restricted to six. The vessel is to be made of a transparent dielectric, so one can observe what is happening inside it. An alternative choice for the vessel is a spherical glass bulb, 10 cm in diameter.

Now on the details of pulse discharge, which is to simulate the lightning bolt. This can be achieved with special, heavy duty pulse capacitors of small inductance equipped with a remotely controlled spark gap. If the vessel of choice is a spherical bulb, we need two capacitors to be discharged synchronously via two horseshoe-like copper bars. The diameter of the horseshoe must be slightly less than the diameter of the bulb. The horseshoe bars shall be located in two vertical planes, parallel to each other, on opposite sides of the bulb in a symmetrical configuration, so that the centers of the horseshoes and the center of the spherical bulb lie on the same horizontal line, while the distance between the centers of the horseshoes is slightly larger than the diameter of the bulb.

The vortex electric field, generated by discharging pulse capacitors, shall be powerful enough to trigger avalanche ionization of the rarefied air inside the bulb. It is important to note here that artificially generated seed electrons might be required to facilitate the launch of the avalanche. Geometrical parameters of the configuration shall be adjusted in such a way that the radial component of the net force (i.e. centrifugal force plus Lorentz force), experienced by free electrons at the stage of current rise, is directed from the center of the bulb, and not to its center. This last condition is of paramount importance: it is this condition that makes creating a set of nested ruptures in the air possible, thereby forming relativistic double layer sheaths and restricting the motion of accelerated electrons in the confines of these sheaths.

## 5. Microwave Oven and Synchrotron Radiation of BL

Now we turn to another puzzling feature of BL, namely, the ability to melt or evaporate certain metal objects—gold jewelry, in particular—on its path. According to our model, BL is accompanied by a synchrotron radiation. What's the frequency range of this radiation? Synchrotron radiation is peaked on the frequency, which is associated with the cyclotron frequency, i.e. with $\nu_0 = c/\pi D$, as [5]

$$\nu_{max} = \frac{3\gamma^3 \nu_0}{2}, \text{ where } \gamma \equiv \frac{E}{m_0 c^2}. \tag{10}$$

As we have noted earlier, by the time fireball gets to casual observer's line of sight, its ultrarelativistic electrons lose part of the energy to synchrotron radiation, retaining at best a few dozen MeV per electron. Let $\gamma$ vary in the range 1–100, i.e., the energy of electrons in the double layer sheath is in the 0.5–50 MeV range. Then, for a typical fireball of 10 cm in diameter, $\nu_{max} = 1.43(10^9 - 10^{15})$ Hz. In other words, synchrotron radiation of BL is roughly in microwave to UV range. So, the assumption that BL is a source of X-rays, which is occasionally made in the literature, is most likely not correct.

Electromagnetic radiation in the microwave oven is of frequency 2.45 GHz and power around 1000 watts. These parameters correspond to that of synchrotron radiation from a fireball of small energy. This invites a curious question. What will happen if we put a metal object in a microwave oven? If BL emits electromagnetic radiation in the microwave range that is powerful enough to melt or evaporate parts of metal objects in an instant, then microwave oven should be able to do the same. With that thought in mind, I took three sewing needles 75 to 80 mm long each and attached them to a cardboard so that a triangle with small gaps at all three vertices is formed. I put all that on the rotating table of the microwave oven in my kitchen and turned it on for 10 s. Around the 7th second, a huge flame flared up and the cardboard caught fire. I turned off the oven as quickly as I could. The result of this experiment was that the sharp tip of one of the needles melted and turned into a ball in a fraction of a second. Thus, it seems that our understanding of the nature of BL can also account for the mysterious ability of fireballs to melt or evaporate parts of metal objects in a blink of an eye.

## 6. Controlled Nuclear Fusion and BL

Provided that we succeed in creating a fireball in a rarefied air inside a glass bulb of 10 cm in diameter, as described above, there is no reason why it cannot be done in a glass sphere of much larger diameter that is filled with heavy hydrogen gas $^2H_2$ at atmospheric pressure or even higher. Then, the positive layer of each double layer sheath would be comprised of deuterium nuclei. We have seen above that ultrarelativistic electrons are the main carriers of energy in BL. Deuterium nuclei, which form the inner positive layers of the sheaths in this case, will also be accelerated by the vortex field, but to considerably lesser degree compared to the electrons, so, in percentage terms, they do not contribute much to the net energy of BL. However, there is a point of larger importance to be made here.

The charge particles, which make up the sheaths, move so fast that the sheaths are perceived as virtual walls: most of the molecules of heavy hydrogen are reflected by the sliding blows of sheath particles, while some of the molecules will penetrate into the "wall" and become targets for direct collisions with the highly energetic deuterons. Figuratively speaking, the two layers of each sheath act as millstones, which grind all the "grains" that end up between them. Now the question is: do the deuterons have enough energy to fuse with the nuclei of atoms of heavy hydrogen molecules that leak into the "wall" and become collision targets? To estimate the energy of deuterons in the sheath, substitute doubled proton mass for the electron mass in formula (2), which we have derived earlier for the energy of electrons accelerated by the vortex field. Let $R = 0.35$ m, $B_a = 1.0$ T, then deuterons get accelerated to 0.73 MeV, while electrons—to 52 MeV.

These estimates are quite encouraging. Firstly, note that electrons stay well below the 100 MeV threshold, above which the losses to synchrotron radiation become unacceptably large. Secondly, the cross section for the two-channel fusion reaction, $^2H + {}^2H \rightarrow {}^3He + n$; $^2H + {}^2H \rightarrow {}^3H + p$, with the energy of striking deuterons at 0.7 MeV is about 0.08 barn, i.e. 80% of its maximum possible value of 0.1 barn. Therefore, if the density of the "target" (which, obviously, grows proportionally with the density of the hydrogen gas, in which the double layer sheaths are immersed) is high enough, an intense fusion reaction is to be expected. In other words, we might have a *nuclear* fusion reactor here, which operates on principles that are quite different from those of the *thermonuclear* fusion reactor.

Is it feasible to control the process of nuclear fusion, which takes place in this reactor, and can it be an efficient source of energy? Time will tell.

## 7. Discussion

Aside from exotic forms of matter (Bose-Einstein condensate, antimatter, etc.), contemporary physics deals with four states of matter only: solid, liquid, gas, and plasma. If we take on faith the characteristic features of BL and the pattern of its motion according to eyewitness reports, the problem with BL is that identifying its substance with any of these four states of matter leads inexorably to violation of one or the other law of physics. This explains why the range of hypothesis underlying speculative models of this phenomenon (the number of which is, probably, way over one hundred by now) is extremely wide, starting from the assumption of optical illusion and all the way to suggestions that fireball is kind of a black hole.

In this study we have shown that the most inscrutable features of BL—including long life, incredibly high density of energy content, low temperature, and the ability to melt or evaporate metal objects—can be accounted for without violating any laws of nature, provided that the substance of BL is not merely a homogeneous, fully ionized gas, but rather an intricate structure comprised of ordinary air plus a number of nested sheaths, each of which is a double electrical layer with voltage drop in the order of 100 kV. The findings indicate that BL is a real phenomenon of nature beyond reasonable doubt, and solving this riddle is likely to have implications, the importance of which can hardly be overestimated. In particular, the possibility of fundamentally new way of looking at the problem of controlled nuclear fusion is noted.

The working hypotheses for the mechanism of BL formation is electromagnetic induction in the wake of a powerful thunderbolt discharge. The future research shall be concentrated on attempts to produce BL in a lab. A fairly detailed experimental schema for carrying out this task in a modestly equipped laboratory has been outlined.

**Funding:** This research received no external funding.

**Conflicts of Interest:** The author declares no conflict of interest.

## Abbreviations

The following abbreviations are used in this manuscript:

BL    ball lightning

## References

1. Kapitsa, P.L. On the nature of ball lightning. *Dokl. Akad. Nauk SSSR* **1955**, *101*, 245–248.
2. Stenhoff, M. *Ball Lightning: An Unsolved Problem in Atmospheric Physics*; Kluwer Academic and Plenum Publishers: New York, NY, USA, 1999.
3. Gulia, N.V. *Amazing Mechanics*; NTS ENAS Publisher: Moscow, Russia, 2006.
4. Butusov, K.P. Electrodynamic model of ball lightning (betatron version). *Probl. Univ. Res.* **2010**, *34*, 61–75. Available online: http://scicom.ru/files/journals/piv/volume34/issue1/piv_vol34_issue1_09.pdf (accessed on 22 July 2019).
5. Ternov, I.M.; Mikhailin, V.V.; Khalilov, V.R. *Synchrotron Radiation and Its Applications*; Moscow University Press: Moscow, Russia, 1980.

© 2019 by the authors. Licensee MDPI, Basel, Switzerland. This article is an open access article distributed under the terms and conditions of the Creative Commons Attribution (CC BY) license (http://creativecommons.org/licenses/by/4.0/).

*Proceedings*

# A Sequential Marginal Likelihood Approximation Using Stochastic Gradients

Scott A. Cameron [1,2,*], Hans C. Eggers [1,2] and Steve Kroon [3]

1. Department of Physics, Stellenbosch University, Stellenbosch 7600, South Africa; eggers@sun.ac.za
2. National Institute for Theoretical Physics, Stellenbosch 7600, South Africa
3. Computer Science Division, Stellenbosch University, Stellenbosch 7600, South Africa; kroon@sun.ac.za
* Correspondence: scott.a.cameron@live.co.uk
† Presented at the 39th International Workshop on Bayesian Inference and Maximum Entropy Methods in Science and Engineering, Garching, Germany, 30 June–5 July 2019.

Published: 3 December 2019

**Abstract:** Existing algorithms like nested sampling and annealed importance sampling are able to produce accurate estimates of the marginal likelihood of a model, but tend to scale poorly to large data sets. This is because these algorithms need to recalculate the log-likelihood for each iteration by summing over the whole data set. Efficient scaling to large data sets requires that algorithms only visit small subsets (mini-batches) of data on each iteration. To this end, we estimate the marginal likelihood via a sequential decomposition into a product of predictive distributions $p(\mathbf{y}_n|\mathbf{y}_{<n})$. Predictive distributions can be approximated efficiently through Bayesian updating using stochastic gradient Hamiltonian Monte Carlo, which approximates likelihood gradients using mini-batches. Since each data point typically contains little information compared to the whole data set, the convergence to each successive posterior only requires a short burn-in phase. This approach can be viewed as a special case of sequential Monte Carlo (SMC) with a single particle, but differs from typical SMC methods in that it uses stochastic gradients. We illustrate how this approach scales favourably to large data sets with some simple models.

**Keywords:** marginal likelihood; Monte Carlo; stochastic gradients

## 1. Introduction

Marginal likelihood (ML), sometimes call evidence, is a quantitative measure of how well a model can describe a particular data set; it is the probability that the data set occurred within that model. Consider a Bayesian model with parameters $\boldsymbol{\theta}$ for a data set $\mathcal{D} = \{\mathbf{y}_n\}_{n=1}^N$. The ML is the integral

$$\mathcal{Z} := p(\mathcal{D}) = \int p(\mathcal{D}|\boldsymbol{\theta}) p(\boldsymbol{\theta}) \, d\boldsymbol{\theta},$$

where $p(\mathcal{D}|\boldsymbol{\theta})$ is the likelihood and $p(\boldsymbol{\theta})$ is the prior. In this paper we consider the case where the data are conditionally independent given the parameters

$$p(\mathcal{D}|\boldsymbol{\theta}) = \prod_n p(\mathbf{y}_n|\boldsymbol{\theta}), \tag{1}$$

as is common in many parametric models. The posterior distribution over a set of models is proportional to their ML and so approximations to it are sometimes used for model comparison, and weighted model averaging [1] (chapter 12). This integral is typically analytically intractable for any but the simplest models, so one must resort to numerical approximation methods. Nested sampling (NS) [2] and annealed importance sampling (AIS) [3] are two algorithms able to produce accurate estimates of the ML. NS accomplishes this by sampling from the prior under constraints of increasing likelihood, and AIS by sampling from a temperature annealed distribution $\propto p(\mathcal{D}|\boldsymbol{\theta})^\beta p(\boldsymbol{\theta})$ and averaging over samples with appropriately calculated importance weights. Although NS and AIS produce accurate estimates of the ML, they tend to scale poorly to large data sets due the fact that they need to repeatedly calculate the likelihood function: for NS this is to ensure staying within the constrained likelihood contour; for AIS the likelihood must be calculated both to sample from the annealed distributions using some Markov chain Monte Carlo (MCMC) method, as well as to calculate the importance weights. Calculation of the likelihood is computationally expensive on large data sets. To combat this problem, various optimization and sampling algorithms rather make use of stochastic approximations of the likelihood by sub-sampling the data set into mini-batches $B \subseteq \mathcal{D}$ [4]. The stochastic log-likelihood approximation is

$$\log p(\mathcal{D}|\boldsymbol{\theta}) \approx \frac{|\mathcal{D}|}{|B|} \sum_{\mathbf{y} \in B} \log p(\mathbf{y}|\boldsymbol{\theta}),$$

with each iteration of these algorithms generally using a different mini-batch. Unfortunately NS and AIS cannot trivially use mini-batching to improve scalability. Using stochastic likelihood approximations changes the statistics of the likelihood contours in NS, allowing particles to occasionally move to lower likelihood instead of higher, violating the basic assumptions of the algorithm. AIS could benefit from using stochastic likelihood gradients during the MCMC steps, but it is not obvious how one would calculate the importance weights in this setting. This work presents an approach for large-scale ML approximations using mini-batches. This is done using a sequential decomposition of the ML into predictive distributions, which can each be approximated using stochastic gradient MCMC methods. The particles sampled from each previous posterior distribution can be reused for efficiency since they will typically be close to the next posterior. This can be viewed as a special case of sequential Monte Carlo with Bayesian updating [5].

We illustrate our approach by calculating ML estimates on three simple models. For these models, we obtain roughly an order of magnitude speedup over nested sampling on data sets with one million observations with negligible loss in accuracy.

## 2. Sequential Marginal Likelihood Estimation

The ML can be decomposed, through the product rule, into a product of predictive distributions of the following form

$$\mathcal{Z} = \prod_n p(\mathbf{y}_n|\mathbf{y}_{<n}),$$

where, from Equation (1),

$$p(\mathbf{y}_n|\mathbf{y}_{<n}) = \int p(\mathbf{y}_n|\boldsymbol{\theta}) p(\boldsymbol{\theta}|\mathbf{y}_{<n}) \, d\boldsymbol{\theta}. \tag{2}$$

Assuming one is able to produce accurate estimates $\hat{p}(\mathbf{y}_n|\mathbf{y}_{<n})$ of the predictive probabilities, the log-ML can be approximated by

$$\log \hat{\mathcal{Z}} = \sum_n \log \hat{p}(\mathbf{y}_n|\mathbf{y}_{<n}). \tag{3}$$

In this way the difficult problem of estimating an integral of an extremely peaked function, $p(\mathcal{D}|\boldsymbol{\theta})$, reduces to the easier problem of estimating many integrals of smoother functions $p(\mathbf{y}_n|\boldsymbol{\theta})$. Note that this approach can more generally be applied using any other sequential decomposition of the data set. We only present derivations using the approach above for notational clarity but we decompose the data into varying-sized chunks of observations during our experiments as discussed in Section 4.

Typical sequential Monte Carlo (SMC) methods use a similar approach, and calculate predictive estimates using a combination of importance resampling and MCMC mutation steps [5]. Generic examples of such algorithms are the bootstrap particle filter [6], which is often used for posterior inference in hidden Markov models and other latent variable sequence models [5], and the "left-to-right" algorithm which is used in [7] to evaluate topic models.

The computational efficiency of using this approach depends on the method of approximating Equation (2). One such estimator is

$$\hat{p}(\mathbf{y}_n|\mathbf{y}_{<n}) = \frac{1}{M} \sum_{i=1}^{M} p(\mathbf{y}_n|\boldsymbol{\theta}_i), \qquad (4)$$

where each $\boldsymbol{\theta}_i$ is drawn from the posterior distribution $p(\boldsymbol{\theta}|\mathbf{y}_{<n})$ using MCMC methods. Again, one might use a chunk of observations for this predictive estimate rather than just one. As described in [8], estimators of this form tend to underestimate the ML, but still converge to the exact ML in the limit $M \to \infty$. Since samples from the previous posterior, $p(\boldsymbol{\theta}|\mathbf{y}_{<n-1})$, would generally be available at each step, we expect only a small number of steps will be needed to accurately sample from the next posterior distribution, $p(\boldsymbol{\theta}|\mathbf{y}_{<n})$. Metropolis-Hastings based MCMC algorithms would have to iterate over the previous $n-1$ data points in order to calculate the acceptance probability for each Markov transition, and so using them to estimate $\log \mathcal{Z}$ in this sequential manner would scale at least quadratically in $N$. The key computational improvement in our approach is instead using stochastic gradient based MCMC algorithms such as stochastic gradient Hamiltonian Monte Carlo [9]. SGHMC utilizes mini-batching, allowing one to efficiently draw samples from the posterior distribution $p(\boldsymbol{\theta}|\mathbf{y}_{<n})$ even when $n$ is large. We now describe how one can use SGHMC to sample from posterior distributions.

## 3. Stochastic Gradient Hamiltonian Monte Carlo

SGHMC [9] simulates a Brownian particle in a potential, by numerically integrating the Langevin equation

$$\begin{aligned} d\boldsymbol{\theta} &= \mathbf{v}\, dt \\ d\mathbf{v} &= -\nabla U(\boldsymbol{\theta})dt - \gamma \mathbf{v} dt + \sqrt{2\gamma} dW, \end{aligned}$$

where $U(\boldsymbol{\theta})$ is the potential energy, $\gamma$ is the friction coefficient and $W$ is the standard Wiener process [10]. That is, each increment $\Delta W$ is independently normally distributed with mean zero and variance $\Delta t$. It can be shown through the use of a Fokker-Planck equation [11], that the above dynamics converge to the stationary distribution

$$p(\boldsymbol{\theta}, \mathbf{v}) \propto \exp\left(-U(\boldsymbol{\theta}) - \frac{\mathbf{v}^2}{2}\right).$$

We can use this to sample from the full data posterior by using a potential energy equal to the negative log-joint $U(\boldsymbol{\theta}) := -\log p(\mathcal{D}, \boldsymbol{\theta})$. The numeric integration is typically discretized [9,12] as follows

$$\begin{aligned} \Delta \boldsymbol{\theta} &= \mathbf{v} \\ \Delta \mathbf{v} &= -\eta \nabla \hat{U}(\boldsymbol{\theta}) - \alpha \mathbf{v} + \epsilon \sqrt{2(\alpha - \hat{\beta})\eta}, \end{aligned} \qquad (5)$$

where $\eta$ is called the learning rate, $1 - \alpha$ is the momentum decay, $\epsilon$ is a standard Gaussian random vector, $\hat{\beta}$ is an optional parameter to offset the variance of the stochastic gradient term and

$$\hat{U}(\boldsymbol{\theta}) = -\frac{|\mathcal{D}|}{|B|} \sum_{\mathbf{y} \in B} \log p(\mathbf{y}|\boldsymbol{\theta}) - \log p(\boldsymbol{\theta})$$

is an unbiased estimate of $U(\boldsymbol{\theta})$. A new mini-batch $B$ is sampled for each iteration of Equation (5). Since $U(\boldsymbol{\theta})$ grows with the size of the data set, a small learning rate $\eta \sim \mathcal{O}(\frac{1}{|\mathcal{D}|})$ is required to minimize the discretization error. The variance of the stochastic gradient term is proportional to $\eta^2$ while the variance of the injected noise is proportional to $\eta$, so in the limit $\eta \to 0$, stochasticity in the gradient estimates becomes negligible and the correct continuum dynamics are recovered, even if one ignores the errors from the stochastic gradient noise and $\hat{\beta} = 0$ is used. We refer the reader to [9,13,14] for an in-depth analysis of the algorithm parameters.

For the purposes of Bayesian updating, we use SGHMC to sample from the $n^{th}$ posterior distribution, $p(\theta|\mathbf{y}_{\leq n})$, using the stochastic potential energy

$$\hat{U}_n(\theta) = -\frac{n-1}{|B|} \sum_{\mathbf{y} \in B} \log p(\mathbf{y}|\theta) - \log p(\mathbf{y}_n|\theta) - \log p(\theta),$$

where the mini-batch is drawn i.i.d. with replacement from the set of all previous data points i.e., $B \subset \{\mathbf{y}_k | k < n\}$. The extra term here is to ensure that the previously unseen data point is always taken into account. If the extra term was not included, there would be some chance that the new data point does not get taken into account during the SGHMC steps. With this potential energy SGHMC still converges to the required posterior distribution, since it is still an unbiased estimator of the negative log-joint.

## 4. Experiments

We use mini-batches of size 500, with the following SGHMC parameters: $\eta = 0.1/n$, $\alpha = 0.2$ and $\hat{\beta} = 0$. Predictive distributions are approximated using $M = 10$ samples and 20 burn-in steps for each new posterior. As mentioned in Section 2, rather than Bayesian updating by adding a single observation at a time, we add chunks of data at a time. In the following experiments we use chunks of 20 data points when $n \leq 80$, chunks of size $\lfloor \frac{n}{4} \rfloor$ when $80 < n < 2000$, and chunks of size 500 thereafter. Our motivation for this is because we expect that smaller chunk sizes will give a lower variance in the estimator Equation (4) when $n$ is small and so the posterior is less peaked, but for large $n$ using larger chunk sizes is more efficient.

We use NS as our reference standard of accuracy. We implement NS with 20 SGHMC steps to sample from the constrained prior. For SGHMC used with NS we used parameters $\eta = 10^{-3}$, $\alpha = 0.1$ and $\hat{\beta} = 0$ because there is no gradient noise when sampling from the prior. Results reported are for two particles; more behave similarly but are slower. We allow NS to run until the additive terms are less than 1% of the current $\hat{Z}$ estimate. This is a popular stopping criterion and is also used in [8]. For more information on the constrained sampler and NS see Appendix A. Our experiments were run on a laptop with an Intel i7 CPU. For fair comparison, all code is single threaded. Multithreading gives a considerable speedup when calculating the likelihood on large data sets but can introduce subtle complexities that are difficult to control for in tests of runtime performance.

We evaluate our approach on the following three models using simulated data sets:

### 4.1. Linear Regression

The data set consists of pairs $(\mathbf{x}, y)$ related by

$$y = \mathbf{w}^T \mathbf{x} + b + \epsilon,$$

where $\epsilon$ is zero mean Gaussian distributed with known variance $\sigma^2$. We do not assume any distribution over $\mathbf{x}$ as it always appears on the right hand side of the conditional.

$$p(y|\mathbf{x}, \mathbf{w}, b) = \frac{1}{\sqrt{2\pi\sigma^2}} \exp\left(-\frac{(y - \mathbf{w}^T \mathbf{x} - b)^2}{2\sigma^2}\right)$$

Parameters are $\mathbf{w}$ and $b$, with standard Gaussian priors. ML can be calculated analytically for this model. We used 5 dimensional vectors $\mathbf{x}$; this model has 6 parameters.

## 4.2. Logistic Regression

The data set consists of pairs $(\mathbf{x}, y)$, where $\mathbf{x}$ is an observation vector which is assigned a class label $y \in \{1, \ldots, K\}$. The labels have a discrete distribution with probabilities given by

$$p(y|\mathbf{x}, \boldsymbol{\theta}) = \frac{\exp(\mathbf{w}_y^T \mathbf{x} + b_y)}{\sum_k \exp(\mathbf{w}_k^T \mathbf{x} + b_k)}.$$

Parameters are $\boldsymbol{\theta} = (\mathbf{w}_{1:K}, b_{1:K})$, with standard Gaussian priors. Again we do not assume any distribution over $\mathbf{x}$ as it always appears on the right hand side of the conditional. We used 10 dimensional vectors $\mathbf{x}$ with 4 classes; this model has 44 parameters.

## 4.3. Gaussian Mixture Model

The data are modeled by a mixture of multivariate Gaussian distributions with diagonal covariance matrices. Mixture weights, means and variances are treated as parameters. This type of model is often treated as a latent variable model, where the mixture component assignments of each data point are the latent variables. Here we marginalize out the latent variables to obtain the following conditional distribution:

$$p(\mathbf{y}|\boldsymbol{\theta}) = \sum_{k=1}^{K} \beta_k \prod_{j=1}^{d} \frac{1}{\sqrt{2\pi\sigma_{k,j}^2}} \exp\left(-\frac{(y_j - \mu_{k,j})^2}{2\sigma_{k,j}^2}\right),$$

$$\boldsymbol{\theta} = (\beta_{1:K}, \mu_{1:K,1:d}, \sigma_{1:K,1:d}^2).$$

Mixture weights $\beta_{1:K}$ are modeled by a Dirichlet prior with $\alpha = 1$; means $\mu_{k,j}$ are modeled by Gaussian priors, centered around zero and with variance $4\sigma_{k,j}^2$; variances $\sigma_{k,j}^2$ are modeled by inverse gamma priors with shape and scale parameters equal to 1. We used 5 Gaussian components and observations were 2 dimensional; this model has 25 parameters with 24 degrees of freedom.

## 5. Results and Discussion

The log-ML typically grows linearly in the number of data points. For this reason, it is natural to measure errors in $\frac{\log \mathcal{Z}}{N}$ rather than $\log \mathcal{Z}$. In [8], the authors suggest that errors in $\frac{\log \mathcal{Z}}{N}$ of 0.1 are acceptable. For each model, we measured the runtime performance of nested sampling and our sequential estimator for various data set sizes up to one million data points. Due to computational constraints we run NS only for data set sizes at logarithmically increasing intervals, while the sequential sampler naturally produces many more intermediate results in a single run. In some of the figures below, the sequential sampler initially underestimates the log-ML. We believe this is due to higher variance of Equation (4) when $n$ is small, and so we also give a hybrid result which replaces the initial terms in Equation (3) with a NS estimate of the ML for the first 100 data points.

### 5.1. Linear Regression

For the linear regression model, the exact ML is available analytically and is shown in Figure 1a for comparison. Both algorithms are able to produce accurate results for this model for all data set sizes. The final error of the sequential sampler on one million data points is only about $10^{-4}N$ (roughly 0.01%). For this model, our method was faster than NS by about a factor of 3 on one million observations.

### 5.2. Logistic Regression

For the largest data set, NS and our sequential sampler produced estimates which differed by $3 \times 10^{-4}N$ (roughly 0.7%), which is negligible. Our sequential sampler was almost a factor 17 faster than the nested sampler on one million observations for this model.

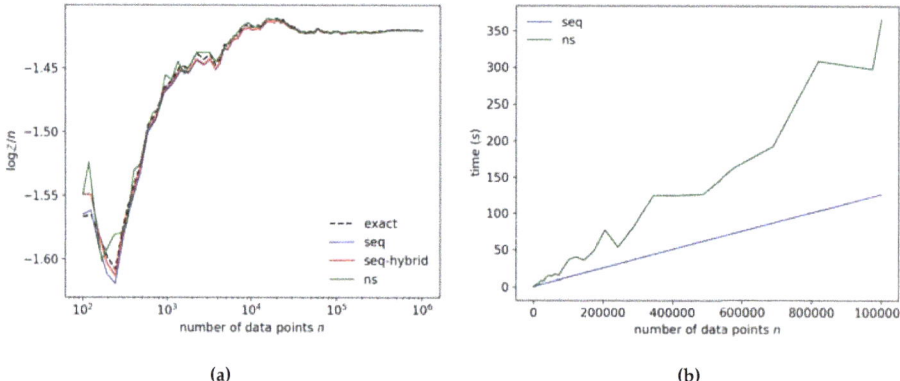

**Figure 1.** Linear regression model. (**a**) shows the accuracy of the sequential ML estimator compared to nested sampling and the exact ML and (**b**) shows the run time of both methods.

*5.3. Gaussian Mixture Model*

The posterior distribution for this model is multimodal. Some modes are due to permutation symmetries; these modes do not have to be explored since each one contains the same information. There are also some local modes which do not necessarily capture meaningful information about the data; for example, fitting a single Gaussian to the whole data set may be a poor local optimum of the likelihood function. If an MCMC walker finds one of these modes it can get trapped. However, we find that by Bayesian updating, the MCMC walkers tend to leave the poor local modes early on, before they become extremely peaked. This is similar to how annealing can help prevent MCMC and optimization algorithms from getting trapped in poor local optima. The estimates produced by NS and our sequential sampler differed on the largest data set by $2 \times 10^{-3} N$ (roughly 0.06%). For this model our sequential sampler was about a factor 11 faster than the nested sampler on one million observations.

In all the experiments our sequential sampler seems to converge to the same result as NS within a negligible error for large $n$. The initial disagreement between NS and our sequential sampler on the first few thousand data points, seen in Figures 2a and 3a, seems as if it can be attributed to the initial terms in Equation (3), since the proposed hybrid approach, replacing early terms in the sequential sampler by estimates based on NS, matches NS closely for all data set sizes.

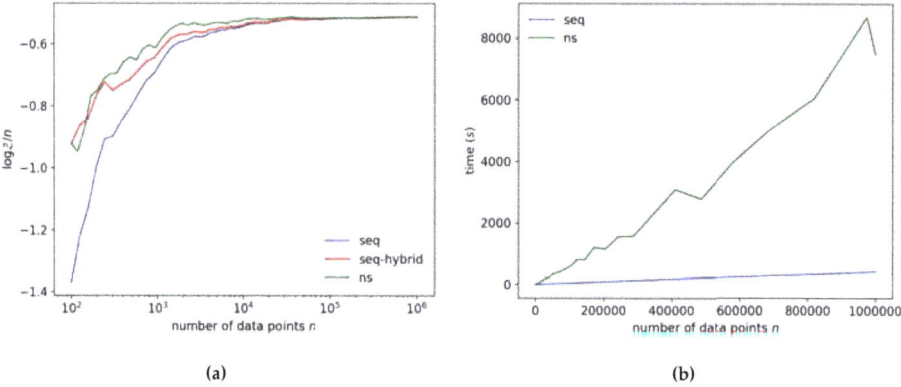

**Figure 2.** Logistic regression model. (**a**) shows the sequential ML estimator compared to nested sampling and (**b**) shows the run time of both methods.

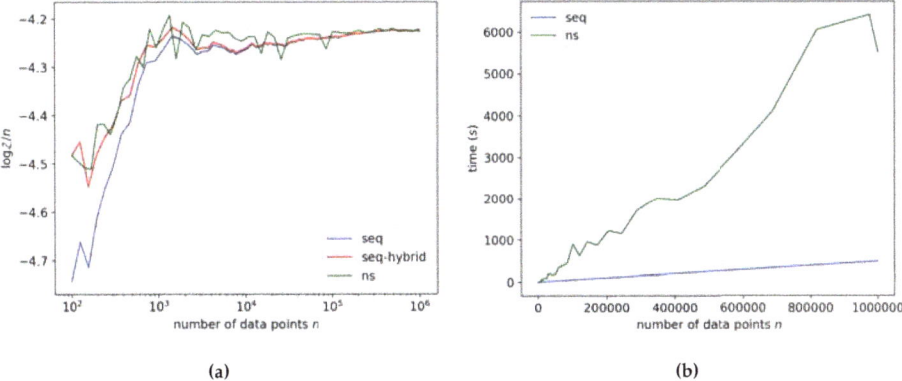

**Figure 3.** Gaussian mixture model. (**a**) shows the sequential ML estimator compared to nested sampling and (**b**) shows the run time of both methods.

## 6. Materials and Methods

Code for this work was implemented using pytorch [15]. The code for our experiments is available at https://gitlab.com/pleased/sequential-evidence.

## 7. Conclusions

We found that our sequential sampler using stochastic gradients was able to produce accurate ML estimates with a speedup over NS. Furthermore, since the marginal cost of updating the ML estimates when new data arrives does not depend on the number of previous data points, this approach may be effective for weighted model averaging in a setting where one periodically gets access to new data, such as in streaming applications. Potential future work may involve a more in-depth analysis of the algorithm parameters, for example automatic tuning of the number of samples, $M$, used for approximating predictive probabilities. One can imagine further exploring the effects of stochastic gradients for ML calculation in the full SMC setting, including latent variable models and resampling steps, with more general stochastic gradient MCMC algorithms such as those in [13].

**Author Contributions:** Conceptualization, methodology, software, formal analysis, investigation, visualization, writing–original draft preparation: S.A.C.; validation, writing–review and editing: S.A.C., H.C.E. and S.K.; resources: S.K.; project administration, supervision: H.C.E. and S.K.; funding acquisition: H.C.E.

**Funding:** S. Cameron received bursary support from the South African National Institute of Theoretical Physics, as well as financial support from both the organizers of MaxEnt 2019 and the National Institute of Theoretical Physics in attending the conference.

**Acknowledgments:** We wish to thank the organizers of MaxEnt 2019 and the South African National Institute of Theoretical Physics for financial support in attending the conference.

**Conflicts of Interest:** The authors declare no conflict of interest.

## Abbreviations

The following abbreviations are used in this manuscript:

| | |
|---|---|
| AIS | Annealed importance sampling |
| MCMC | Markov chain Monte Carlo |
| ML | Marginal likelihood |
| NS | Nested sampling |
| SGHMC | Stochastic gradient Hamiltonian Monte Carlo |

## Appendix A

Nested sampling requires one to sample from the prior under increasing likelihood constraints. Sampling under constraints is, in general, a difficult problem. This sampling can be accomplished by repeatedly sampling from the prior until the constraint is satisfied. This method of rejection sampling scales extremely poorly with both the size of the data set and the dimension of the parameter space, and is not feasible on any but the smallest problems. However NS using this rejection sampling method is a theoretically perfect implementation of NS, as it satisfies the assumptions of perfect i.i.d. sampling upon which NS is based. Further more it is extremely easy to implement correctly and so is a useful tool for testing the correctness of other NS implementation.

Our implementation of NS is based on SGHMC, simply because the code for SGHMC was already written. In order to sample under constraints we use a similar strategy to Galilean Monte Carlo as described in [16], where the particle reflects off of the boundary of the likelihood contour by updating the momentum as follows:

$$\Delta \mathbf{v} = -2\mathbf{n}(\mathbf{v} \cdot \mathbf{n}).$$

Here $\mathbf{n}$ is a unit vector parallel to the likelihood gradient $\nabla_\theta p(\mathcal{D}|\theta)$. While it is not the focus of our paper, we note that we have not previously encountered the idea of applying SGHMC to sampling under constraints; our implementation allows a fairly direct approach to implementing NS in a variety of contexts, and the technique may also potentially be of value in other constrained sampling contexts.

We found that, due to discretization error, the constrained sampler tends to undersample slightly at the boundaries of the constrained region; however the undersampled volume is of the order of the learning rate $\eta$ and so can be neglected if a small enough learning rate is used.

We tested the correctness of our constrained sampler against NS with rejection sampling for up to 250 observations. We found that the slight undersampling at the constraint boundaries tends to make the NS estimate slightly higher that that of the rejection sampler—see Figure A1—but the difference was within the acceptable range of $0.1N$, and appears to be decreasing with the number of observations.

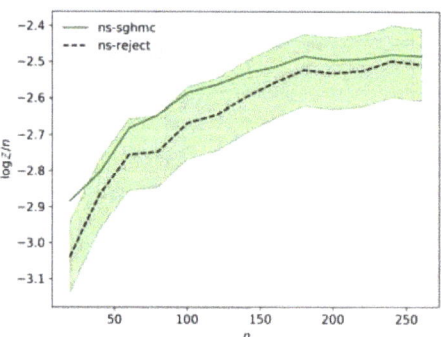

**Figure A1.** Comparison of NS with SGHMC to NS with rejection sampling for a 1-dimensional Gaussian mixture model with 9 parameters (8 degrees of freedom). The size of the shaded band is $\pm 0.1N$ around the rejection sampling implementation.

## References

1. Barber, D. *Bayesian Reasoning and Machine Learning*; Cambridge University Press: New York, NY, USA, 2012.
2. Skilling, J. Nested sampling for general Bayesian computation. *Bayesian Anal.* **2006**, *1*, 833–859, doi:10.1214/06-BA127.
3. Neal, R.M. Annealed Importance Sampling. *arXiv* **1998**, arXiv:physics/9803008.

4. Welling, M.; Teh, Y.W. Bayesian Learning via Stochastic Gradient Langevin Dynamics. In Proceedings of the 28th International Conference on Machine Learning, ICML 2011, Bellevue, WA, USA, 28 June–2 July 2011; Getoor, L., Scheffer, T., Eds.; Omnipress: Madison, WI, USA, 2011; pp. 681–688.
5. Naesseth, C.A.; Lindsten, F.; Schön, T.B. Elements of Sequential Monte Carlo. *arXiv* **2019**, arXiv:1903.04797.
6. Gordon, N.J.; Salmond, D.J.; Smith, A.F.M. Novel approach to nonlinear/non-Gaussian Bayesian state estimation. *IEE Proc. F Radar Signal Process.* **1993**, *140*, 107–113, doi:10.1049/ip-f-2.1993.0015.
7. Wallach, H.M.; Murray, I.; Salakhutdinov, R.; Mimno, D. Evaluation Methods for Topic Models. In Proceedings of the 26th Annual International Conference on Machine Learning, Montreal, QC, Canada, 14–18 June 2009; ACM: New York, NY, USA, 2009; pp. 1105–1112, doi:10.1145/1553374.1553515.
8. Grosse, R.B.; Ghahramani, Z.; Adams, R.P. Sandwiching the marginal likelihood using bidirectional Monte Carlo. *arXiv* **2015**, arXiv:1511.02543.
9. Chen, T.; Fox, E.; Guestrin, C. Stochastic Gradient Hamiltonian Monte Carlo. In Proceedings of the 31st International Conference on Machine Learning, Beijing, China, 21–26 June 2014; Volume 5.
10. Durrett, R. *Stochastic Calculus: A Practical Introduction*; Probability and Stochastics Series; CRC Press: Boca Raton, FL, USA, 1996; pp. 177–207
11. Gardiner, C.W. *Handbook of Stochastic Methods for Physics, Chemistry and the Natural Sciences*, 3rd ed.; Volume 13, Springer Series in Synergetics; Springer: Berlin, Germany, 2004; p. xviii+415.
12. Zhang, R.; Li, C.; Zhang, J.; Chen, C.; Wilson, A.G. Cyclical Stochastic Gradient MCMC for Bayesian Deep Learning. *arXiv* **2019**, arXiv:1902.03932.
13. Ma, Y.A.; Chen, T.; Fox, E.B. A Complete Recipe for Stochastic Gradient MCMC. *arXiv* **2015**, arXiv:1506.04696.
14. Springenberg, J.T.; Klein, A.; Falkner, S.; Hutter, F. Bayesian Optimization with Robust Bayesian Neural Networks. In *Advances in Neural Information Processing Systems 29*; Lee, D.D., Sugiyama, M., Luxburg, U.V., Guyon, I., Garnett, R., Eds.; Curran Associates, Inc.: Barcelona, Spain, 2016; pp. 4134–4142.
15. Paszke, A.; Gross, S.; Chintala, S.; Chanan, G.; Yang, E.; DeVito, Z.; Lin, Z.; Desmaison, A.; Antiga, L.; Lerer, A. Automatic Differentiation in PyTorch. Available online: https://openreview.net/pdJsrmfCZ (accessed on 24 June 2019)
16. Skilling, J. Bayesian Computation in Big Spaces-Nested Sampling and Galilean Monte Carlo. *AIP Conf. Proc.* **2012**, *1443*, 145–156, doi:10.1063/1.3703630.

© 2019 by the authors. Licensee MDPI, Basel, Switzerland. This article is an open access article distributed under the terms and conditions of the Creative Commons Attribution (CC BY) license (http://creativecommons.org/licenses/by/4.0/).

*Proceedings*

# Quantum Trajectories in Entropic Dynamics [†]

Nicholas Carrara

Department of Physics Graduate Students, University at Albany, Albany, NY 12222, USA; ncarrara@albany.edu
† Presented at the 39th International Workshop on Bayesian Inference and Maximum Entropy Methods in Science and Engineering, Garching, Germany, 30 June–5 July 2019.

Published: 13 December 2019

**Abstract:** Entropic Dynamics is a framework for deriving the laws of physics from entropic inference. In an (ED) of particles, the central assumption is that particles have definite yet unknown positions. By appealing to certain symmetries, one can derive a quantum mechanics of scalar particles and particles with spin, in which the trajectories of the particles are given by a stochastic equation. This is much like Nelson's stochastic mechanics which also assumes a fluctuating particle as the basis of the microstates. The uniqueness of ED as an entropic inference of particles allows one to continuously transition between fluctuating particles and the smooth trajectories assumed in Bohmian mechanics. In this work we explore the consequences of the ED framework by studying the trajectories of particles in the continuum between stochastic and Bohmian limits in the context of a few physical examples, which include the double slit and Stern-Gerlach experiments.

## 1. Introduction

Entropic Dynamics (ED) [1] is a unique approach to foundational quantum mechanics with its emphasis on entropic inference. It is argued, simply, that physics cannot be an exception to the rules of inductive reasoning; physics is constrained to be consistent with the rules for inference. (ED) is an exercise in deriving physical laws from inductive inference. The main assumption in (ED) is that particles have definite yet unknown positions, and that these positions determine entirely the *ontic* elements of the theory (One might argue for assuming that momentum, in place of position, is the ontic quantity, however as we will see the momentum in (ED) is not necessarily defined, except for certain classes of constraints. Momentum is also dependent on position, it is a coordinate in a derived manifold from the manifold of positions and so is not as fundamental). All other *observable* quantities, such as momentum, spin, electric charge, etc., are necessarily epistemic. This is a slight departure from the Copenhagen interpretation, which claims that particles have *no* properties until they are measured. Other foundational approaches, such as the Bohmian [2] (or causal interpretation) and Nelson's stochastic mechanics [3], also assume ontic positions for particles. These approaches however also give onticity to the macroscopic variables, such as the wave function $\psi(x)$, and the probability distribution $\rho(x) = |\psi(x)|^2$. In (ED) the macroscopic variables are also necessarily epistemic.

In the causal approach particles are assumed to follow smooth trajectories whose velocities are determined by the probability flow [4]. In this way it is a deterministic theory with respect to particle positions; given initial conditions the trajectory of the particle is known exactly. The uncertainty in positions can therefore only be blamed on not knowing the initial conditions, or not knowing the proper Hamiltonian. While entirely consistent with quantum mechanics, it is impossible to determine whether the Bohmian interpretation is general enough with respect to particle trajectories, since we cannot set up experiments in which the Hamiltonian is known exactly. Therefore, fluctuations can always be blamed on a lack of this information and not, necessarily, on some sub-quantum effect that (BM) has failed to include. Nelson's stochastic mechanics (NSM) is more general in this regard, since it begins with a stochastic equation for the motion of particles and proceeds to derive the dynamics of the macroscopic variables from these assumptions. While in this way (NSM) is more general than (BM), it singles out a particular sub-quantum dynamics for particles which is a Brownian motion. Much like the causal picture, (NSM) gives ontic privilege to the macroscopic variables which is part of the reason for its downfall [5]. While (NSM) can obtain the Bohmian limit, simply by sending the fluctuations to zero, they cannot necessarily motivate the generalized dynamics offered by (ED)

Entropic Dynamics allows for a more generalized sub-quantum dynamics which includes the (NSM) and (BM) limits as special cases. Particle trajectories are derived from the principle of maximum entropy by incorporating uncertainty in their motion for small steps $\Delta t$. Once we specify the relevant constraints in the problem, we can find the transition probability for these small steps $P(x'|x)$. The Lagrange multipliers, or equivalently the constraints, provides a freedom to specify the sub-quantum dynamics. The family of possible sub-quantum dynamics which reproduce the Schrödinger equation is potentially infinite, however experiments may constrain these theories once a proper understanding of quantum gravity is achieved.

## 2. Entropic Dynamics

In any application of entropic inference, we must supply three pieces of information. The first of these is the subject matter, the microstates, which is discussed in the next section. We will then have to supply a prior and any relevant constraints for the problem.

### 2.1. The Microstates

In general treatments of (ED) we consider the positions, $x \in \mathbf{X}$, of $N$ particles in configuration space, $\mathbf{X} = \mathbf{X}_1 \times \cdots \times \mathbf{X}_N$ which are definite yet *unknown*. Their unknown values are quantified by a probability density $\rho(x)$. We also make another assumption, that the particles follow continuous trajectories; the particles move in short steps [1]. The inference framework allows us to find a large change by iterating over many small steps, and thus we only need to find the transition probability for a short step. The principle of maximum entropy tells us that such a probability should maximize the relative entropy,

$$S[P,Q] = -\int dx' \, P(x'|x) \log \frac{P(x'|x)}{Q(x'|x)} \tag{1}$$

subject to constraints, however first we must specify the prior $Q(x'|x)$.

### 2.2. The Prior

To incorporate our ignorance about the motion of the particles, we can choose a prior that includes the symmetries in the problem. For $N$ particles, such a prior is a Gaussian,

$$Q(x'|x) \propto \exp\left[-\frac{1}{2}\sum_n \alpha_n \delta_{ab} \Delta x_n^a \Delta x_n^b\right] \tag{2}$$

where $a = \{1,2,3\}$ are spatial coordinates, $n = \{1,\ldots,N\}$ denotes the $n$th particle and where $\alpha_n$ is some particle dependent constant for which we can take the limit $\alpha_n \to \infty$ to impose short steps. Such a prior quantifies the rotational symmetries present in the problem. In order to break the symmetry, we impose a family of constraints.

### 2.3. The Constraints

Depending on the subject matter, we impose a family of different constraints that incorporate the information that is relevant to the problem. There are two main classes of problems that we will discuss here, although such a list is not exhaustive. The first concerns *scalar particles*, or particles without spin, while the second concerns particles *with* spin, and hence the second kind requires additional constraints. Special cases of either approach concern the study of a single particle [6], which we will mainly focus on in this paper. The microstates for a single particle is simply three dimensional space, $\mathbf{X} \subset \mathbb{R}^3$.

The Local $U(1)$ Constraint-

The constraints for particles, whether of the scalar or spin variety, also incorporate symmetry information (much like the prior). The main symmetry group for scalar particles is $U(1)$, the unitary group in infinite dimensions. This corresponds to a local gauge symmetry at each position $x \in X$ and is represented by the following constraint,

$$\langle \Delta x^a \rangle [\partial_a \phi(x) - \beta A_a(x)] = \kappa(x) \tag{3}$$

where $\phi(x)$ is a field that has the topological properties of an angle (Identifying $\phi(x)$ as an angle field may seem strange, but there is a deeper reason for this. It will become clear once one introduces spin into the picture [6], that angle fields are a natural set of constraints for describing rotations), $\vec{A}$ is a connection field that sets the zero of $\phi(x)$ at each $x$ and $\kappa(x)$ is some position dependent constant. The factor $\beta$ is identified with electric charge [7].

The $SU(2)$ Constraint -

In order to capture the appropriate rotational properties of the system, we incorporate an additional set of constraints on the motion of the particle (The results of this section are from joint work with A. Caticha that will appear in [6]). A useful representation of rotations in $\mathbb{R}^3$ is a frame field $\vec{s}_k(x)$ at each point in space, the dynamics of which will be coupled to the particle motion (The use of frame fields for describing spin has been used throughout the literature [8–11]). Just like the fields $\phi(x)$ and $\vec{A}(x)$, the field $\vec{s}_k(x)$ is entirely epistemic; it is merely a convenient representation of our information about the motion of the particle, there is no assumption that the field $\vec{s}_k(x)$ is "real".

A frame $\vec{s}_k(x)$ at a point $x \in X$ is a triad, $\vec{s}_k(x) = \{\vec{s}_1(x), \vec{s}_2(x), \vec{s}_3(x)\}$, whose individual components span $\mathbb{R}^3$. Each frame at $x \in X$ can be constructed by rotating the lab frame, which we denote with the basis vectors $\vec{e}_k = \{\vec{e}_1, \vec{e}_2, \vec{e}_3\}$, through three Euler angles $\{\chi(x), \theta(x), \varphi(x)\}$ which depend on position. This is performed through the action of a rotor $U(\chi, \theta, \varphi)$,

$$\vec{s}_k(x) = U(x)\vec{e}_k U^\dagger(x) = U(\chi, \theta, \varphi)\vec{e}_k U^\dagger(\chi, \theta, \varphi) \tag{4}$$

where

$$U(\chi, \theta, \varphi) = U_z(\varphi) U_y(\theta) U_z(\chi) = e^{-i\vec{e}_3 \varphi/2} e^{-i\vec{e}_2 \theta/2} e^{-i\vec{e}_3 \chi/2} \tag{5}$$

The frame is said to be oriented along the $\vec{s}_3$ direction with constant magnitude; i.e. $\vec{s}(x) = |\vec{s}|\vec{s}_3(x)$ and $|\vec{s}(x)| = |\vec{s}| = $ const.. Since the frames can take arbitrary orientation at each $x$, we would like to know how the frame changes its direction from $x$ to $x'$. In the same way that the constraint (3) involves the displacement being directed along the gradient of an angle, we incorporate the spin by also coupling the displacement to the gradient of an angle $\zeta_3(x)$

$$\langle \Delta x^a \rangle \partial_a \zeta_3 = \langle \Delta x^a \rangle (\vec{\omega}_a \cdot \vec{s}_3) = \langle \Delta x^a \rangle (\partial_a \chi + \cos\theta \partial_a \varphi) = \kappa'(x) \tag{6}$$

which is a combination of gradients along the polar angle $\chi(x)$ and the precession angle $\varphi(x)$ (While the derivatives of the angles $\zeta_k(x)$ are well defined, their solutions are in general not integrable) given in the frame velocity $\vec{\omega}_a$. Since the motion of the particle is being directed along the $\vec{s}_3$ direction, there is an arbitrariness in the setting of the zero angle of the $\chi(x)$. This suggests that the $\chi(x)$ in the spin constraint (6) plays the same role of a gauge field as the constraint for scalar particles (3), and we will see that it is only their joint dynamics that contributes to the evolution of the system. Thus in cases of a single particle with spin, the constraints (3) and (6) can be combined into a single constraint which is gauge invariant.

## 2.4. The Transition Probability

Maximizing the relative entropy (1) subject to the constraints (3) and (6) leads to the transition probability

$$P(x'|x) \propto \exp\left[-\frac{\alpha}{2}\delta_{ab}\Delta x^a \Delta x^b + \left(\alpha'(\partial_a \phi - \beta A_a) + \gamma(\vec{\omega}_a \cdot \vec{s}_3)\right)\Delta x^a\right] \quad (7)$$

with Lagrange multipliers $\alpha'$ and $\gamma$. This distribution is Gaussian, and a generic displacement $\Delta x^a$ can be written

$$\Delta x^a = \langle \Delta x^a \rangle + \Delta w^a \quad (8)$$

where the expected displacement $\langle \Delta x^a \rangle$ is given by

$$\langle \Delta x^a \rangle = \frac{1}{\alpha}\delta^{ab}\left(\alpha'(\partial_b \phi - \beta A_b) + \gamma(\vec{\omega}_b \cdot \vec{s}_3)\right) \quad (9)$$

and the fluctuations obey

$$\langle \Delta w^a \rangle = 0 \quad \text{and} \quad \langle \Delta w^a \Delta w^b \rangle = \frac{1}{\alpha}\delta^{ab} \quad (10)$$

The Lagrange multiplier $\alpha'$ plays the role of controlling the relative strength of the fluctuations [12]. In the theory of spin the value of $\gamma = 1/2$, while $\beta = e/c$ is proportional to the electric charge (The (ED) framework offers a unique argument for the quantization of electric charge which is a consequence of the circulation conditions of the spin frame $\vec{s}(x)$ and the single-valuedness of the wave function [6,7]). An important quantity is the ratio of the Lagrange multipliers,

$$\frac{\alpha'}{\alpha} \propto \frac{\gamma}{\alpha} \propto \frac{\hbar}{m}\Delta t \quad (11)$$

The form of the Lagrange multipliers determines a *class* of motions,

$$\alpha' = \frac{1}{\eta(\Delta t)^n}, \quad \gamma = \frac{1}{\xi(\Delta t)^n} \quad \text{and}$$
$$\langle \delta_{ab}\Delta w^a \Delta w^b \rangle = \frac{\hbar}{m}\eta(\Delta t)^{n+1}, \quad |\Delta w| \propto \sqrt{\frac{\hbar}{m}}\eta(\Delta t)^{(n+1)/2} \quad (12)$$

for some integer $n$ and constants $\eta$ and $\xi$, which control the relative strength of the constraints to the fluctuations. For $n = 0$, the particles follow Brownian trajectories, which in the limit of $\eta \to 0$ and $\xi \to 0$ recovers the smooth Bohmian trajectories [12].

## 2.5. Entropic Time

At this point Entropic Dynamics describes a theory of particles which undergo a particular class of motion depending on the choice of constraints (3) and (6). The next step is to define an *entropic time* [13] by associating to the equation,

$$\rho(x') = \int dx\, P(x'|x)\rho(x) \quad (13)$$

a notion of *duration*, supplied by the fluctuations ($\Delta t$). The distribution $\rho(x)$ *becomes* the distribution $\rho(x')$, and the procedure in (13) has an implicit direction as demonstrated by Bayes' rule (The update provided by marginalizing over the transition probability $P(x'|x)$ is not necessarily symmetric. Updating in reverse is constrained by Bayes' rule, $P(x|x') = P(x)P(x'|x)/P(x')$). Much in the way that time is defined in classical mechanics by the free particle—the free particle moves equal distances in equal time—entropic time is defined by the free quantum particle; the free quantum particle undergoes equal fluctuations in equal entropic time.

It's often easier to work with the differential form of the integral in Equation (13), which can be found to be,

$$\partial_t \rho = -\partial_a(v^a \rho) \quad (14)$$

where the current velocity $v^a$ depends on the class of motion determined by the constraints and $(\eta, n)$. For Brownian trajectories, $n = 0$, the velocity is,

$$v^a = \frac{\hbar}{m}\delta^{ab}\left(\alpha'(\partial_b\phi - \beta A_b) + \gamma(\vec{\omega}_b \cdot \vec{s}_3) - \partial_b \log \rho^{1/2}\right) \quad (15)$$

and Equation (14) is the Fokker-Plank equation, which includes the appearance of the osmotic term $\log \rho^{1/2}$. For the smoother trajectories, $n = 1$, the osmotic term disappears and the velocity becomes $v^a = \langle \Delta x^a \rangle / \Delta t$. The Equation (14) is a diffusion equation which can be rewritten as a functional derivative, $\partial_t \rho = \delta \tilde{H} / \delta \Phi$ for some functional $\tilde{H}$ which we eventually identify as a Hamiltonian.

From here the discussion extends to the symplectic and information geometry of the statistical manifold $\Delta$ from which we can derive a Hamiltonian and Hamilton's equations by identifying certain symmetries [1]. These symmetries form the group $Sp(2n) \cap O(2n) = U(n)$ which are the intersection of the symplectic group and orthogonal group in $(2n)$ dimensions, of which the constraints (3) and (6) are a subset. The group $U(n)$ also leads to another important consequence, the appearance of a complex structure. The complex structure allows one to use complex coordinates, which we identify as wave functions,

$$\psi_\pm = \rho_\pm^{1/2} e^{i\Phi_\pm/\hbar} \quad \text{and} \quad i\hbar\psi_\pm^\dagger = i\hbar \rho_\pm^{1/2} e^{-i\Phi_\pm/\hbar} \quad (16)$$

where $\rho_\pm = (1/2)(\rho \pm \rho_s)$ and $\Phi_\pm = (\Phi \pm \Phi_s)$. The conjugate momenta to $\rho(x)$ ends up being the phase [7], which for the Brownian case is $\Phi(x)/\hbar = \gamma \chi + \alpha' \phi - \log \rho^{1/2}$, and where $\rho_s = \rho \cos \theta$ and $\Phi_s / \hbar = \gamma \varphi$ are conjugate variables incorporating the extra spin degrees of freedom. The Hamiltonian is,

$$\tilde{H}[\psi_\pm, i\hbar\psi_\pm^*] = \int dx \left(-\frac{\hbar^2}{2m}\psi_\pm^*(\partial_a - (i/2)A_a)^2 \psi_\pm + \psi_\pm^* V \psi_\pm + \psi_\pm^*(B_x \mp iB_y)\psi_\mp \pm \psi_\pm^* B_z \psi_\pm \right) \quad (17)$$

and the associated Hamilton's equation,

$$i\hbar \partial_t \psi_\pm = \frac{\delta \tilde{H}}{\delta \psi_\pm^*} = -\frac{\hbar^2}{2m}(\partial_a - (i/2))^2 \psi_\pm + V\psi_\pm + (B_x \mp iB_y)\psi_\mp \pm B_z \psi_\pm \quad (18)$$

is the Schrödinger equation for the ($\pm$) components of the Pauli equation. In the limit that the variables $\theta, \varphi$ are not dynamical, the Pauli equation reduces to the Schrödinger equation for a scalar particle. While we will not go into further detail on these aspects of (ED), for a more detailed discussion see [1,6].

## 3. Entropic Trajectories

Entropic trajectories are a generalization of the trajectories assumed by (BM) and (NSM). In Bohmian mechanics these trajectories are smooth, with well defined velocities, that are also constrained to never cross. They are determined from the probability flow, which for scalar particles is given by

$$\frac{d\vec{x}_n}{dt} = \vec{v}_n, \quad \text{where} \quad \vec{v}_n = i\frac{\hbar}{2m_n}\left(\frac{\vec{\nabla}_n \psi}{\psi} - \frac{\vec{\nabla}_n \psi^*}{\psi^*}\right) \quad (19)$$

where $\vec{x}_n$ is the position of the $n$th-particle. The Bohmian velocity $\vec{v}_n$ is equivalent to the drift velocity $\vec{b}$ in (ED). In (NSM) the equation of motion for the particles is given by the stochastic equation,

$$d\vec{x} = b(\vec{x}, t)dt + dw(t) \quad (20)$$

The velocity from (19) is not defined in (20) in the standard limit calculus sense and hence one can only evaluate finite differences. In Entropic Dynamics, it is the displacement (8) which determines the motion of the particles. The displacement contains the fluctuation term, which is stochastic, hence the limit in (19) is not always defined. While one can evaluate the limit using stochastic calculus, we will relegate that discussion to a future paper. For the collection of simulations in this paper, we will simply use a unit fluctuation $\Delta \tilde{w}$ in place of the Wiener process $\Delta \bar{w}$, which simulates a random walk on the unit sphere. A finite time step is simulated by providing a duration $\Delta t$ and some prescribed values of $n$ and $\eta$. The displacement for Brownian motion is then found from,

$$\Delta x^a = b^a \Delta t + \sqrt{\frac{\hbar}{m} \eta} (\Delta t)^{1/2} \Delta \tilde{w}^a \tag{21}$$

where $b^a$ is given from the Bohmian limit. In the examples below, Equation (21) is integrated using the standard 4th-order Runge Kutta method.

### 3.1. The Double-Slit Experiment

The double slit experiment (DS) [14,15] is a special case where the wave function can be solved exactly by assuming that each slit produces a Gaussian wave packet with a width equal to the width of the slit, $\sigma_0$, and that the total wave function is represented by a super-position of each packet,

$$\psi_i(x,y,t) = (2\pi\sigma^2)^{-1/4} \exp\left[ -\frac{(y_i - d - \hbar k_y t/m)^2}{4\sigma\sigma_0} + i\left\{ \left(k_y(y_i - d) - \frac{\hbar k_y^2 t}{2m}\right) + \left(k_x x - \frac{\hbar k_x^2 t}{2m}\right)\right\}\right] \tag{22}$$

where $\hbar k_y = mv_y$, $y_a = y$, $y_b = -y$, and $2d$ is the distance between the slits. The factor $\sigma$ is $\sigma = \sigma_0 \left(1 + \frac{i\hbar t}{2m\sigma_0^2}\right)$. Each wave function $\psi_i(x,y,t)$, is found from integrating the Schrödinger equation for a free particle, $i\hbar \partial_t \psi_i = -(\hbar^2/2m)\vec{\nabla}^2 \psi_i$ with an initial Gaussian wave function. The total wavefunction is the superposition, $\Psi(x,y,t) = N[\psi_a(x,y,t) + \psi_b(x,y,t)]$.

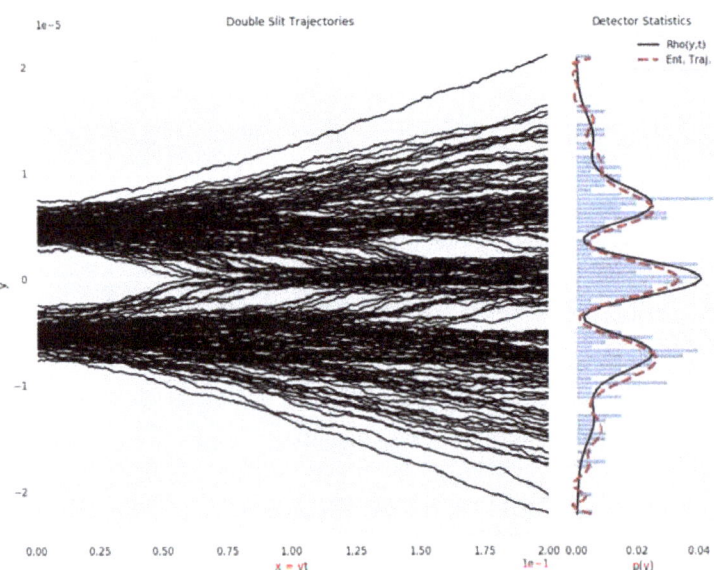

**Figure 1.** Cont.

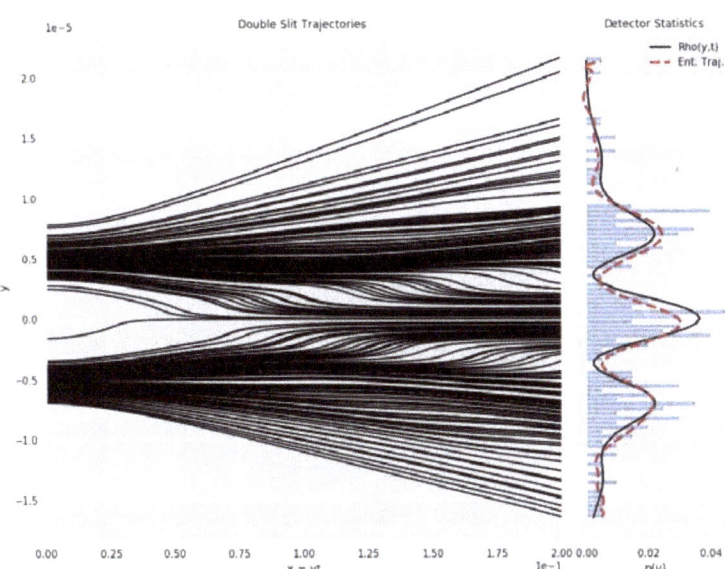

**Figure 1.** Entorpic trajectories for the double slit experiment with $n = 0$ and $\eta = 1,0$ for $N = 200$ particles. The black curve ($Rho(y,t)$) is the probability distribution determined from the wave function at the detector screen, while the red curve (Ent. Traj.) is the interpolated distribution from the detector statistics using a fitting polynomial of order 15 to show the shape of the distribution.

We simulated the trajectories of electrons, $m = m_e$, with an initial velocity in the $x$ direction of $2 \times 10^6$ m/s and random initial positions along the $y$ direction sampled according to the initial Gaussian distribution with standard deviation equal to the slit width, $\sigma_0 = 10^{-6}$ m and with distance between the slits $d = 5\sigma_0 = 5 \times 10^{-6}$ m. The initial velocity in the $y$ direction is set to zero and the distance to the screen is $x_f = 0.2$ m. One can see that the value of $\eta = 1$ generates fluctuations which give rise to similar statistics as the Bohmian limit (Figure 1).

### 3.2. The Stern-Gerlach Experiment

In a similar way to the Double-slit experiment (DS), we can solve the Pauli equation in the case of the Stern-Gerlach experiment (SG) [16] by making a few approximating assumptions [17]. Following the arguments in [4,18,19], we assume that the Stern-Gerlach magnet produces a magnetic field, $\vec{B} = (B_0 + zB_0')\hat{z}$, within a region $\Delta x$ and is assumed to be zero outside this region. Given an initial particle velocity $v_x$ along the $x$ direction, the particle remains in the magnetic field for a time $\Delta t = \Delta x / v_x$. After the particle leaves the magnetic field, the spinor wave function breaks up into two packets which can be solved for all $t$ (For detailed calculations see [15,17,20,21]) as,

$$\Psi(z, t + \Delta t) = (2\pi\sigma_0)^{-1/2} \begin{pmatrix} \cos\frac{\theta_0}{2} \exp\left[ -\frac{(z-\Delta_z-ut)^2}{4\sigma_0^2} + \frac{i}{\hbar}(muz + \hbar\varphi_+) \right] \\ \sin\frac{\theta_0}{2} \exp\left[ -\frac{(z+\Delta_z+ut)^2}{4\sigma_0^2} - \frac{i}{\hbar}(muz - \hbar\varphi_-) \right] \end{pmatrix} \qquad (23)$$

where $\theta_0$ is the initial azimuthal angle for $\vec{s}_3$ with respect to the $z$ axis, $u$ is the packet velocity in the $z$ direction, $\Delta_z = \mu_B B'_0 (\Delta t)^2 / 2m$ and $\varphi_\pm = \pm \varphi_0 / 2 \mp \mu_B B_0 \Delta t / \hbar - \mu_B^2 (B'_0)^2 (\Delta t)^3 / 6m\hbar$, where $\mu_B$ is the Bohr magneton. The width $\sigma_0$ of the initial packet is set to the (SG) device opening of $\sigma_0 = 10^{-4}$ m.

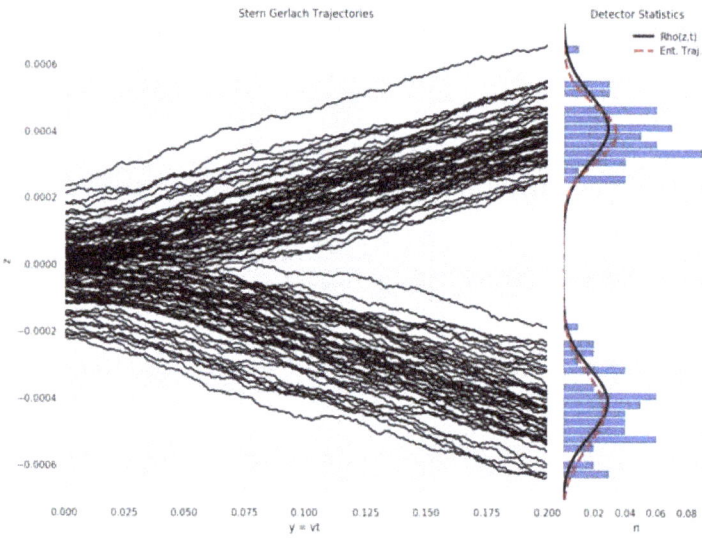

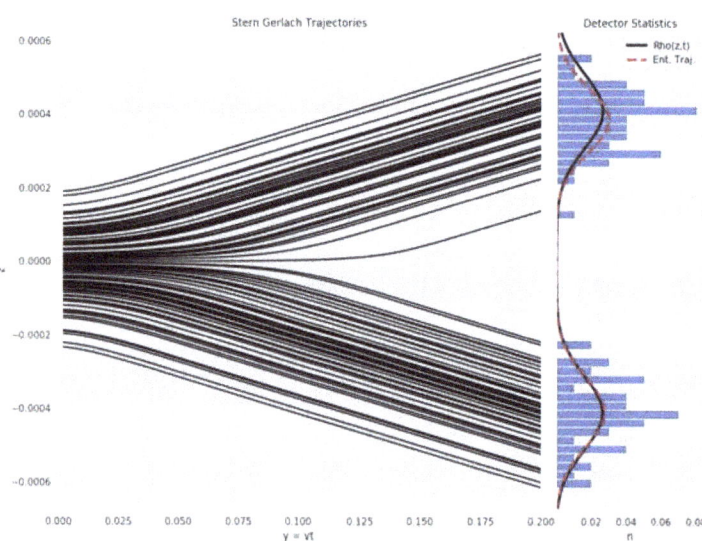

**Figure 2.** Entropic trajectories for the Stern-Gerlach experiment with $n = 0$ and $\eta = 0, 10^5$ for $N = 100$ particles.

We simulated the trajectories of silver atoms with mass $m \approx 1.8e - 25$ and an initial velocity along the $x$ direction of 500 m/s sampled according to the initial wave packet. The magnetic field parameters are set to $B_0 = 5$ T and $B'_0 = 10^3$ T/m. Assuming the particle remains in the magnetic field for a time $\Delta t = 2 \times 10^{-5}$ s, the factors $\Delta z = 10^{-5}$ m and $u = 1$ m/s.

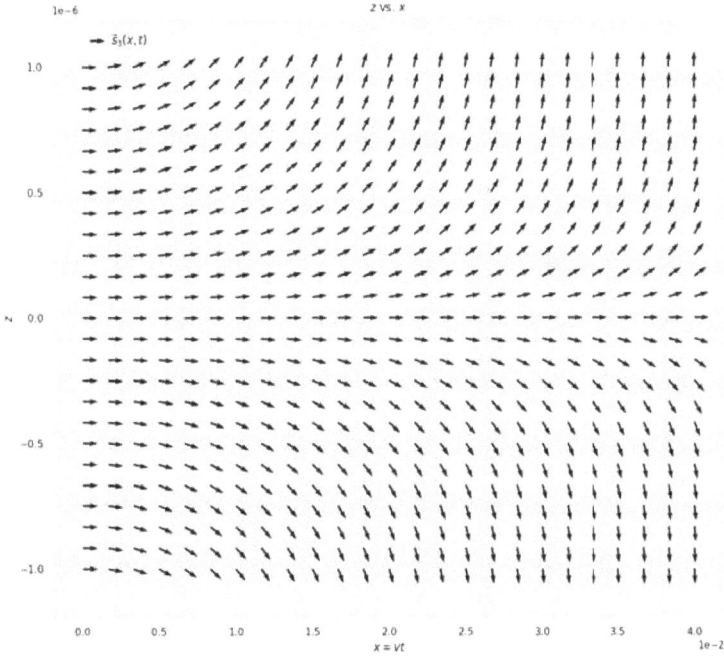

**Figure 3.** Starting with the initial condition $\theta_0 = \pi/2$, we show the evolution of the direction of the spin frame over $x$ and $t$ with respect to the xz-plane.

Unlike in the (DS) experiment, the fluctuations are suppressed in this example since the mass of silver is so much larger than electrons, hence the need for $\eta \propto 10^5$ before we start to see Brownian motion (Figure 2).

As we've stated in the introduction and throughout, the spin is entirely epistemic and is not assumed to be a *property* of the particle, but rather a property of its motion. Much like in [4], the above example shows how the epistemic spin frame evolves over space and time (Figure 3). The two-valuedness of spin measurements is not the same type of quantization that is attributed to the particle, but rather just a consequence of measurement, as can be seen from the trajectories. By measuring the particle up or down on the screen, we then assume that the spin must have been up or down at the magnet. From fig. 3 however, the up and down trajectories are *created* by the (SG) magnet and initially the spin is only up in the $x$ direction.

## 4. Discussion

The (ED) formalism allows for generalized particle trajectories which are not a priori realizable in other foundational approaches. This freedom is granted by (ED)'s foundation in entropic inference, which requires us to supply information about the symmetries in the problem through constraints. As we have seen, the Bohmian and Brownian limits are easily attainable, and both give consistent results with respect to experiment. It still remains an open question as to what classes of sub-quantum dynamics are allowable in quantum mechanics, and ultimately in quantum gravity. While at the moment we cannot answer the latter, we will address the former question in longer paper which will extrapolate further on the discussion from section three.

**Acknowledgments:** We would like to thank A. Caticha, J. Ernst, S. Ipek, P. Pessoa and K. Vanslette for insightful conversations.

## References

1. Caticha, A. Entropic Dynamics: Quantum Mechanics from Entropy and Information Geometry. *arXiv* **2017**, arXiv:1711.02538.
2. Bohm, D. A Suggested Interpretation of the Quantum Theory in Terms of 'Hidden Variables' I. *Phys. Rev.* **1952**, *85*, 166–179.
3. Nelson, E. *Quantum Fluctuations*; Princeton University Press: Princeton, NJ, USA, 1985.
4. C. Dewdney, P.H.; Kypianidis, A. What happens in a spin measurement? *Phys. Lett. A* **1986**, *119*, 259–267.
5. Wallstrom, T. Inequivalence between the Schr??dinger equation and the Madelung hydrodynamic equations. *Phys. Rev. A* **1994**, *49*, 1613.
6. Caticha, A.; Carrara, N. The Entropic Dynamics Of Spin. Forthcoming.
7. Carrara, N.; Caticha, A. Quantum phases in entropic dynamics. *arXiv* **2017**, arXiv:1708.08977.
8. Takabayasi, T. The Vector Representation of Spinning Particle in the Quantum Theory, I*. *Prog. Theor. Phys.* **1955**, *14*, 283–302. doi:10.1143/PTP.14.283.
9. Takabayasi, T. Vortex, Spin and Triad for Quantum Mechanics of Spinning Particle.I: General Theory. *Prog. Theor. Phys.* **1983**, *70*, 1–17. doi:10.1143/PTP.70.1.
10. Hestenes, D. Spin and Uncertainty in the Interpretation of Quantum Mechanics. *Am. J. Phys.* **1979**, *47*, 399–415.
11. Gurtler, R.; Hestenes, D. Consistency in the Formulation of the Dirac, Pauli and Schrödinger Theories. *J. Math. Phys.* **1975**, *16*, 573–584.
12. Bartolomeo, D.; Caticha, A. Trading drift and fluctuations in entropic dynamics: Quantum dynamics as an emergent universality class. *J. Phys. Conf. Ser.* **2016**, *701*, 012009. doi:10.1088/1742-6596/701/1/012009.
13. Caticha, A. Entropic Time. *Am. Inst. Phys. Conf. Ser.* **2011**, *1305*, 200–207. doi:10.1063/1.3573617.
14. Jönsson, C. Elektroneninterferenzen an mehreren k??nstlich hergestellten Feinspalten. *Z. Phys.* **1961**, *161*, 454–474.
15. Gondran, M.; Gondran, A. Numerical Simulation of the Double Slit Interference with Ultracold Atoms. *arXiv* **2007**, arXiv:0712.0841.
16. Gerlach, W.; Stern, O. Der Experimentelle Nachweisdes Magnetischen Moments des Silberatoms. *Z. Phys.* **1922**, *8*, 110–111.
17. Gondran, M.; Gondran, A. A complete analysis of the Stern-Gerlach experiment using Pauli spinors. *arXiv* **2005**, arXiv:quant-ph/0511276.
18. Bohm, D.; Schiller, R. A causal interpretation of the pauli equation (A). *Nuovo Cimento* **1955**, *1*, 48–66.
19. Bohm, D.; Schiller, R. A causal interpretation of the pauli equation (B). *Nuovo Cimento* **1955**, *1*, 467–491.
20. Gondran, M.; Gondran, A. Measurement in the de Broglie-Bohm interpretation: Double-slit, Stern-Gerlach and EPR-B. *arXiv* **2013**, arXiv:1309.4757.
21. Carrara, N. Entropic Trajectories. Forthcoming.

 © 2019 by the authors. Licensee MDPI, Basel, Switzerland. This article is an open access article distributed under the terms and conditions of the Creative Commons Attribution (CC BY) license (http://creativecommons.org/licenses/by/4.0/).

*Proceedings*

# On the Estimation of Mutual Information

Nicholas Carrara *[ID] and Jesse Ernst

Physics department, University at Albany, 1400 Washington Ave, Albany, NY 12222, USA; jae@albany.edu
* Correspondence: ncarrara@albany.edu
† Presented at the 39th International Workshop on Bayesian Inference and Maximum Entropy Methods in Science and Engineering, Garching, Germany, 30 June–5 July 2019.

Published: 15 January 2020

**Abstract:** In this paper we focus on the estimation of mutual information from finite samples ($\mathcal{X} \times \mathcal{Y}$). The main concern with estimations of mutual information (MI) is their robustness under the class of transformations for which it remains invariant: i.e., type I (coordinate transformations), III (marginalizations) and special cases of type IV (embeddings, products). Estimators which fail to meet these standards are not *robust* in their general applicability. Since most machine learning tasks employ transformations which belong to the classes referenced in part I, the mutual information can tell us which transformations are most optimal. There are several classes of estimation methods in the literature, such as non-parametric estimators like the one developed by Kraskov et al., and its improved versions. These estimators are extremely useful, since they rely only on the geometry of the underlying sample, and circumvent estimating the probability distribution itself. We explore the robustness of this family of estimators in the context of our design criteria.

**Keywords:** mutual information; non-parametric entropy estimation; dimension reduction; machine learning

## 1. Introduction

Interpretting mutual information (MI) as a measure of correlation has gained considerable attention over the past couple of decades for it's application in both machine learning [1–4] and in dimensionality reduction [5], although it has a rich history in Communication Theory, especially in applications of Rate-Distortion theory [6] (While MI has been exploited in these examples, it has only recently been derived from first principles [7]). MI has several useful properties, such as having a lower bound of $I_{\min} = 0$ for variables which are uncorrelated. Mostly we are interested in how the MI changes whenever we pass one set of variables through a function, which according to the data processing inequality can only ever destroy correlations and not increase them. Thus, the MI for any set of variables $\mathbf{X} \times \mathbf{Y}$ is an upper bound for any transformation $f(\mathbf{X}) \times g(\mathbf{Y})$. This feature makes MI a good measure of performance for any machine learning (ML) task, which is why has gained so much attention recently [1–4].

The main challenge with using MI in any inference task is computing it when one only has a sample $\mathcal{X} \times \mathcal{Y} \subset \mathbf{X} \times \mathbf{Y}$. The MI one estimates from the sample is highly dependent on the assumptions about the underlying joint distribution $p(x,y)$; effectively, one estimates MI by estimating the density $p(x,y)$. The most popular method for estimating MI is by using the class of non-parametric estimators built on the method derived by Kraskov et al. [8] (KSG). The KSG estimator uses local geometric information about the sample to approximate the density $p(x_i, y_i)$ at each point $(x_i, y_i)$ and then calculates a local estimate of MI from it. While this approach has been very successful, there are some weaknesses which we will discuss in this paper. Specifically, the use of local geometric information causes the estimator to not be coordinate invariant in general, which is a violation of the basic properties of MI. What's worse, is it's inability to see through useless information, i.e., noise. This is also a consequence of using local information without regard to the overall global structure of the space. When combined, these two problems cause unwanted behavior in even the simplest of situations.

There have been some improvements to KSG [9] which we will discuss. Most often studies of estimators are concerned with its effectiveness with small numbers of samples in large dimension, while here we will be mostly concerned with its robustness under coordinate transformations, redundancy and noise. We will define redundancy and noise more precisely in a later section, but one can also check [7] for a more rigorous definition. As was shown in [1] KSG handles redundant information well, which we will reiterate in a later section. It is KSG's inability to handle noise that diminishes it's effectiveness in real data sets. In the next section we will briefly discuss the basic properties of MI, and then discuss the ideas behind non-parametric entropy estimators. We will then examine the robustness of KSG and it's improvements in section IV. We end with a discussion.

## 2. Mutual Information

We will reiterate some the basic properties of mutual information. Consider two spaces of propositions, $\mathbf{X}$ and $\mathbf{Y}$, whose joint space is given by $\mathbf{X} \times \mathbf{Y}$ (The spaces $\mathbf{X}, \mathbf{Y}$ can be either discrete/categorical or continuous.). The *global* correlations present between the two spaces is determined from the mutual information,

$$I[\mathbf{X};\mathbf{Y}] = \int dxdy\, p(x,y) \log \frac{p(x,y)}{p(x)p(y)} \tag{1}$$

where $p(x,y)$ is the joint probability density and,

$$p(x) = \int dy\, p(x,y) = \int dy\, p(y)p(x|y) \quad \text{and} \quad p(y) = \int dx\, p(x,y) = \int dx\, p(x)p(y|x) \tag{2}$$

are the marginals. The product marginal $p(x)p(y)$ can be interpreted as an independent prior and the MI gives a *ranking* of joint distributions $p(x,y)$ according to their *amount* of correlation; joint distributions with *more* correlation have higher values of MI. The MI is bounded from below by $I_{\min} = 0$ whenever the spaces $\mathbf{X}$ and $\mathbf{Y}$ are uncorrelated; i.e., $p(x,y) = p(x)p(y)$, and is typically unbounded from above (except in cases of discrete distributions).

One immediate consequence of the functional form of (1) is its invariance under coordinate transformations. Since the probabilities,

$$p(x,y)dxdy = p(x',y')dx'dy' \tag{3}$$

are equivalent, then $I[\mathbf{X};\mathbf{Y}] = I[\mathbf{X}';\mathbf{Y}']$. While this fact is somewhat trivial on its own, when combined with other types of transformations it can be quite powerful. In this paper we will study three main types of transformations, the first being coordinate transformations. The second kind of transformation of interest is *marginalization*, and the third is *products*. Marginalization is simply the the projecting out of some variables, which according to the design criteria of MI [7], can only ever decrease the correlations present. On the other hand, products of spaces can increase correlations when the new variables provide new information. The most trivial type of product is an embedding.

One advantage of MI is its invariance under the inclusion of redundant information. For example, consider adding to the space $\mathbf{X}$ another space which is simply a function of $\mathbf{X}$, i.e., $\mathbf{X} \to \mathbf{X} \times f(\mathbf{X})$. The joint probability distribution becomes,

$$p(x,f(x),y) = p(x,f(x))p(y|x,f(x)) = p(x)\delta(f(x)-f)p(y|x) \tag{4}$$

and hence the MI is,

$$I[\mathbf{X} \times f(\mathbf{X}); \mathbf{Y}] = \int dxdydf\, p(x)\delta(f(x)-f)p(y|x) \log \frac{p(x)p(y|x)}{p(x)p(y)} = I[\mathbf{X}; \mathbf{Y}] \quad (5)$$

The map $\mathbf{X} \to \mathbf{X} \times f(\mathbf{X})$ is an embedding of $\mathbf{X}$ into a higher dimensional space. Such a transformation does not increase the intrinsic dimension of $\mathbf{X}$. Machine learning algorithms exploit this type of transformation in conjunction with coordinate transformations and marginalizations.

Much like in Equation (4), MI is also invariant under the addition of *noise*, which are defined as variables, $\mathbf{Z}$, that are uncorrelated to both $\mathbf{X}$ and $\mathbf{Y}$,

$$p(x,z,y) = p(x,z)p(y|x,z) = p(x)p(z)p(y|x) = p(z)p(x,y) \quad (6)$$

And like (5) the mutual information is invariant,

$$I[\mathbf{X} \times \mathbf{Z}; \mathbf{Y}] = I[\mathbf{X}; \mathbf{Y} \times \mathbf{Z}] = \int dxdydz\, p(z)p(x,y) \log \frac{p(x,y)}{p(x)p(y)} = I[\mathbf{X}; \mathbf{Y}] \quad (7)$$

Unlike with redundancy, noise variables necessarily increase the dimension of the underlying space.

## 3. Non-Parametric Estimation

Non parametric entropy estimators attempt to utilize the geometry of the underlying sample to estimate the *local* density and hence the *local* entropy. A popular estimator is the one developed by Kozachenko and Leonenko (KL) [10], which we will briefly motivate. Consider the task of estimating the entropy of a sample $\mathcal{X}$ from an underlying space $\mathbf{X}$. Our goal is to find an unbiased estimator of the form $\hat{H}[\mathcal{X}] = N^{-1} \sum_{i=1}^{N} \log p(x_i)$, which converges to the *true* Shannon entropy as $N \to \infty$ (We highlight the word true here, since the underlying probability distribution is not known and hence our *estimation* depends on our assumptions about its form.). To find an approximation of $\log p(x_i)$, consider the following probability distribution,

$$P_\varepsilon(x_i)d\varepsilon = \frac{(N_1)!}{1!(k-1)!(N-k-1)!} p_i^{k-1}(1-p_i)^{N-k-1} \frac{dp_i}{d\varepsilon} d\varepsilon \quad (8)$$

which is the probability that the *k*th-nearest neighbor of the point $x_i$ exists within the small spherical shell of radius $\varepsilon/2$ and that there are $k-1$ points at $r_i < \varepsilon/2$ and $N-k-1$ points at $r_i > \varepsilon/2 + d\varepsilon$. This distribution is of course properly normalized, and upon evaluating the expected value of the logarithm of $p_i$, we find,

$$\langle \log p_i \rangle = \int d\varepsilon\, P_\varepsilon(x_i) \log p_i = \psi(k) - \psi(N) \quad (9)$$

where $\psi(k)$ is the digamma function. From here one can determine an approximation for the logarithm of the *true* distribution by assuming something about the local behavior of $p(x_i)$ with respect to the probability mass $p_i$. In the KL approximation (and as well in the KSG approximation), it is assumed that the probability within the region defined by $p_i$ is uniform with respect to the *true* distribution at the point $x_i$,

$$p_i \approx c_d \varepsilon^d p(x_i) \quad (10)$$

where $d$ is the dimension of the space and $c_d$ is the volume of the unit $d$-ball (The form of $c_d$ depends on the choice of metric for the space $\mathbf{X}$. As we will see a useful choice for MI estimation is the $L^\infty$ norm.). Putting (10) into the unbiased estimator one arrives at,

$$\hat{H}[\mathcal{X}] = \psi(N) - \psi(k) + \log c_d + \frac{d}{N} \sum_{i=1}^{N} \log(\varepsilon_i) \quad (11)$$

## 3.1. The Vanilla KSG Estimator

The KSG estimator of the first kind is derived by taking the expression in (11) and applying it to the decomposition,

$$\hat{I}[\mathcal{X};\mathcal{Y}] = \hat{H}[\mathcal{X}] + \hat{H}[\mathcal{Y}] - \hat{H}[\mathcal{X},\mathcal{Y}] \quad (12)$$

where $\hat{H}[\mathcal{X},\mathcal{Y}]$ is the entropy over the joint distribution $p(x,y)$. As has been shown and argued by KSG, the approximation above is slightly naive since the local densities in the joint and marginal spaces can be different, leading to errors in the terms involving $\log(\varepsilon_i)$ which don't necessarily cancel. As a neat trick, KSG suggests using the same density found in the joint space in the marginal spaces, so that the factors $(d_x/N)\sum_{i=1}^{N}\log(\varepsilon_i^x)$,$(d_y/N)\sum_{i=1}^{N}\log(\varepsilon_i^y)$ and $((d_x+d_y)/N)\sum_{i=1}^{N}\log(\varepsilon_i^{xy})$ will cancel. Choosing the same density for fixed $k$ in the joint space causes the $k$ values in the marginal spaces to vary, and hence we arrive at the expression,

$$\hat{I}^1[\mathcal{X};\mathcal{Y}] = \psi(k) + \psi(N) - \langle\psi(n_x+1)\rangle - \langle\psi(n_y+1)\rangle \quad (13)$$

where $n_x$ and $n_y$ are the number of points which land in the $d_x$ and $d_y$ balls of radius $\varepsilon/2$ in the marginal spaces.

One unfortunate consequence of the KSG estimator of the first kind is its reliance on the $L^\infty$ norm for finding neighbors. As has been pointed out by others[9], such a choice can lead to problems in regions where the probability varies greatly, which can easily happen in spaces of large dimension. Unless the density of samples increases exponentially with respect to the dimension of the space, the errors in choosing $L^\infty$ will compound quickly.

## 3.2. The LNC Correction to KSG

As an attempted correction to KSG's problem with using the $L^\infty$ box, S. Gao et al. proposed the *local non-uniform correction* (LNC) technique. This technique adjusts the unbiased estimator for MI by replacing the $L^\infty$ volume in the joint space with a volume computed from a PCA analysis. The basic idea is the following. Consider a point $x_i$ whose $k$th-neighbor is $x_k$. With the collection of $k+1$ points including $x_i, x_k$ and all points closer than $x_k$, construct the correlation matrix $C_{ij}$ and find its eigenvectors. By then projecting each point along the maximal eigenvectors, we can find a PCA bounding box, which is rotated and skewed with respect to the $L^\infty$ box. The assumption in this case is that the rotated PCA box is a much better representation of the region of uniform probability around $x_i$. Once each volume is found, the MI is given by,

$$\hat{I}_{LNC} = \hat{I}_{KSG} - \frac{1}{N}\sum_{i=1}^{N}\log\frac{\tilde{V}_i}{V_i} \quad (14)$$

where $\tilde{V}_i$ is the PCA volume and $V_i$ is the $L^\infty$ volume. Such an estimator has shown to give vast improvement to the naive KSG method, however current results are limited to two dimensional problems. The reason for this is its inability to deal with redundant information. To see this, consider a two-dimensional problem in which the variables $\mathbf{X} \times \mathbf{Y}$ have some non-trivial correlations. If we add to $\mathbf{X}$ a redundant copy, $\mathbf{X} \to \mathbf{X} \times f(\mathbf{X})$, then we expect the MI to be invariant. If one naively uses the LNC method, one will find that the MI increases. This is because the volumes $\tilde{V}_i$ will most often decrease when computed in the redundant scenario and hence the LNC correction term will most often increase.

A possible fix to this problem is to not only correct the volume in the joint space, but to fix the volumes in the marginal spaces as well, leading to the LNC correction of the second kind,

$$\hat{I}_{LNC^2} = \hat{I}_{KSG} - \frac{1}{N}\sum_{i=1}^{N}\log\frac{\tilde{V}_i^{xy}}{\tilde{V}_i^x \tilde{V}_i^y} \quad (15)$$

In investigating the efficacy of such a method, we discovered that it's not very robust. This is mainly due to the fact that like the original method in (14), (15) is not coordinate invariant in general, and while the volume supplied by redundant variables can in principle be canceled in the denominator of (15), the volumes themselves will be computed with respect to spaces of different dimension, and will therefore not exactly cancel. The effect is to still increase MI under the influence of redundant variables, which is undesirable since vanilla KSG is most successful in this domain. Thus while LNC fixes one aspect of KSG, it reduces its efficacy in another aspect.

## 4. Robustness Tests of NP Estimators

We will access the robustness of the KSG estimator and its variants with respect to the three types of transformations outlined in section two (coordinate transformations, redundancy and noise). Most tests in this section will use a multivariate normal distribution, $\mathcal{N}_k = ((2\pi)^k|\Sigma|)^{-1/2} \exp\left[-\frac{1}{2}(\mathbf{x}-\mu)^T\Sigma^{-1}(\mathbf{x}-\mu)\right]$ where $\Sigma$ is the covariance matrix. The mutual information between two sets of variables $\mathbf{X}_n$ and $\mathbf{X}_m$, where $n+m=k$, is given by,

$$I_{\mathcal{N}_k}[\mathbf{X}_n; \mathbf{X}_m] = -\frac{1}{2} \log \left( \frac{|\Sigma_k|}{|\Sigma_n||\Sigma_m|} \right) \quad (16)$$

where $|\Sigma_k|$ is the determinant of the covariance matrix $\Sigma_k$. For computing the KSG estimate, we will use a python package called (NPEET) [11] developed by G. ver Steeg et al. For the LNC correction we use a similar package [9].

### 4.1. Coordinate Transformations

Since KSG uses the $L^\infty$ norm to define the region of uniform probability for the estimate of $p_i$, this automatically presents a problem with coordinate invariance. KSG will not even be invariant under linear scalings of the data, let alone arbitrary coordinate transformations (This is likely the motivation for KSG's estimator of the second kind, which gave different weight to each of the variables, however they were unable to derive a closed form expression.). Essentially, if one variable, say $z$, is scaled by a large order of magnitude with respect to the other variables, then the side lengths of the $L^\infty$ box will get chosen to be the length of the $k$th nearest neighbor in the direction of $z$. While this will not necessarily cause a problem with the values of $\langle \psi(x_z + 1) \rangle$, it will cause the counts for the other variables to be much larger than they necessarily should be. As an example, consider the following bivariate normal case.

As one can see from the figure, scaling one variable of the bivariate normal by $10^5$ renders the KSG method useless. While one can always argue for a heuristic scaling method, any robust method for computing MI should be invariant under arbitrary scalings. Below is the same experiment using the LNC method.

While the LNC method corrects the behavior of the curve in the highly correlated region, it still fails to capture the correct value overall due to KSG's inability to handle arbitrary coordinate transformations. The effects of arbitrary coordinate transformations are even worse in higher dimensional situations. Consider the eight-dimensional multivariate Gaussian below.

As you can see, multiplying all the variables by a factor of 10 greatly reduces the accuracy of the KSG algorithm. To see this effect happening more gradually, we focus on one particular value of the correlation matrix (where all correlation coefficients are equal to 1/2) and dial up the transformation on the four variables in one set. The results for 10,000 and 100,000 points are below.

As you can see, the KSG approximation begins to deteriorate very quickly under linear transformations when the dimensionality is high. Increasing the number of points by a factor of ten seems to do little to help this. While this is certainly a flaw in the method, it isn't as dire as the others we will explore in the next section. For now, one can adopt a strategy in which each variable is scaled in a way that gives equal weight to each of them. Proposed methods for this were given in [1].

## 4.2. Redundancy vs. Noise

Another simple test we can perform is to see how MI in high dimensional situations handles redundant and noisy variables. Specifically, we will look at how the MI changes as we dial up the noise present in redundant variables by randomizing their values with respect to the other variables. As we saw in [1], KSG does quite well with redundant variables, however noisy variables still present a problem.

We have studied the ability of vanilla KSG to calculate MI under the presence of redundant variables in [1]. Here we will briefly discuss the highlights. We examined a binary decision problem in which two distributions (signal and background) are separated by a certain amount with respect to their means. We showed in [1] that the neural network transformation leaves the MI unchanged which is expected according to its design criteria [7]. We tested this claim on a more general data set which was generated as part of a machine learning analysis on a mock SUSY search [12]. The SUSY data set contains eight low-level variables and ten high-level variables which are functions of the low-level ones. From the tests in [1], we again see that KSG performs well under the addition of the high-level variables when they are redundant.

However when the variables are shuffled so that they are independent of both $X$ and $\Theta$, i.e., when they are noise, then KSG starts to deteriorate very quickly. We can see this effect in the case where the MI is known exactly. Using the multi-variate Gaussian with eight variables from the previous tests, we added three copies of the last variable on one side, essentially taking $x_4 \in X$ and creating $X' = X \cup x_4 \cup x_4 \cup x_4$. We then dialed up the randomness in the three variables to 100%. What we mean here is that we rearranged the points in the set $(x_4 \cup x_4 \cup x_4)$ so that a particular value $x_i \in (x_4 \cup x_4 \cup x_4) \neq f(X)$ and $x_i \in (x_4 \cup x_4 \cup x_4) \neq f(Y)$.

This shows that noisy variables cause the KSG estimate to go down as a function of their uselessness. This problem compounds quickly when the number of useless variables increases, making KSG's ability to determine MI in high-dimensional cases problematic.

## 5. Discussion

We've shown in practical examples that KSG is not robust under coordinate transformations (Figures 1–4) and noise (Figures 5 and 6). While the effect of including noise is not as drastic in cases of simple distributions (Figure 3), it is must more dramatic in cases where the the distribution is not simple (Figure 5).

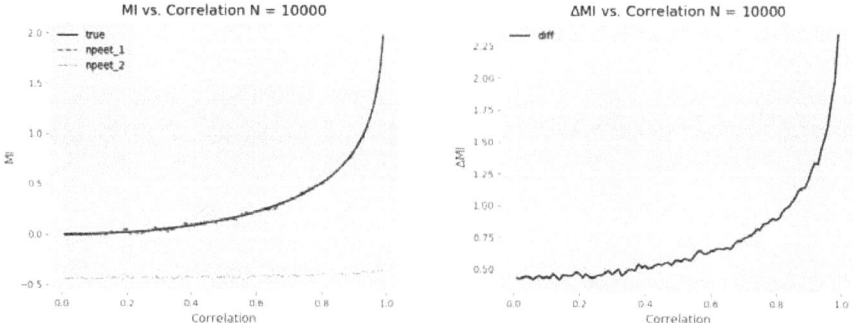

**Figure 1.** Comparison of mutual information estimates KSG for a bivariate normal distribution before (npeet_1) and after (npeet_2) a linear transformation of one variable by a factor of $10^5$. The second plot shows the difference between (npeet_1) and (npeet_2).

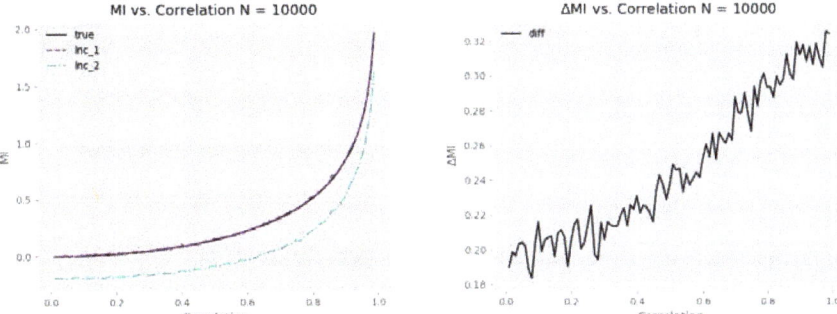

**Figure 2.** Comparison of mutual information estimates LNC for a bivariate normal distribution before (lnc_1) and after (lnc_2) a linear transformation of one variable by a factor of $10^5$. The second plot shows the difference between (lnc_1) and (lnc_2).

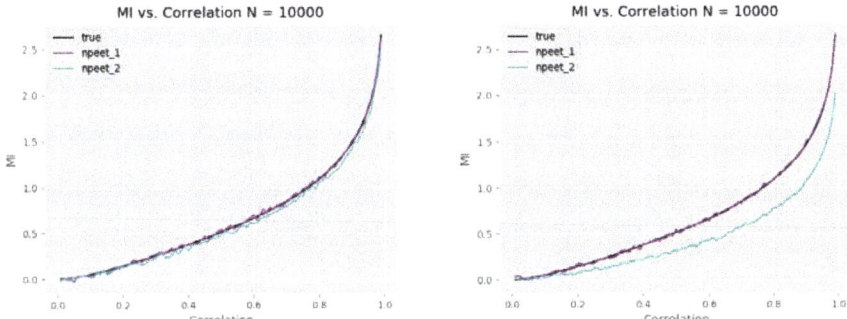

**Figure 3.** Comparison of mutual information estimates KSG for a multivariate normal distribution with equal correlation coefficients $\rho_{ij}$ before (npeet_1) and after (npeet_2) a linear transformation of one variable by a factor of 10. The second plot compares the KSG estimators after four variables are multiplied by a factor of 10.

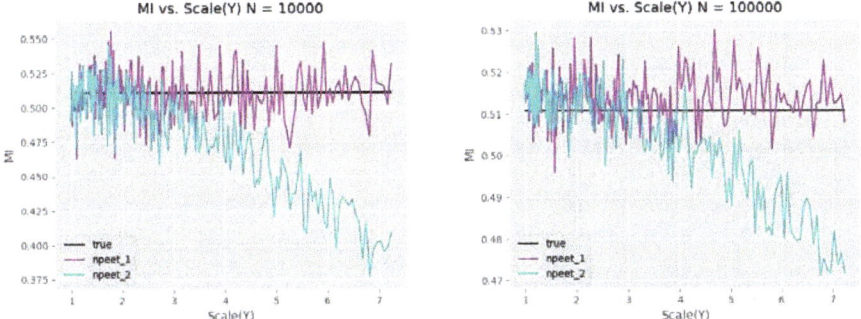

**Figure 4.** Comparison of true mutual information between the eight variables of a multivariate normal distribution where all correlation coefficients are equal to $1/2$ (black line) and the value estimated from $N = 10,000$ and $N = 100,000$ samples before (npeet_1) and after (npeet_2) a linear transformation is applied to one variable $x \to x' = 10x$.

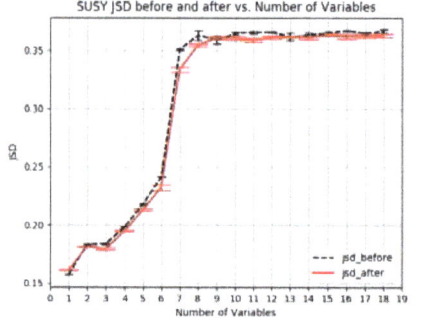 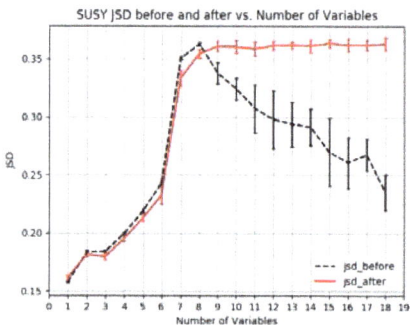

**Figure 5.** Comparison of MI values for increasing additions of discriminating variables for the SUSY data set. The first eight variables are low-level and the last ten are functions of the first eight. The first plot shows a comparison of (NN) performance when the last ten variables are redundant while the second shows how KSG's accuracy deteriorates when the high-level variables are shuffled.

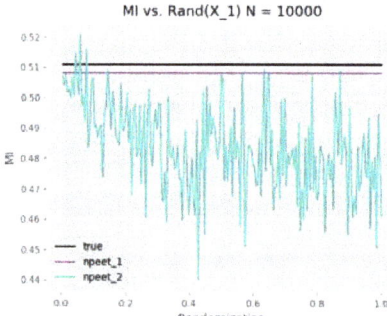 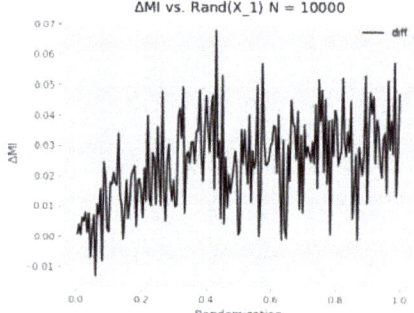

**Figure 6.** Comparison of true mutual information between the eight variables of a multivariate normal distribution where all correlation coefficients are equal to $1/2$ (black line) and the value estimated from $N = 10,000$ samples before (npeet_1) and after (npeet_2) a randomization is applied to the set of redundant variables.

**Acknowledgments:** We would like to thank A. Caticha, K. Knuth, G. ver Steeg and K. Vanslette for insightful conversations.

## References

1. Carrara, N.; Ernst, J.A. On the Upper Limit of Separability. *arXiv* **2017**, arXiv:1708.09449.
2. Tishby, N.; Pereira, F.C.; Bialek, W. The information bottleneck method. *arXiv* **2000**, arXiv:physics/0004057.
3. Tishby, N.; Zaslavsky, N. Deep Learning and the Information Bottleneck Principle. *arXiv* **2015**, arXiv:1503.02406.
4. Chen, X.; Duan, Y.; Houthooft, R.; Schulman, J.; Sutskever, I.; Abbeel, P. InfoGAN: Interpretable Representation Learning by Information Maximizing Generative Adversarial Nets. *arXiv* **2016**, arXiv:1606.03657.
5. Ver Steeg, G.; Galstyan, A. The Information Sieve. *arXiv* **2015**, arXiv:1507.02284.
6. Cover, T.M.; Thomas, J.A. *Elements of Information Theory*; Wiley-Interscience: Hoboken, NJ, USA, 2006.
7. Carrara, N.; Vanslette, K. The Inferential Foundations of Mutual Information. *arXiv* **2019**, arXiv:1907.06992.
8. Kraskov, A.; Stögbauer, H.; Grassberger, P. Estimating mutual information *Phys. Rev. E* **2004**, *69*, 066138.
9. Gao, S.; Ver Steeg, G.; Galstyan, A. Efficient Estimation of Mutual Information for Strongly Dependent Variables. *arXiv* **2014**, arXiv:1411.2003.

10. Kozachenko, L.F.; Leonenko, N.N. A statistical estimate for the entropy of a random vector. *Probl. PeredachiInformats* **1987**, *2*, 9–16.
11. Ver Steeg, G. Non-parametric Entropy Estimation Toolbox. Available online: https://github.com/gregversteeg/NPEET (accessed on 14 January 2020).
12. Baldi, P.; Sadowski, P.; Whiteson, D. Searching for exotic particles in high-energy physics with deep learning. *Nat. Commun.* **2014**, *5*, 4308.

© 2020 by the authors. Licensee MDPI, Basel, Switzerland. This article is an open access article distributed under the terms and conditions of the Creative Commons Attribution (CC BY) license (http://creativecommons.org/licenses/by/4.0/).

*Proceedings*

# The Information Geometry of Space-Time [†]

### Ariel Caticha

Physics Department, University at Albany-SUNY, Albany, NY 12222, USA; ariel@albany.edu
† Presented at the 39th International Workshop on Bayesian Inference and Maximum Entropy Methods in Science and Engineering, Garching, Germany, 30 June–5 July 2019.

Published: 28 November 2019

**Abstract:** The method of maximum entropy is used to model curved physical space in terms of points defined with a finite resolution. Such a blurred space is automatically endowed with a metric given by information geometry. The corresponding space-time is such that the geometry of any embedded spacelike surface is given by its information geometry. The dynamics of blurred space, its *geometrodynamics*, is constructed by requiring that as space undergoes the deformations associated with evolution in local time, it sweeps a four-dimensional space-time. This reproduces Einstein's equations for vacuum gravity. We conclude with brief comments on some of the peculiar properties of blurred space: There is a minimum length and blurred points have a finite volume. There is a relativistic "blur dilation". The volume of space is a measure of its entropy.

**Keywords:** information geometry; general relativity; geometrodynamics

## 1. Introduction

The problem of reconciling quantum theory (QT) and general relativity (GR) has most commonly been addressed by preserving the framework of QT essentially unchanged while modifying the structure and dynamics of space-time. This is not unreasonable. Einstein's equation, $G_{\mu\nu} = 8\pi G\, T_{\mu\nu}$, relates geometry on the left to matter on the right. Since our best theories for the matter right hand side are QTs it is natural to try to construct a theory in which the geometrical left hand side is also of quantum mechanical origin [1–3].

Further thought however shows that this move carries a considerable risk, particularly because the old process of quantization involves ad hoc rules which, however successful in the past, have led to conceptual difficulties that would immediately spread and also infect the gravitational field. One example is the old quantum measurement problem and its closely related cousin the problem of macroscopic superpositions. Do quantum superpositions of space-times even make sense? In what direction would the future be? Another example is the cosmological constant problem. Does the zero point energy of quantum fields gravitate? Why does it not give rise to unacceptably large space-time curvatures? Considerations such as these suggest that the issue of whether and how to quantize gravity hinges on a deeper understanding of the foundations of QT and also on a deeper understanding of GR and of geometry itself—what, after all, is *distance*? Why are QT and GR framed in such different languages? Recent developments indicate that they might be closer than previously thought—the link is entropy. Indeed, in the entropic dynamics approach [4–6] QT is derived as an application of entropic methods of inference [7] with a central role assigned to concepts of information geometry [8–16]. And, on the GR side, the link between gravity and entropy has been recognized from the early work of Bekenstein and Hawking and further reaffirmed in more recent thermodynamic approaches to GR [17–22].

In a previous paper [23] we used the method of maximum entropy to construct a model of physical space in which points are blurred; they are defined with a finite resolution. Such a blurred space is a statistical manifold and therefore it is automatically endowed with a Riemannian metric given by information geometry. Our goal here is to further close the gap between QT and GR by formulating the corresponding Lorentzian geometry of space-time.

The extension from space to space-time is not just a simple matter of applying information geometry to four dimensions rather than three. The problem is that information geometry leads to metrics that are positive—statistical manifolds are inevitably Riemannian—which cannot reproduce the light-cone structure of space-time. Some additional ingredient is needed. We do not model space-time as a statistical manifold. Instead, space-time is modelled as a four-dimensional manifold such that the geometry of all space-like embedded surfaces is given by information geometry. We find that in the limit of a flat space-time our model coincides with a stochastic model of space-time proposed long ago by Ingraham by following a very different line of argument [24].

Blurred space is a curious hybrid: some features are typical of discrete spaces while other features are typical of continuous manifolds [25,26]. For example, there is a minimum length and blurred points have a finite volume. The volume of a region of space is a measure of the number of blurred points within it, and it is also a measure of its bulk entropy. Under Lorentz transformations the minimum length suffers a dilation which is more analogous to the relativistic time dilation than to the familiar length contraction.

The dynamics of blurred space, its *geometrodynamics*, is constructed by requiring that as three-dimensional space undergoes the deformations associated with time evolution it sweeps a four-dimensional space-time. As shown in a remarkable paper by Hojman, Kuchař, and Teitelboim [27] in the context of the familiar sharp space-time this requirement is sufficient to determine the dynamics. Exactly the same argument can be deployed here. The result is that in the absence of matter the geometrodynamics of four-dimensional blurred space-time is given by Einstein's equations. The coupling of gravity to matter will not be addressed in this work.

## 2. The Information Geometry of Blurred Space

To set the stage we recall the model of blurred space as a smooth three-dimensional manifold **X** the points of which are defined with a finite resolution [23]. It is noteworthy that, unlike the very rough space-time foams expected in some models of quantum gravity, one expects blurred space to be very smooth because irregularities at scales smaller than the local uncertainty are suppressed. Blurriness is implemented as follows: when we say that a test particle is located at $x \in \mathbf{X}$ (with coordinates $x^a$, $a = 1, 2, 3$) it turns out that it is actually located at some unknown neighboring $x'$. The probability that $x'$ lies within $d^3x'$ is $p(x'|x)d^3x'$. Since to each point $x \in \mathbf{X}$ one associates a distribution $p(x'|x)$ the space **X** is a statistical manifold automatically endowed with a metric. Indeed, when points are blurred one cannot fully distinguish the point at $x$ described by the distribution $p(x'|x)$ from another point at $x + dx$ described by $p(x'|x + dx)$. The quantitative measure of distinguishability [7,11] is the *information distance*,

$$d\ell^2 = g_{ab}(x)dx^a dx^b, \tag{1}$$

where the metric tensor $g_{ab}$—the *information metric*—is given by,

$$g_{ab}(x) = \int dx' \, p(x'|x) \, \partial_a \log p(x'|x) \, \partial_b \log p(x'|x). \tag{2}$$

(We adopt the standard notation $\partial_a = \partial/\partial x^a$ and $dx' = d^3x'$.) Thus, in a blurred space *distance is distinguishability*.

In Section 4 we will briefly address the physical/geometrical interpretation of $d\ell$. For now we merely state [23] that $d\ell$ measures the distance between two neighboring points in units of the local uncertainty defined by the distribution $p(x'|x)$, that is, information length is measured in units of the local blur.

In order to completely define the information geometry of **X** which will allow us to introduce notions of parallel transport, curvature, and so on, one must specify a connection or covariant derivative

∇. The natural choice is the Levi-Civita connection, defined so that $\nabla_a g_{bc} = 0$. Indeed, as argued in [28], the Levi-Civita connection is to be preferred because, unlike the other $\alpha$-connections [11], it does not require imposing any additional structure on the Hilbert space of functions $(p)^{1/2}$.

The next step is to use the method of maximum entropy to assign the blur distribution $p(x'|x)$. The challenge is to identify the constraints that capture the physically relevant information. One might be tempted to consider imposing constraints on the expected values of $\langle x'^a - x^a \rangle$ and $\langle (x'^a - x^a)(x'^b - x^b) \rangle$ but this does not work because in a curved space neither of these constraints is covariant. This technical difficulty is evaded by maximizing entropy on the flat space $\mathbf{T}_P$ that is tangent to $\mathbf{X}$ at $P$ and then using the *exponential map* (see [23]) to "project" the distribution from the flat $\mathbf{T}_P$ to the curved space $\mathbf{X}$. It is important to emphasize that the validity of this construction rests on the assumption that the normal neighborhood of every point $x$—the region about $x$ where the exponential map is 1-1—is sufficiently large. The assumption is justified provided the scale of the blur is much smaller than the scale over which curvature effects are appreciable.

Consider a point $P \in \mathbf{X}$ with generic coordinates $x^a$ and a positive definite tensor field $\gamma^{ab}(x)$. The components of $y \in \mathbf{T}_P$ are $y^a$. The distribution $\hat{p}(y|P)$ on $\mathbf{T}_P$ is assigned on the basis of information about the expectation $\langle y^a \rangle_P$ and the variance-covariance matrix $\langle y^a y^b \rangle_P$,

$$\langle y^a \rangle_P = 0 \quad \text{and} \quad \langle y^a y^b \rangle_P = \gamma^{ab}(P) \, . \tag{3}$$

On $\mathbf{X}$ it is always possible to transform to new coordinates

$$x^i = X^i(x^a) \, , \tag{4}$$

such that

$$\gamma^{ij}(P) = \delta^{ij} \quad \text{and} \quad \partial_k \gamma^{ij}(P) = 0 \, , \tag{5}$$

where $i, j, \ldots = 1, 2, 3$. If $\gamma^{ab}$ were a metric tensor the new coordinates would be called Riemann Normal Coordinates at $P$ (RNC$_P$). The new components of $y$ are

$$y^i = X_a^i y^a \quad \text{where} \quad X_a^i = \frac{\partial x^i}{\partial x^a} \, , \tag{6}$$

and the constraints (3) take the simpler form,

$$\langle y^i \rangle_P = 0 \quad \text{and} \quad \langle y^i y^j \rangle_P = \delta^{ij} \, . \tag{7}$$

We can now maximize the entropy

$$S[\hat{p}, q] = -\int d^3y \, \hat{p}(y|P) \log \frac{\hat{p}(y|P)}{\hat{q}(y)} \tag{8}$$

relative to the measure $\hat{q}(y)$ subject to (7) and normalization. Since $\mathbf{T}_P$ is flat we can take $\hat{q}(y)$ to be constant and we may ignore it. The result in RNC$_P$ is

$$\hat{p}(y^i|P) = \frac{1}{(2\pi)^{3/2}} \exp\left[-\frac{1}{2}\delta_{ij} y^i y^j\right] \, . \tag{9}$$

Using the inverse of Equation (6) we can transform back to the original coordinates $y^a$,

$$y^a = X_i^a y^i \quad \text{and} \quad \gamma_{ab} = X_a^i X_b^j \delta_{ij} \, . \tag{10}$$

The resulting distribution is also Gaussian,

$$\hat{p}(y^a|P) = \frac{(\det \gamma_{ab})^{1/2}}{(2\pi)^{3/2}} \exp\left[-\frac{1}{2}\gamma_{ab}y^a y^b\right], \tag{11}$$

and the matrix $\gamma_{ab}$ of Lagrange multipliers turns out to be the inverse of the correlation matrix $\gamma^{ab}$, $\gamma_{ab}\gamma^{bc} = \delta_a^c$.

Next we use the exponential map to project $y^i$ coordinates on the flat $\mathbf{T}_P$ to the RNC$_P$ coordinates on the curved $\mathbf{X}$,

$$x'^i = x^i(P) + y^i. \tag{12}$$

The corresponding distribution $p(x'^i|P)$ induced on $\mathbf{X}$ by $\hat{p}(y^i|P)$ on $\mathbf{T}_P$ is

$$p(x'^i|P)d^3x' = \hat{p}(y^i|P)d^3y, \tag{13}$$

or

$$p(x'^i|x^i) = \frac{1}{(2\pi)^{3/2}} \exp\left[-\frac{1}{2}\delta_{ij}(x'^i - x^i)(x'^j - x^j)\right]. \tag{14}$$

Thus, in RNC$_P$ the distribution $p(x'^i|x^i)$ retains the Gaussian form. We can now invert (4) and transform back to the original generic frame of coordinates $x^a$ and define $p(x'^a|x^a)$ by

$$p(x'^a|x^a)d^3x'^a = p(x'^i|x^i)d^3x'^i, \tag{15}$$

which is an identity between scalars and holds in all coordinate systems. In the original $x^a$ coordinates the distribution $p(x'^a|x^a)$ will not, in general, be Gaussian,

$$p(x'^a|x^a) = \frac{(\det \gamma_{ab})^{1/2}}{(2\pi)^{3/2}} \exp\left[-\frac{1}{2}\delta_{ij}\left(X^i(x'^a) - X^i(x^a)\right)\left(X^j(x'^a) - X^j(x^a)\right)\right]. \tag{16}$$

Finally we substitute (16) into (2) to calculate the information metric $g_{ab}$. (The integral is easily handled in RNC$_P$.) The result is deceptively simple,

$$g_{ab} = X_a^i X_b^j \delta_{ij} = \gamma_{ab}. \tag{17}$$

The main result of [23] was to show that the metric $g_{ab}$ of a blurred space is a statistical concept that measures the "degree of distinguishability" between neighboring points. The metric is given by the Lagrange multipliers $\gamma_{ab}$ associated to the covariance tensor $\gamma^{ab}$ that describes the blurriness of space.

## 3. Space-Time and the Geometrodynamics of Pure Gravity

The constraint that determines the dynamics is the requirement that blurred space be a three-dimensional spacelike "surface" embedded in four-dimensional space-time. As shown in [27] the reason this condition is so constraining is that when evolving from an initial to a final surface every intermediate surface must also be embeddable in the same space-time and, furthermore, the sequence of intermediate surfaces—the path or foliation—is not unique. Such a "foliation invariance", which amounts to the local relativity of simultaneity, is a requirement of consistency: if there are two alternative paths to evolve from an initial to a final state, then the two paths must lead to the same result.

Space-time is foliated by a sequence of space-like surfaces $\{\Sigma\}$. Points on the surface $\Sigma$ are labeled by coordinates $x^a$ ($a = 1, 2, 3$) and space-time events are labeled by space-time coordinates $X^\mu$ ($\mu = 0, 1, 2, 3$). The embedding of $\Sigma$ within space-time is defined by four functions $X^\mu = X^\mu(x)$.

An infinitesimal deformation of $\Sigma$ to a neighboring $\Sigma'$ is specified by $X^\mu(x) \to X^\mu(x) + \delta X^\mu(x)$. The deformation vector $\delta X^\mu(x)$ is decomposed into normal and tangential components,

$$\delta X^\mu = \delta X^\perp n^\mu + \delta X^a X_a^\mu, \tag{18}$$

where $n^\mu$ is the unit normal to the surface and the three vectors $X_a^\mu = \partial X^\mu / \partial x^a$ are tangent to the coordinate lines $x^a$ ($n_\mu n^\mu = -1$, $n_\mu X_a^\mu = 0$).

We assume a phase space endowed with a symplectic structure: the basic dynamical variables are the surface metric $g_{ab}(x)$ and its canonically conjugate momentum $\pi^{ab}(x)$. This leads to a Hamiltonian dynamics where the super-Hamiltonian $H_\perp(x)[g,\pi]$ and the super-momentum $H_a(x)[g,\pi]$ generate normal and tangential deformations respectively. In order for the dynamics to be consistent with the kinematics of deformations the Poisson brackets of $H_\perp$ and $H_a$ must obey two sets of conditions [29,30]. First, they must close in the same way as the "group" of deformations, that is, they must provide a representation of the "algebra" of deformations [31],

$$[H_\perp(x), H_\perp(x')] = \left(g^{ab}(x) H_b(x) + g^{ab}(x') H_b(x')\right) \partial_{ax} \delta(x,x'), \tag{19}$$

$$[H_a(x), H_\perp(x')] = H_\perp(x) \partial_{ax} \delta(x,x'), \tag{20}$$

$$[H_a(x), H_b(x')] = H_a(x') \partial_b \delta(x,x') + H_b(x) \partial_a \delta(x,x'). \tag{21}$$

And second, the initial values of the variables $g_{ab}$ and $\pi^{ab}$ must be restricted to obey the weak constraints

$$H_\perp(x) \approx 0 \quad \text{and} \quad H_a(x) \approx 0. \tag{22}$$

A remarkable feature of the resulting dynamics is that once the constraints (22) are imposed on one initial surface $\Sigma$ they will be satisfied automatically on all subsequent surfaces. As shown in [27] the generators that satisfy (19)–(21) are

$$H_a = -2 \nabla_b \pi_a^b, \tag{23}$$

$$H_\perp = 2\kappa G_{abcd} \pi^{ab} \pi^{cd} - \frac{1}{2\kappa} g^{1/2} (R - 2\Lambda), \tag{24}$$

$$G_{abcd} = \frac{1}{2} g^{-1/2} \left(g_{ac} g_{bd} + g_{ad} g_{bc} - g_{ab} g_{cd}\right), \tag{25}$$

where $\kappa$ and $\Lambda$ are constants which, once the coupling to matter is introduced, can be related to Newton's constant $G = c^4 \kappa / 8\pi$ and to the cosmological constant $\Lambda$. Equations (22)–(25) are known to be equivalent to Einstein's equations in vacuum.

To summarize: (a) Space-time is constructed so that the geometry of any embedded spacelike surface is given by information geometry. (b) The geometrodynamics of blurred space is given by Einstein's equations. These are the main conclusions of this paper.

## 4. Discussion

**Dimensionless Distance?**—As with any information geometry the distance $d\ell$ given in Equations (1) and (2) turns out to be dimensionless. The interpretation [23] is that an information distance is measured distances in units of the local uncertainty—the blur. To make this explicit we write the distribution (14) that describes a blurred point in RNC$_P$ in the form

$$p(x'^i | x^i) = \frac{1}{(2\pi \ell_0^2)^{3/2}} \exp\left[-\frac{1}{2\ell_0^2} \delta_{ij} (x'^i - x^i)(x'^j - x^j)\right], \tag{26}$$

so that the information distance between two neighboring points is

$$d\ell^2 = \frac{1}{\ell_0^2}\delta_{ij}dx^i dx^j .\tag{27}$$

Since the blur $\ell_0$ is the only unit of length available to us (there are no external rulers) it follows that $\ell_0 = 1$ but it is nevertheless useful to write our equations showing $\ell_0$ explicitly. In (26) the two points $x$ and $x'$ are meant to be simultaneous.

**Minimum Length**—To explore the geometry of blurred space it helps to distinguish the abstract "mathematical" points that are sharply defined by the coordinates $x$ from the more "physical" blurred points. We shall call them c-points and b-points respectively. In RNC$_P$ the distance between two c-points located at $x$ and at $x + \Delta x$ is given by (27). To find the corresponding distance $\Delta\lambda$ between two b-points located at $x$ and at $x + \Delta x$ we recall that when we say a test particle is at $x$ it is actually located at $x' = x + y$ so that

$$\Delta\lambda^2 = \frac{1}{\ell_0^2}\delta_{ij}(\Delta x^i + \Delta y^i)(\Delta x^j + \Delta y^j) .\tag{28}$$

Taking the expectation over $y$ with the probability (26)—use $\langle y^i \rangle = 0$ and $\langle y^i y^j \rangle = \ell_0^2 \delta^{ij}$—we find

$$\langle \Delta\lambda^2 \rangle = \frac{1}{\ell_0^2}\delta_{ij}\langle(\Delta x^i + \Delta y^i)(\Delta x^j + \Delta y^j)\rangle = \Delta\ell^2 + 6 .\tag{29}$$

We see that even as $\Delta x \to 0$ and the two b-points coincide we still expect a minimum *rms* distance of $\sqrt{6}\ell_0$.

**Blur Dilation**—The size of the blur of space is a length but it does not behave as the length of a rod. When referred to a moving frame it does not undergo a Lorentz contraction. It is more analogous to time dilation: just as a clock marks time by ticking along the time axis, so are lengths measured by ticking $\ell_0$s along them. By the principle of relativity all inertial observers measure the same blur in their own rest frames — the proper blur $\ell_0$. Relative to another inertial frame the blur is dilated to $\gamma\ell_0$ where $\gamma$ is the usual relativistic factor. This implies the proper blur $\ell_0$ is indeed the minimum attainable.

**The Volume of a Blurred Point: Is Space Continuous or Discrete?**—A b-point is smeared over the whole of space but we can still define a useful measure of its volume by adding all volume elements $g^{1/2}(x')d^3x'$ weighed by the scalar density $p(x'|x)/g^{1/2}(x')$. Therefore in $\ell_0$ units a blurred point has unit volume. This means that we can measure the volume of a finite region of space by *counting* the number of b-points it contains. It also means that the number of distinguishable b-points within a region of finite volume is finite which is a property one would normally associate to discrete spaces. In this sense blurred space is both continuous and discrete. (See also [26].)

**The Entropy of Space**—The statistical state of blurred space is the joint distribution of all the $y_x$ variables associated to every b-point $x$. We assume that the $y_x$ variables at different $x$s are independent, and therefore their joint distribution is a product,

$$\hat{P}[y] = \prod_x \hat{p}(y_x|x) .\tag{30}$$

From (11) and (17) the distribution $\hat{p}(y_x^a|x)$ in the tangent space $\mathbf{T}_x$ is Gaussian,

$$\hat{p}(y_x|x) = \frac{(\det g_x)^{1/2}}{(2\pi)^{3/2}} \exp\left[-\frac{1}{2}g_{ab}(x)y_x^a y_x^b\right] ,\tag{31}$$

which shows explicitly how the information metric $g_{ab}$ determines the statistical state of space.

Next we calculate the total entropy of space,

$$S[\hat{P}, \hat{Q}] = - \int Dy\, \hat{P}[y] \log \frac{\hat{P}[y]}{\hat{Q}[y]} \stackrel{\text{def}}{=} S[g] \tag{32}$$

relative to the uniform distribution

$$\hat{Q}[y|g] = \Pi_x g^{1/2}(x), \tag{33}$$

which is independent of $y$ — a constant. Since the $y$'s in Equation (30) are independent variables the entropy is additive, $S[g] = \sum_x S(x)$, and we only need to calculate the entropy $S(x)$ associated to a b-point at a generic location $x$,

$$S(x) = - \int d^3y\, \hat{p}(y|x) \log \frac{\hat{p}(y|x)}{g^{1/2}(x)} = \frac{3}{2} \log 2\pi e = s_0. \tag{34}$$

Thus, the entropy per b-point is a numerical constant $s_0$ and the entropy of any region $R$ of space, $S_R[g]$, is just its volume,

$$S_R[g] = \sum_{x \in R} S(x) = s_0 \int_R d^3x\, g^{1/2}(x). \tag{35}$$

Thus, the entropy of a region of space is proportional to the number of b-points within it and is proportional to its volume.

**Canonical Quantization of Gravity?**—The picture of space as a smooth blurred statistical manifold stands in sharp contrast to ideas inspired from various models of quantized gravity in which the short distance structure of space is dominated by extreme fluctuations. From our perspective it is not surprising that attempts to quantize gravity by imposing commutation relations on the metric tensor $g_{ab}$ have not been successful. The information geometry approach suggests a reason why: quantizing the Lagrange multipliers $g_{ab} = \gamma_{ab}$ would be just as misguided as formulating a quantum theory of fluids by imposing commutation relations on those Lagrange multipliers like temperature, pressure, or chemical potential, that define the thermodynamic macrostate.

**Physical Consequences of a Minimum Length?**—A minimum length will eliminate the short wavelength divergences in QFT. This in turn will most likely illuminate our understanding of the cosmological constant and affect the scale dependence of running coupling constants. One also expects that QFT effects that are mediated by short wavelength excitations should be suppressed. For example, the lifetime of the proton ought to be longer than predicted by grand-unified theories formulated in Minkowski space-time. The nonlocality implicit in a minimum length might lead to possible violations of CPT symmetry with new insights into matter-antimatter asymmetry. Of particular interest would be the early universe cosmology where inflation might amplify minimum-length effects possibly making them observable.

**Acknowledgments:** I would like to thank N. Carrara, N. Caticha, S. Ipek, and P. Pessoa, for valuable discussions.

## References and Notes

1. For an introduction to the extensive literature on canonical quantization of gravity, loop quantum gravity, string theory, and causal sets see e.g., [2,3].
2. Kiefer, C. *Quantum Gravity*; Oxford U.P.: Oxford, UK, 2007.
3. Ashtekar, A.; Berger, B.; Isenberg, J.; MacCallum, M. (Eds.) *General Relativity and Gravitation*; Cambridge U.P.: Cambridge, UK, 2015.
4. Caticha, A. The Entropic Dynamics approach to Quantum Mechanics. *Entropy* **2019**, *21*, 943; doi:10.3390/e21100943, arXiv **2019** arXiv:1908.04693.
5. Ipek, S.; Abedi, M.; Caticha, A. Entropic Dynamics: Reconstructing Quantum Field Theory in Curved Spacetime. *Class. Quantum Grav.* **2019**, *36*, 205013, arXiv **2018**, arXiv:1803.07493.
6. Ipek, S.; Caticha, A. An Entropic Dynamics approach to Geometrodynamics. *arXiv* **2019**, arXiv:1910.01188.

7. Caticha, A. *Entropic Inference and the Foundations of Physics*; International Society for Bayesian Analysis-ISBrA: Sao Paulo, Brazil, 2012. Available online: http://www.albany.edu/physics/ACaticha-EIFP-book.pdf (accessed on 20 September 2019).
8. The subject of information geometry was introduced in statistics by Fisher [9] and Rao [10] with important later contributions by other authors [11–14]. Important aspects were also independently discovered in thermodynamics [15,16].
9. Fisher, R.A. Theory of statistical estimation. *Math. Proc. Camb. Philos. Soc.* **1925**, *22*, 700–725.
10. Rao, C.R. Information and the accuracy attainable in the estimation of statistical parameters. *Bull. Calcutta Math. Soc.* **1945**, *37*, 81.
11. Amari, S. *Differential-Geometrical Methods in Statistics*; Springer: Berlin, Germany, 1985.
12. Čencov, N.N. *Statistical Decision Rules and Optimal Inference*; American Mathematical Soc.: Providence, RI, USA, 1981; Volume 53.
13. Rodríguez, C.C. The metrics generated by the Kullback number. In *Maximum Entropy and Bayesian Methods*; Skilling, J., Ed.; Kluwer: Dordrecht, The Netherlands, 1989.
14. Ay, N.; Jost, J.; Vân Lê, H.; Schwanchhöfer, L. *Information Geometry*; Springer: Berlin, Germany, 2017.
15. Weinhold, F. Metric geometry of equilibrium thermodynamics. *J. Chem. Phys.* **1975**, *63*, 2479.
16. Ruppeiner, G. Thermodynamics: A Riemannian geometric model. *Phys. Rev. A* **1979**, *20*, 1608.
17. Bekenstein, J.D. Black holes and entropy. *Phys. Rev. D* **1973**, *7*, 2333.
18. Hawking, S. Black Holes and Thermodynamics. *Phys. Rev. D* **1976**, *13*, 191.
19. Jacobson, T. Thermodynamics of space-time: the Einstein equation of state. *Phys. Rev. Lett.* **1995**, *75*, 1260.
20. Padmanabhan, T. Thermodynamical aspects of gravity: New insights. *Rep. Prog. Phys.* **2010**, *73*, 046901.
21. Verlinde, E.P. On the origin of gravity and the laws of Newton. *J. High Energy Phys.* **2011**, *2011*, 29.
22. Jacobson, T. Entanglement equilibrium and the Einstein equation. *Phys. Rev. Lett.* **2016**, *116*, 201101.
23. Caticha, A. Geometry from Information Geometry. In *Bayesian Inference and Maximum Entropy Methods in Science and Engineering*; Giffin, A., Knuth, K., Eds.; AIP American Institute of Physics: College Park, MD, USA, 2016; Volume 1757, p. 030001, *arXiv* **2015**, arXiv:1512.09076.
24. Ingraham, R.L. Stochastic Space-time. *Nuovo Cimento* **1964**, *34*, 182.
25. It is possible that there is some connection with ideas proposed by Kempf [26] expressed in the language of spectral geometry. This is a topic for future research.
26. Kempf, A. Information-theoretic natural ultraviolet cutoff for spacetime. *Phys. Rev. Lett.* **2009**, *103*, 231301..
27. Hojman, S.A.; Kuchař, K.; Teitelboim, C. Geometrodynamics Regained. *Ann. Phys.* **1976**, *96*, 88.
28. Brodie, D.J.; Hughston, L.P. Statistical Geometry in Quantum Mechanics. *Proc. R. Soc. Lond. Ser. A* **1998**, *454*, 2445–2475.
29. Teitelboim, C. How Commutators of Constraints Reflect the Spacetime Structure. *Ann. Phys.* **1973**, *79*, 542.
30. Kuchař, K. Canonical Quantization of Gravity. In *Relativity, Astrophysics, and Cosmology*; Israel, W., Ed.; Reidel: Dordrecht, The Netherlands, 1973; pp. 237–288.
31. The quotes in "group" and "algebra" are a reminder that the set of deformations do not form a group. The composition of two successive deformations is itself a deformation but it depends on the surface to which the first deformation is applied.

© 2019 by the authors. Licensee MDPI, Basel, Switzerland. This article is an open access article distributed under the terms and conditions of the Creative Commons Attribution (CC BY) license (http://creativecommons.org/licenses/by/4.0/).

*Proceedings*

# Entropic Dynamics for Learning in Neural Networks and the Renormalization Group [†]

**Nestor Caticha**

Instituto de Fisica, Universidade de Sao Paulo, 05508-090 Sao Paulo, SP, Brazil; ncaticha@usp.br

† Presented at the 39th International Workshop on Bayesian Inference and Maximum Entropy Methods in Science and Engineering, Garching, Germany, 30 June–5 July 2019.

Published: 25 November 2019

**Abstract:** We study the dynamics of information processing in the continuous depth limit of deep feed-forward Neural Networks (NN) and find that it can be described in language similar to the Renormalization Group (RG). The association of concepts to patterns by NN is analogous to the identification of the few variables that characterize the thermodynamic state obtained by the RG from microstates. We encode the information about the weights of a NN in a Maxent family of distributions. The location hyper-parameters represent the weights estimates. Bayesian learning of new examples determine new constraints on the generators of the family, yielding a new pdf and in the ensuing entropic dynamics of learning, hyper-parameters change along the gradient of the evidence. For a feed-forward architecture the evidence can be written recursively from the evidence up to the previous layer convoluted with an aggregation kernel. The continuum limit leads to a diffusion-like PDE analogous to Wilson's RG but with an aggregation kernel that depends on the the weights of the NN, different from those that integrate out ultraviolet degrees of freedom. Approximations to the evidence can be obtained from solutions of the RG equation. Its derivatives with respect to the hyper-parameters, generate examples of Entropic Dynamics in Neural Networks Architectures (EDNNA) learning algorithms. For simple architectures, these algorithms can be shown to yield optimal generalization in student- teacher scenarios.

**Keywords:** Neural Networks; Renormalization Group; Entropic Dynamics; learning algorithms

---

## 1. Introduction

Neural networks are information processing systems that learn from examples. Loosely inspired in biological neural systems, they have been used for several types of problems such as classification, regression, dimensional reduction and clustering [1]. Biological systems selection is based on a measure of performance that combines not only accuracy but also ease of computation and implementation. Predictions based on expectations over posterior Bayesian distributions may lead to saturating bounds for optimal accuracy learning but will typically lack in ease of computation and speed in reaching a result. Neural networks are parametric models and if we don't address the determination of the architecture, which we don't in this paper, the problem of learning from examples is reduced to obtaining fast estimates of the weights or parameters, avoiding the integration over large dimensional spaces. The spectacular explosion of applications in several areas is witness to the fact that several training methods and large data sets are available. Despite these victories, the mechanisms of information dynamics processing remain obscure and despite several decades of theoretical analysis using methods of Statistical Mechanics, much remains to be understood. Here we study on-line learning in feed-forward architectures, where (input,output) examples are presented one at a time. Theoretical analysis is easier than for batch or off-line learning where the cost function depends on a large number of example pairs, however on-line accuracy performance remains high. This is in part due to the fact that since the cost function changes from example to example, the local minima of the cost function that plague off-line learning are not so important. Local stationary points of the learning dynamics are still a problem, but good performances are possible. An important problem to be addressed is what cost function is the most appropriate. If an algorithm is going to be successful it has to approach Bayesian estimates for the available information. But any Bayes algorithm leads to high, even in the millions, dimensional integrals. Monte Carlo strategies cannot be used if simplicity is a requirement. The strategy to determine optimized algorithms for on-line learning has been studied in the past for restricted scenarios and architectures. We present a more general approach, with the following strategy. We are in a situation of incomplete information, thus a probability distribution represents, at a given point in the dynamics, what is known about the parameters. We have to commit to a family of distributions and we choose a Maxent family. Location hyperparameters give the current estimate of the weights. A new (input,output) example pair arrives and Bayes rule permits an update. The choice of the likelihood is a reflection of what we know about the architecture of the NN. In general it is not conjugated to the chosen family.

Still, the Bayes posterior, while not in the family, points to a unique member of the family, since it imposes new constraints on the expected values of the generators.

The resulting learning algorithm is the entropic dynamics imposed by the arrival of information in the examples that induces a change of the hyperparameters of the family. It turns out that changes in the weights are in the direction of decreasing the model Bayesian evidence and it is a stochastic gradient descent algorithm, where the cost function is the log evidence of the model.

The denominator of the Bayes update can be interpreted either as the evidence of the model or alternatively as the predictive probability distribution of the output conditioned on the input and the weights. Once it is written as the marginalization over the internal representation, i.e. the activation values of the internal units, of the joint distribution of activities of the whole network, and under the supposition that the information flows only from one layer to the next, a Markov chain structure follows. Recursion relations of the partial evidence up to a given internal layer are obtained and in the continuous depth limit (CDL) a Fokker-Planck parabolic partial differential equation is obtained. It generalizes Wilson's Renormalization Group [2] diffusion equation for general kernels. The usual, e.g., majority rule that eliminates high frequency degrees of freedom are replaced by the weights of the NN. The RG dynamics can be seen as a classifier of Statistical Mechanics microstates into thermodynamics states. A NN extracts the relevant degrees of freedom that describe the macroscopic concept onto which an input pattern is to be assigned. The first authors to relate the RG and NN were [3] and [4] generating a large flow of ideas into the possible connections between these two areas [5–7].

## 2. Maxent Distributions and Bayesian Learning

Let $f_a(w)$, for $a = 1, ...K$, $w \in \mathbb{R}^N$, be the generators of a family $\mathcal{Q}$ of distributions $Q(w|\lambda)$. If information about $w$ is given in the form of constraints $\mathbb{E}_Q(f_a) = F_a$, for the set of numbers $\{F_a\}_{a=1,K}$, the Maxent distribution is

$$Q(w|\lambda) = \frac{1}{z} \exp\left(-\sum_{i=1}^{K} \lambda_a f_a(w)\right), \tag{1}$$

where $z$ ensures normalization. Then

$$\frac{\partial \ln z}{\partial \lambda_a} = -F_a \text{ and } \frac{\partial Q(w|\lambda)}{\partial \lambda_a} = (-f_a + F_a) Q(w|\lambda). \tag{2}$$

Now consider a system learning a map from inputs $x$ to outputs $y$, and the model is a known function which depends on a parameter array $w$: $y = T(x; w)$. The aim of learning is to obtain the parameters from the information in the learning set $\mathcal{D}_n = \{(x_i, y_i)\}_{i=1,n}$. We want to obtain a distribution for the parameters and consider that up to $n-1$ examples the information is coded in a member of the $\mathcal{Q}$

family: $Q(w|\lambda_{n-1}) = Q_{n-1}$. Calling the likelihood of the problem $L_n = P(y_n|x_n, w)$, the product rule permits the Bayesian updating

$$P_n = P(w|\mathcal{D}_n) = \frac{Q_{n-1}L_n}{Z_n}, \qquad (3)$$

where the partition function or the evidence is $Z(y_n|x_n, \lambda_{n-1}) = \int Q_{n-1}L_n dw = P(y_n|x_n\lambda_{n-1})$. The Bayes posterior given by eq. 3 in general doesn't belong to the $Q$ family. We have to choose the member of the family that is closest to the Bayes posterior. This is the Maxent posterior. The way to proceed is based on the fact that a member of the $Q$ family is determined solely by the values of the constraints $\{F_a\}$. The Bayes posterior defines a set of values for the constraints $\{\langle f_a \rangle\}$. It points in a unique way to the Maxent posterior $Q_n$ within the family $Q$, obtained at the extreme of

$$S[Q_n||Q_{n-1}] = -\int Q_n \log \frac{Q_n}{Q_{n-1}} dw - \Delta\lambda_a \left(\mathbb{E}_n(f_a) - \langle f_a \rangle\right), \qquad (4)$$

subject to the only possible constraints on its expected values $\mathbb{E}_n(f_a)$ which are taken to be the Bayes posterior expected values $\langle f_a \rangle$. Then for every generator

$$\mathbb{E}_{Q_n}(f_a) = \int \frac{Q_{n-1}L_n}{Z_n} f_a(w) dw = \mathbb{E}_{P_n}(f_a) = F_a^n. \qquad (5)$$

Subtract from both sides $F_a^{n-1}$, and use equation 2, then

$$F_a^n - F_a^{n-1} = -\frac{\partial \ln Z}{\partial \lambda_a^{n-1}} \qquad (6)$$

since the likelihood is independent of the Lagrange multiplier. This learning dynamics is deduced from entropy maximization and thus will be called Entropic dynamics. Learning occurs along the gradient of the log evidence. It will turn out that the sign is such that typically the evidence for the new model is higher than before learning. These equations hold for any family, but it is interesting to consider the case that will be most likely to be useful in practice, where the family is determined by the functions $f_0 = 1$, $f_i = w_i$ and $f_{ij} = w_i w_j$, for $i, j = 1, N$. The constraints after $n$ examples are the normalization, $\mathbb{E}(w_i) = \hat{w}_{ni}$ and $\mathbb{E}(w_i w_j) = (C_n)_{ij} + \hat{w}_{ni}\hat{w}_{nj}$. The result is the gaussian family $Q \propto \exp(-\lambda_0 - \sum_i \lambda_i w_i - \sum_{ij} \lambda_{ij} w_i w_j)$. The entropic dynamics update equations, driven by the arrival of the $n^{th}$ example are

$$\hat{w}_n = \hat{w}_{n-1} + C_{n-1}.\nabla_{\hat{w}_{n-1}} \log Z_n, \qquad (7)$$
$$C_n = C_{n-1} + C_{n-1}.\nabla^2_{\hat{w}_{n-1}} \log Z_n.C_{n-1}. \qquad (8)$$

For a layered network, these are the equations associated to the update of the weights afferent to a particular unit in layer $d$ from unit $i$ in layer $d-1$ and of the component of the covariance matrix describing the correlation between weights coming from units $i$ and $j$. The update equations, induced by a maximum entropy approximation to Bayesian learning is the learning algorithm of the neural network which implements the map $y = T(x; \hat{w})$.

An approximation to this scheme was found for simple networks with no hidden units using a variational procedure ([8]) and applied to several architectures [9–13]. Then Opper [14] showed the Bayesian connection, explored elsewhere [15]. Recently it has been applied to societies of interacting neural networks [16–19]. While [12] attacked the neural network with a hidden layer, the challenge remains to study networks with deep architectures.

## 3. Deep Multilayer Perceptron

In this section we show that the evidence for a multilayer feedforward neural network can be written recursively as a map. Actually we will get two maps that are essentially the same. This type of map is typical of Renormalization Group transformations and in a continuous limit representation of the neural network as a field theory, we will show that the map leads to a partial differential equation analogous to Wilson's diffusion-like RG equation.

We fix our attention at the $n^{th}$ example, and hence don't write temporal (lower) indices anymore. A layer (upper) index now appears and $x^d$ is the internal representation at the the unit layer $d$. Layers start with $d = 0$ and the depth of the network is $D$. Layer $d$ weights are collectively denoted $w^d$ and individually $w^d_{ij}$ is the weight connecting unit $i$ at layer $d-1$ to unit $j$ at layer $d$. The data pair used for the learning step are $X_0$ and $y$. The distributions of the representation at the input is $\delta(x^0 - X^0)$ and at the output $\delta(x^D - y)$. The partition function $Z(y_n|x_n, \lambda_{n-1})$ in Equation (3) is $Z(X^D|x^0, \lambda) = \int Q(w|\lambda) L dw$, where $Q(w|\lambda)$ is the prior joint distribution of the weights over all the layers. We will eventually take this to be a product over layers, $Q(w|\lambda) = \prod_{d=1}^{D-1} Q(w^d|\lambda_d)$, which will permit a simpler analytical treatment, but it is not a necessity at this moment. To obtain the likelihood we marginalize the joint distribution of the internal representations $P(x^D, x^{D-1}....x^1|x^0, w^1, ...w^D)$ over all internal representations at the hidden units doing the same trick that leads to the Chapman-Kolmogorov equation

$$L = P(x^D|x^0 = X^0, w^1, ...w^D) = \int P(x^D, x^{D-1}, ...x^1|x^0 = X^0, w^1, ...w^D) \prod_{d=1}^{D-1} dx^d. \tag{9}$$

The evidence can be written as

$$Z_D(x^D|X^0, \lambda) = \int Q^T(x^D, x^{D-1}....x^1|x^0 = X^0, \lambda) \prod_{d=1}^{D-1} dx^d. \tag{10}$$

where

$$Q^T(x^D, x^{D-1}....x^1|x^0 = X^0, \lambda) = \int P(x^D, x^{D-1}....x^1|x^0 = X^0, w^1, ...w^D)$$
$$\times \prod_{d=1}^{D-1} Q(w^d|\lambda^d) dw^d \tag{11}$$

is the joint transition distribution. Define the partially integrated $Z_d$ for any $d = 1....D$

$$Z_d(x^D, x^{D-1}, ....x^d|x^0, \lambda) = \int Q^T(x^D, x^{D-1}....x^1|x^0 = X^0, \lambda) \prod_{d'=1}^{d-1} dx^{d'}. \tag{12}$$

It satisfies the recursion

$$Z_d = \int Z_{d-1} dx^{d-1}. \tag{13}$$

and the evidence is

$$Z_D = \int Z_d \prod_{d'=d}^{D-1} dx^{d'} \tag{14}$$

At this point this is analogous to a Statistical Mechanics (SM) or euclidean field theory (EFT) partition function in which all field configurations with momentum components above a cutoff have been integrated out. The equivalent of the effective action of the EFT, or the renormalized hamiltonian in the SM is $-\log Z_d$.

Now we get a similar map, where the renormalization group transformation of the internal representations can be seen. Recall the likelihood in equation 9 and use the product rule

$$L = P(x^D|x^0, w^1, ... w^D) = \int P(x^D|x^{D-1}w_D)P(x^{D-1}....x^1|x^0, w^1, ... w^D) \prod_{d=1}^{D-1} dx^d$$

and finally

$$L = P(x^D|x^0, w^1, ... w^D) = \int \prod_{d=1}^{D-1} P(x^{d+1}|x^d, w^{d+1}) dx^d$$

Since the prior is also a product, then the partition function $Z_D = Z_D(x^D = y|x^0 = X^0, \{\lambda^d\})$ is given by

$$Z_D = \int \prod_{d=1}^{D} Q_d(w^d|\lambda^d) P(x^d|x^{d-1}, w^d) \prod_{d=1}^{D} dx^{d-1} dw^d \quad (15)$$

We integrate over $x_0$ and $x^D$ with the constraints that their distribution are deltas at the input $X^0$ and output $y$.

$$Z_D = \prod_{d=1}^{D} \int dw^d \left[ \int dx^{d-1} Q_d(w^d|\lambda^d) P(x^d|x^{d-1}, w^d) \right]$$

Define the evidence up to a given layer $\rho(x^d)$, with initial condition $\rho(x^0) = \delta(x^0 - X^0)$ and the map

$$\rho(x^{d+1}) = \int \rho(x^d) P(x^{d+1}|x^d w^{d+1}) Q_{d+1}(w^{d+1}|\lambda^{d+1}) dx^d dw^{d+1} \quad (16)$$

The last step for the map of a network of depth $D$ is for $x^D = y$ leading to the evidence of the model defined by the architecture of the network with weight and hyperparameters given by the set of $\lambda_d$:

$$Z_D(y) = \rho(x^D) = \int \rho(x^{D-1}) P(x^D|x^{D-1} w^D) Q_D(w^D|\lambda^D) dx^{D-1} dw^D \quad (17)$$

Define a layer to layer transition distribution

$$Q_{d-1}^T(x^d|x^{d-1}\lambda^d) = \int P(x^d|x^{d-1}, w^d) Q_d(w^d|\lambda^d) dw^d \quad (18)$$

$$\quad (19)$$

then, we have a map that gives the evidence after $d$ layers as an integral over internal representations at layer $d-1$ of the evidence at layer $d-1$ with a kernel $Q^T$ that implements an aggregation RG-like step:

$$\rho(x^d) = \int dx^{d-1} \rho(x^{d-1}) Q_{d-1}^T(x^d|x^{d-1}, \lambda^d) \quad (20)$$

We have obtained two RG-like maps, Equations (13) and (20). $Z_d$ depends on all internal representations from layer $d$ to $D$ and on all the hyperparameters $\lambda$. The simpler $\rho_d$ only depends on the internal representation at layer $d$ and on the hyperparameters of the previous layers. The map for $Z_d$ is simpler and the map for $\rho_d$ requires, at each step the input on the transition distribution $Q^T(x^d|x^{d-1}, \lambda^d)$. The transition distribution describes the renormalization group like transformation implemented by the neural network that takes the internal representation at one layer to the next. It is simple to see that

$$Z_d = \rho(x^d) \prod_{d' \geq d}^{D} Q^T(x^{d'+1}|x^{d'} \lambda^d) \quad (21)$$

## 3.1. Generalized RG Differential Equation of a Neural Network in the Continuous Depth Limit

The layer index is obviously discrete, but we can take the continuous limit, where now layers are represented by a time like $\tau$ variable. A discrete variable $i$ still labels the units. The evidence at depth $\tau$ is related to the evidence at depth $\tau_0$ by a generalization of Equation (20):

$$\rho(x,\tau) = \int Q^T(x(\tau)|x'(\tau_0),\lambda)\rho(x',\tau_0)Dx', \qquad (22)$$

where the integration measure $Dx = \prod_i dx_i$. The distribution $Q^T(x(\tau)|x'(\tau_0),\lambda)$ is the probability, that a network with parameters $\lambda$, conditional on being in state $x'$ at $\tau_0$ has an internal representation $x$ at depth $\tau$. It must satisfy the composition law

$$Q^T(x(\tau+\Delta\tau)|x'(\tau_0),\lambda) = \int Q^T(x(\tau+\Delta\tau)|z(\tau),\lambda)Q^T(z(\tau)|x'(\tau_0),\lambda)Dz$$

For a deterministic neural network, conditional on the weights $w$, the evolution of the internal representation is given by the transfer function. To obtain a well behaved limit it is supposed to vary slowly:

$$x_i(\tau+\Delta\tau) = T_i(x(\tau),w) = x_i(\tau) + \Delta\tau \tilde{b}_i(x(\tau),w), \qquad (23)$$

so that interpretation of $\tilde{b}$ is the gradient of the transfer function. The transition distribution is

$$Q^T(x|\tau,x',\tau_0,\lambda) = \int \prod_{\tau'\in[\tau_0,\tau]} \delta\left(x(\tau+\Delta\tau) - T(x'(\tau),w)\right) Q(w|\lambda,\tau)dw_{\tau'}, \qquad (24)$$

obtained by integrating over all configuration of the weights in the slice. We have chosen a Gaussian family to represent the informational state of the network, which now takes the form of a product of Gaussians for all $\tau$ slices:

$$Q(w|\lambda,\tau) \propto \prod_\tau \exp -\frac{1}{2}\{\Delta w \cdot C_\tau^{-1} \cdot \Delta w\}$$

where $\Delta w = w - \hat{w}_\tau$ and $\lambda = \{\hat{w}_\tau,C_\tau\}$ for all values of $\tau$, but only the hyperparameters of the particular slice under consideration matters. To define the continuous limit we impose that the limits below exit:

$$\lim_{\Delta\tau\downarrow 0}\frac{1}{\Delta\tau}\int Q^T(x|\tau+\Delta\tau,x',\tau,\lambda)(x-x')Dx = E_w[\tilde{b}(x(\tau),w] = b(x',\tau,\lambda),$$

$$\lim_{\Delta\tau\downarrow 0}\frac{1}{\Delta\tau}\int Q^T(x|\tau+\Delta\tau,x',\tau,\lambda)(x_i-x'_i)(x_j-x'_j)Dx = E_w[\tilde{b}_i(x(\tau),w)\tilde{b}_j(x(\tau),w)] = B_{ij}(x',\tau,\lambda). \qquad (25)$$

At each layer the drift vector $b(x',\tau,\lambda)$ is the expected value of the change in internal representation and the diffusion matrix $B_{ij}(x',\tau,\lambda)$ to the expectation of quadratic change, which are related to the expected values of the gradient and Hessian of the transfer function respectively. As usual, take the time derivative of the expected value, with respect to $Q^T(x|x',\lambda)$ of a well behaved test function $g(x)$. Taylor expand $g(x)$ around $x'$ and integrate by parts, use that $g(x)$ is arbitrary and obtain that $Q^T$ satisfies a parabolic PDE and so does the evidence (see Equation (22))

$$\frac{\partial\rho(x,\tau)}{\partial\tau} = -\frac{\partial}{\partial x_i}(b_i(x,\tau,\lambda)\rho(x,\tau)) + \frac{1}{2}\frac{\partial^2}{\partial x_i\partial x_j}(B_{ij}(x,\tau,\lambda)\rho(x,\tau)). \qquad (26)$$

The long time limit of Equation (26) is the predictive distribution $\rho(y,\tau = D) = P(y|x_0,\lambda)$. Equation (26) is a generalization of an analogous diffusion equation which appears in Wilson's incomplete integration formulation of the renormalization group (e.g., [2]). It extends the type of transformation by permitting that the transformations that leads from $\tau$ to $\tau + d\tau$ are not a simple spatial average, which would eliminate high spatial frequency components. Instead, the transformations are mediated by the weights $\hat{w}$. It differs from the usual statistical mechanics or field theories also in the following sense. In those approaches, the transformation $\hat{w}$ is known and uniform and the aim is to obtain the final $\rho_D$, which describes the infrared limit or the thermodynamics of the theory. In supervised learning in neural networks, the starting point, defined by the input $X^0$ and the output $Y$ are given. The problem is to find the correct set of weights $\hat{w}$ that implements the correct input-output association. There are two regimes for the neural network. In the learning phase the set of examples is a set of microscopic-macroscopic variables that describe a task. The aim of learning is to determine the appropriate generalized RG transformation that maps from the microscopic description to the macroscopic. After learning, the network is used to find out, for the current RG transformation, the unknown macroscopic generalized thermodynamics or infrared properties associated to the microstate. The next step is to derive optimized learning algorithms, from the solutions of Equation (26) and the EDNNA learning described by (7) and (8).

## References

1. Goodfellow, I.; Bengio, Y.; Courville, A. *Deep Learning*; MIT Press: Cambridge, MA, USA, 2016. Available online: http://www.deeplearningbook.org (accessed on 19 November 2019).
2. Wilson, K.G.; Kogut, J. The renormalization group and the $\epsilon$ expansion. *Phys. Rep.* **1974**, *12*, 75–199. doi:doi:10.1016/0370-1573(74)90023-4.
3. Bény, C. Deep Learning and the Renormalization Group. Available online: https://arxiv.org/abs/1301.3124 (accessed on 19 November 2019).
4. Mehta, P.; Schwab, D.J. An exact mapping between the Variational Renormalization Group and Deep Learning. *arXiv* **2014**, arXiv:1410.3831. [arXiv:1410.3831].
5. Koch-Janusz, M.; Ringel, Z. Mutual information, neural networks and the renormalization group. *Nat. Phys.* **2018**, *14*, 578–582. doi:10.1038/s41567-018-0081-4.
6. Li, S.H.; Wang, L. Neural Network Renormalization Group. *Phys. Rev. Lett.* **2018**, *121*, 260601. doi:10.1103/PhysRevLett.121.260601.
7. Lin, H.W.; Tegmark, M.; Rolnick, D. Why Does Deep and Cheap Learning Work So Well? *J. Stat. Phys.* **2017**, *168*, 1223–1247. doi:10.1007/s10955-017-1836-5.
8. Kinouchi, O.; Caticha, N. Optimal generalization in perceptrons. *J. Phys. A* **1992**, *25*, 6243.
9. Biehl, M.; Riegler, P. On-Line Learning with a Preceptron. *Europhys. Lett.* **1994**, *28*, 525.
10. Kinouchi, O.; Caticha, N. Lower Bounds for Generalization with Drifting Rules. *J. Phys. A* **1993**, *26*, 6161.
11. Copelli, M.; Caticha, N. On-line learning in the Committee Machine. *J. Phys. A* **1995**, *28*, 1615.
12. Vicente, R.; Caticha, N. Functional optimization of online algorithms in multilayer neural networks. *J. Phys. A Gen. Phys.* **1997**, *30*. doi:10.1088/0305-4470/30/17/002.
13. Caticha, N.; de Oliveira, E. Gradient descent learning in and out of equilibrium. *Phys. Rev. E* **2001**, *63*, 061905.
14. Opper, M. *A Bayesian Approach to Online Learning in On-line Learning in Neural Networks*; Saad, D., Ed.; Cambridge University Press: Cambridge, UK, 1998.
15. Solla, S.A.; Winther, O. Optimal online learning: A Bayesian approach. *Comput. Phys. Commun.* **1999**, *121–122*, 94–97.
16. Caticha, N.; Vicente, R. Agent-based Social Psychology: From Neurocognitive Processes to Social Data. *Adv. Complex Syst.* **2011**, *14*, 711–731.
17. Vicente, R.; Susemihl, A.; Jerico, J.; Caticha, N. Moral foundations in an interacting neural networks society: A statistical mechanics analysis. *Phys. A Stat. Mech. Appl.* **2014**, *400*, 124–138. doi:doi:10.1016/j.physa.2014.01.013.

18. Caticha, N.; Cesar, J.; Vicente, R. For whom will the Bayesian agents vote? *Front. Phys.* **2015**, *3*. doi:10.3389/fphy.2015.00025.
19. Caticha, N.; Alves, F. Trust, law and ideology in a NN agent model of the US Appellate Courts. In *ESANN 2019 Proceedings, Proceedings of the European Symposium on Artificial Neural Networks, Computational Intelligence and Machine Learning, Bruges, Belgium, 24–26 April 2019*; pp. 511–516. ISBN 978-287-587-065-0. Available online: https://www.elen.ucl.ac.be/Proceedings/esann/esannpdf/es2019-72.pdf (accessed on 19 November 2019).

© 2019 by the authors. Licensee MDPI, Basel, Switzerland. This article is an open access article distributed under the terms and conditions of the Creative Commons Attribution (CC BY) license (http://creativecommons.org/licenses/by/4.0/).

*Proceedings*

# Variational Bayesian Approach in Model-Based Iterative Reconstruction for 3D X-Ray Computed Tomography with Gauss-Markov-Potts Prior [†]

**Camille Chapdelaine [1,*], Ali Mohammad-Djafari [2], Nicolas Gac [2] and Estelle Parra [1]**

[1] SAFRAN SA, Safran Tech, Pôle Technologie du Signal et de l'Information, 78772 Magny-Les-Hameaux, France; estelle.parra@safrangroup.com
[2] Laboratoire des signaux et systèmes, CNRS, CentraleSupélec-Université Paris-Saclay, 91190 Gif-sur-Yvette, France; ali.mohammad-djafari@l2s.centralesupelec.fr (A.M.-D.); nicolas.gac@l2s.centralesupelec.fr (N.G.)
* Correspondence: camille.chapdelaine@safrangroup.com
[†] Presented at the 39th International Workshop on Bayesian Inference and Maximum Entropy Methods in Science and Engineering, Garching, Germany, 30 June–5 July 2019.

Published: 21 November 2019

**Abstract:** 3D X-ray Computed Tomography (CT) is used in medicine and non-destructive testing (NDT) for industry to visualize the interior of a volume and control its healthiness. Compared to analytical reconstruction methods, model-based iterative reconstruction (MBIR) methods obtain high-quality reconstructions while reducing the dose. Nevertheless, usual Maximum-A-Posteriori (MAP) estimation does not enable to quantify the uncertainties on the reconstruction, which can be useful for the control performed afterwards. Herein, we propose to estimate these uncertainties jointly with the reconstruction by computing Posterior Mean (PM) thanks to Variational Bayesian Approach (VBA). We present our reconstruction algorithm using a Gauss-Markov-Potts prior model on the volume to reconstruct. For PM calculation in VBA, the uncertainties on the reconstruction are given by the variances of the posterior distribution of the volume. To estimate these variances in our algorithm, we need to compute diagonal coefficients of the posterior covariance matrix. Since this matrix is not available in 3D X-ray CT, we propose an efficient solution to tackle this difficulty, based on the use of a matched pair of projector and backprojector. In our simulations using the Separable Footprint (SF) pair, we compare our PM estimation with MAP estimation. Perspectives for this work are applications to real data as improvement of our GPU implementation of SF pair.

**Keywords:** Computed Tomography, Gauss-Markov-Potts, variational Bayesian approach, Separable Footprint

---

## 1. Introduction

In 3D X-ray CT, MBIR methods enforce a prior model on the volume to image, so the reconstruction quality is enhanced compared to filtered backprojection (FBP) methods [1], and the dose can be reduced [2]. Smoothing and edge-preserving priors, such as total variation regularization [3,4], Gauss-Markov-Potts prior model [5] or sparsity-inducing priors in a wavelet or learnt transform domain [6–8], have provided promising results for the development of MBIR methods in medicine and NDT for industry. Due to the high dimension and to the fact that the reconstruction problem is ill-posed [9], exact estimation of the unknown volume is not available [10]. As a consequence, uncertainties on the estimation are a desirable tool for the analysis of the reconstructed volume.

After the reconstruction has been performed, an iterative method to estimate the uncertainties is proposed in [10]. Nevertheless, its high computational cost makes it only applicable to a few voxels of interest [10]. Since MBIR methods mostly estimate the maximum of the posterior distribution of the unknowns (MAP), confidence regions can be computed following the reconstruction [11] but this procedure is difficult to apply for discrete-continuous channels estimation, such as joint reconstruction and segmentation [5]. For this reason, in this paper, we propose to compute Posterior Mean (PM) rather than MAP. For PM estimator, the uncertainties on the reconstruction correspond to the variances. Our algorithm estimates these variances jointly with the reconstruction based on variational Bayesian approach (VBA) [12,13].

In the following, we first present our reconstruction algorithm based on VBA, applied with a Gauss-Markov-Potts prior model on the volume to reconstruct [5]. To implement this algorithm, the main difficulty is the computation of diagonal coefficients of the posterior covariance matrix, which are linked to projection and backprojection operators (P/BP) : we solve this problem thanks to the use of a matched pair which is here the Separable Footprint (SF) [14]. We present simulation results and compare the obtained reconstruction with the one given by joint maximization a posteriori (JMAP) [5,15]. To the best of our knowledge, this work is the first attempt to apply VBA to a very general 3D inverse problem such as 3D X-ray CT.

## 2. Variational Bayesian Approach

We consider a cone-beam acquisition process : X-rays are sent from a source through the object to control and hit a flat detector which measures the decrease of intensity they have undergone inside the volume. Several perspectives of the volume are acquired by rotating the object around its vertical axis. The $M$ collected measurements $g$ are called the projections and are connected to volume $f$, of size $N$, by the linear forward model taking uncertainties into account [16]

$$g = Hf + \zeta \quad (1)$$

where $H$ is called the projection operator. Its adjoint $H^T$ is the backprojection operator [14]. Since both the data and the volume are huge, matrix $H$, which is size $M \times N$, is not storable in memory. Consequently, successive projections and backprojections in MBIR methods are computed on-the-fly [14,15]. Uncertainties $\zeta$ are zero-mean Gaussian [16]

$$p(\zeta_i | \rho_{\zeta_i}) = \mathcal{N}(\zeta_i | 0, \rho_{\zeta_i}^{-1}), \forall i \in \{1, \ldots, M\}. \quad (2)$$

Precisions $\rho_\zeta = (\rho_{\zeta_i})_i$ are assigned Gamma conjugate prior [5] :

$$p(\rho_{\zeta_i} | \alpha_{\zeta_0}, \beta_{\zeta_0}) = \mathcal{G}(\rho_{\zeta_i} | \alpha_{\zeta_0}, \beta_{\zeta_0}), \forall i. \quad (3)$$

The prior model on the volume is a Gauss-Markov-Potts prior which consists in labelling each voxel $j$ according to its material $z_j = k \in \{1, \ldots, K\}$, where $K$ is the number of materials. Then, the distribution of value $f_j$ of voxel $j$ depends on its material $z_j$ :

$$f_j \sim \mathcal{N}(m_k, \rho_k^{-1}) \text{ if } z_j = k, \forall j \in \{1, \ldots, N\}. \quad (4)$$

Means $m = (m_k)_k$ and inverses $\rho = (\rho_k)_k$ of variances of the classes have to be estimated and are assigned conjugate priors [5] :

$$\begin{cases} p(m_k | m_0, v_0) = \mathcal{N}(m_k | m_0, v_0) \\ p(\rho_k | \alpha_0, \beta_0) = \mathcal{G}(\rho_k | \alpha_0, \beta_0) \end{cases}, \forall k. \quad (5)$$

A Potts model is assigned to labels $z$ in order to favour compact regions in the volume [5] : denoting by $\mathcal{V}(j)$ the neighbourhood of voxel $j$, we have, according to Hammersley-Clifford theorem [17],

$$p(z|\alpha,\gamma_0) \propto \exp\left[\sum_{j=1}^{N}\left(\sum_{k=1}^{K}\alpha_k\delta(z_j-k) + \gamma_0 \sum_{i\in\mathcal{V}(j)}\delta(z_j-z_i)\right)\right]. \qquad (6)$$

From our prior model $\mathcal{M}$, the posterior distribution of unknowns $\psi = (f, z, \rho_\zeta, m, \rho)$ is given by Bayes' rule [5]

$$p(f,z,\rho_\zeta,m,\rho|g;\mathcal{M}) \propto p(g|f,\rho_\zeta)p(f|z,m,\rho)p(z|\alpha,\gamma_0)p(\rho_\zeta|\alpha_{\zeta_0},\beta_{\zeta_0})p(m|m_0,v_0)p(\rho|\alpha_0,\beta_0), \qquad (7)$$

where $\alpha = (\alpha_k)_k$. Based on this distribution, JMAP can be performed [5] but does not provide uncertainties on the result. MCMC methods for joint computation of the means and the variances of the posterior distribution are too computationally costly for 3D applications [5,18]. For this reason, we apply VBA which consists in approximating the true posterior distribution $p$ by a simpler distribution $q$ on which posterior means and variances can be easily estimated. Approximating distribution $q$ minimizes Kullback-Leibler (KL) divergence $KL(q||p)$ on a chosen set of simple distributions [12]. The choice we make for $q$ is a factorizable approximation, which only preserves a dependence between value $f_j$ of voxel $j$ and its label [19] :

$$q(f,z,\rho_\zeta,m,\rho) = \prod_{j=1}^{N} q_{f_j}(f_j|z_j) \times \prod_{j=1}^{N} q_{z_j}(z_j) \times \prod_{i=1}^{M} q_{\rho_{\zeta_i}}(\rho_{\zeta_i}) \times \prod_{k=1}^{K} q_{m_k}(m_k) \times \prod_{k=1}^{K} q_{\rho_k}(\rho_k). \qquad (8)$$

Minimizing KL divergence with respect to each factor while fixing the others leads to [13,19]

$$\begin{cases} q_{f_j}(f_j|z_j = k) = \mathcal{N}(f_j|\tilde{m}_{jk}, \tilde{v}_{jk}) \\ q_{z_j}(k) \propto \exp\left[\tilde{\alpha}_{jk} + \gamma_0 \sum_{i\in\mathcal{V}(j)} q_{z_i}(k)\right], \forall k \\ q_{\rho_{\zeta_i}}(\rho_{\zeta_i}) = \mathcal{G}(\rho_{\zeta_i}|\tilde{\alpha}_{\zeta_{0_i}}, \tilde{\beta}_{\zeta_{0_i}}) \\ q_{m_k}(m_k) = \mathcal{N}(m_k|\tilde{m}_{0_k}, \tilde{v}_{0_k}) \\ q_{\rho_k}(\rho_k) = \mathcal{G}(\rho_k|\tilde{\alpha}_{0_k}, \tilde{\beta}_{0_k}) \end{cases} \qquad (9)$$

The VBA algorithm turns into the iterative updating of the parameters of these distributions with respect to the others. The updating formulae and the order of their applications are given in [13]. In particular, at iteration $t$, the variances of the approximating distribution for the volume are updated by

$$\tilde{v}_{jk}^{(t)} = \left(\frac{\tilde{\alpha}_{0_k}^{(t-1)}}{\tilde{\beta}_{0_k}^{(t-1)}} + \left[H^T \tilde{V}_\zeta^{-1} H\right]_{jj}\right)^{-1} \qquad (10)$$

where $\tilde{V}_\zeta = \text{diag}\,[\tilde{v}_\zeta]$ and $\tilde{v}_{\zeta_i} = \frac{\tilde{\beta}_{\zeta_{0_i}}^{(t-1)}}{\tilde{\alpha}_{\zeta_{0_i}}^{(t-1)}}, \forall i$ [13]. Moreover, the updating formula for intensity parameter of the approximating Gamma distribution for $\rho_{\zeta_i}$ is [13]

$$\tilde{\beta}_{\zeta_{0_i}}^{(t)} = \beta_{\zeta_0} + \frac{1}{2}\left((g_i - [H\tilde{m}]_i)^2 + \left[H\tilde{V}H^T\right]_{ii}\right) \qquad (11)$$

where $\tilde{V} = \text{diag}\,[v]$ and

$$\begin{cases} \tilde{m}_j = \sum_{k=1}^{K} \tilde{m}_{jk}^{(t)} q_{z_j}^{(t)}(k) \\ \tilde{v}_j = \sum_{k=1}^{K} \left[\tilde{v}_{jk}^{(t)} + \left(\tilde{m}_{jk}^{(t)} - \tilde{m}_j\right)^2\right] q_{z_j}^{(t)}(k) \end{cases}. \qquad (12)$$

To compute approximate posterior variances, formula (10) needs the computation of diagonal coefficients of $H^T \tilde{V}_\zeta^{-1} H$, while formula (11) needs diagonal coefficients of $H\tilde{V} H^T$. Both of these matrices imply projector and backprojector which are not in memory, contrary to 2D applications [19]. Therefore, in order to implement VBA for 3D X-ray CT, we need to find a way to compute diagonal coefficients in formulae (10) and (11) efficiently. We propose a strategy which is detailed in the next section.

## 3. Computation of diagonal coefficients

At one iteration of the algorithm, for any voxel $j$, diagonal coefficient used to compute $v_{jk}$ by (10) is

$$d_{v_j} = \left[ H^T \tilde{V}_\zeta^{-1} H \right]_{jj} = \| H e^{(j)} \|^2_{\tilde{V}_\zeta} \tag{13}$$

where $e_i^{(j)} = \delta(j-i), \forall i$. As $d_v = (d_{v_j})_j$ has the size of a volume, formula (13) implies to compute $N$ projections, which is very long, even if the projector implemented on GPU is very fast. We calculated that, if we have to reconstruct a volume of size $N = 256^3$ voxels from 64 projections of size $256^2$ pixels, and if one projection takes only 10 milliseconds, computing all dialgonal coefficients $d_{v_j}, \forall j$, for only one iteration of proposed VBA algorithm [13], would require more than 40 hours. Due to this huge computational cost, we prefer to consider the algebraic formula :

$$d_{v_j} = \left[ H^T \tilde{V}_\zeta^{-1} H \right]_{jj} = \sum_{i=1}^{M} H_{ij}^2 \tilde{v}_{\zeta_i}^{-1}, \forall j. \tag{14}$$

From this formula, diagonal coefficients $d_v$ appear to be similar to a backprojection of $\tilde{v}_\zeta^{-1} = (\tilde{v}_{\zeta_i}^{-1})_i$, except that coefficients $H_{ij}$ are replaced by their squares $H_{ij}^2, \forall i,j$. Similarly, diagonal coefficients

$$d_{\zeta_i} = \left[ H \tilde{V} H^T \right]_{ii} = \sum_{j=1}^{N} H_{ij}^2 \tilde{v}_j, \forall i, \tag{15}$$

appear like a projection of volume $\tilde{v}$, with $H_{ij}^2$ instead of $H_{ij}$. Given formulae (14) and (15), we implement a *squared-projector* $H^{(2)}$ such that $H_{ij}^{(2)} = H_{ij}^2, \forall i,j$, and a *squared-backprojector* $(H^{(2)})^T$. Both are implemented exactly like the projector and the backprojector respectively. In order to ensure the validity of formulae (14) and (15), and therefore the convergence of our algorithm, we use a matched P/BP pair, which is here the Separable Footprint (SF) pair [14]. This pair is implemented on GPU as described in [15]. The same implementation is used for $H^{(2)}$ and $(H^{(2)})^T$.

Thanks to these new operators, in one iteration of our algorithm, diagonal coefficients $d_{v_j}, \forall j$, are simultaneously computed by applying $(H^{(2)})^T$, which is very fast because it takes exactly the same time as a backprojection, instead of $N$ projections. Similarly, diagonal coefficients $d_{\zeta_i}, \forall i$, are simultaneously computed by applying $H^{(2)}$, as fast as one projection, instead of $M$ backprojections.

Figure 1 shows diagonal coefficients of $HH^T$ and $H^T H$, computed by $H^{(2)}$ and $(H^{(2)})^T$ respectively. Diagonal coefficients of $HH^T$ have the size of projections and are shown as it in Figure 1, while those of $H^T H$ are shown as a volume. We now apply our VBA algorithm to simulated data, and compare the estimated PM with JMAP. JMAP algorithm is described in [5] and applied with SF pair as we did in [15].

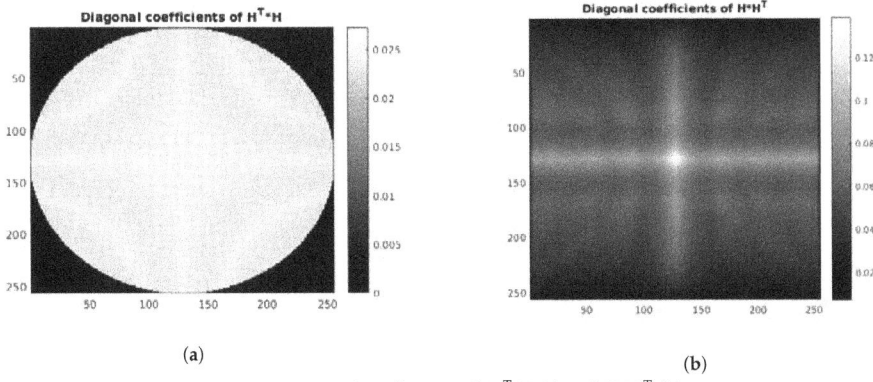

**Figure 1.** Diagonal coefficients of $H^T H$ (a) and $H H^T$ (b).

## 4. Results

The simulated phantom is of size $256^3$ voxels and contains $K = 5$ classes. It is shown in Figure 2. We reconstruct this volume from 64 projections of size $256^2$ pixels, uniformly distributed over $[0, 2\pi]$. These projections are noisy with SNR equal to 20 db.

Parameters $(\alpha_{\zeta_0}, \beta_{\zeta_0}, \alpha_0, \beta_0)$ are fixed near Jeffreys' prior as in [13,19]. The strategies to fix other parameters $\alpha$, $\gamma_0$, $m_0$ and $v_0$ are explained in [13]. The values of the parameters for VBA are given in Table 1, excepted $m_0$ and $\alpha$ which are fixed automatically as in [5]. For our comparison, the parameters are the same for JMAP.

**Table 1.** Parameters for JMAP and VBA algorithms.

| Parameters | K | $\gamma_0$ | $v_0$ | $\alpha_{\zeta_0}$ | $\beta_{\zeta_0}$ | $\alpha_0$ | $\beta_0$ |
|---|---|---|---|---|---|---|---|
| Values | 5 | 6 | 1 | $10^{-4}$ | $10^{-2}$ | $10^{-6}$ | $10^{-2}$ |

The initialization of approximating distributions for VBA is described in [13]. This initialization requires initial volume and segmentation, obtained as explained in [13]. The same initialization is used for JMAP.

Figures 3 and 4 show the reconstructions obtained by JMAP and VBA respectively. They are compared with total-variation (TV) regularization. For TV, the reconstruction, shown in Figure 5, is obtained thanks to Primal-Dual Frank-Wolfe algorithm (PDFW) [20]. Thanks to the use of Gauss-Markov-Potts prior model, JMAP and VBA reconstructions have compact and well-distinguishable regions, while contours are slightly blurred for TV. VBA reconstruction has smoother contours than JMAP.

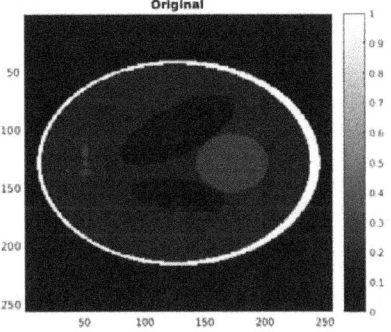

**Figure 2.** Original phantom.

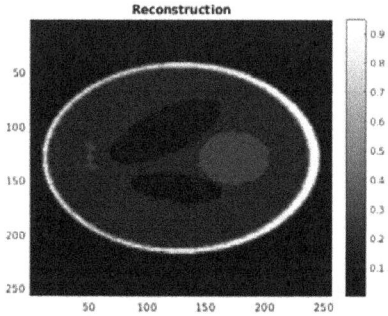

**Figure 3.** Reconstruction by JMAP.

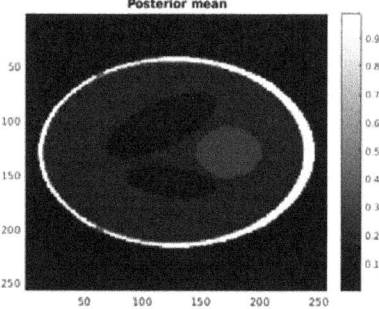

**Figure 4.** Reconstruction by VBA.

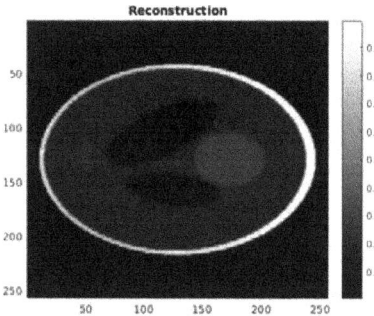

**Figure 5.** Reconstruction by PDFW.

For each reconstruction, the $\mathcal{L}_2$-relative error with respect to the original phantom is shown in Table 2. As we see in Figure 4, details are lost by VBA because of the factorized approximating distribution. Consequently, VBA has the highest error, while it is roughly the same for PDFW and JMAP. The variances of the posterior distribution of the volume estimated by VBA are shown in Figure 6. Unsurprisingly, the highest variances are on the thinest part of the phantom which is the bone. Nevertheless, the loss of details in the reconstruction is not highlighted by posterior variances. Indeed, uncertainties are known to be under-estimated in VBA when considering divergence $KL(q||p)$ [12]. The stop criterion for PDFW is given in [20] and is minimized, while those for JMAP and VBA are maximized and given in [5,13] respectively. For each algorithm, the evolution of stop criterion is shown in Figures 7–9 respectively. One iteration of JMAP contains 20 sub-iterations and few sub-iterations for segmentation step [5], while VBA and PDFW do not have sub-iterations [13,20]. Consequently, in Table 2, the computation time of VBA is much less than the one of JMAP and quite similar to the one of PDFW. Furthermore, during our experiments, we have noticed that, compared to JMAP, VBA has a higher sensitivity to the choice of the parameters, as to the number of iterations. Indeed, for a too large number of iterations of VBA, the reconstruction is over-regularized. This is a drawback of VBA compared to JMAP.

Moreover, the memory cost of VBA is much higher than the one of JMAP and PDFW. This makes VBA only applicable to small regions-of-interest (ROI), typically of size $256^3$. Based on a reconstruction of high quality (for instance, obtained by JMAP [5]), the reconstruction of ROI can be performed following the method of [21], as done for other MBIR methods [14]. This point will be covered in future works.

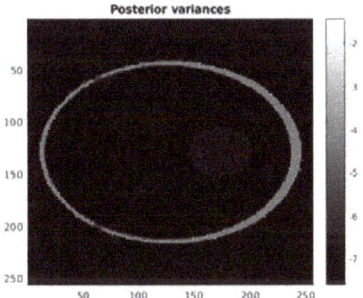

**Figure 6.** Variances (log) obtained by VBA

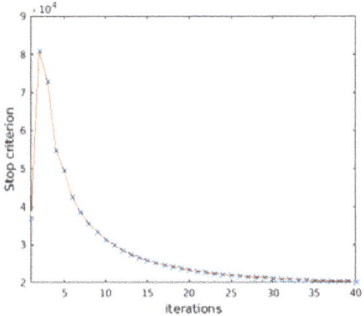

**Figure 7.** Convergence of PDFW

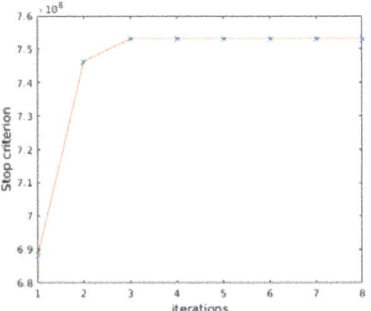

**Figure 8.** Convergence of JMAP

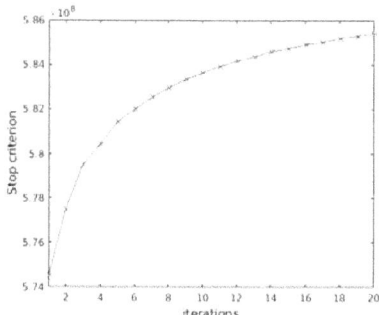

**Figure 9.** Convergence of VBA

**Table 2.** Comparaison of PDFW, JMAP and VBA algorithms.

| Algorithm | $\mathcal{L}_2$-Relative Error | Computation Time |
|---|---|---|
| PDFW | 6.0 % | 126.3 s |
| JMAP | 9.1 % | 751.6 s |
| VBA | 13.5 % | 150.0 s |

## 5. Conclusions and Perspectives

In this paper, we have presented an application for 3D X-ray CT of variational Bayesian approach (VBA) with Gauss-Markov-Potts prior model. By computing posterior mean (PM) thanks to VBA, we have been able to jointly perform the reconstruction and the estimation of the posterior variances, which give the uncertainties on the reconstruction. To compute these variances, we have seen that the huge dimension in 3D X-ray CT hinders to easily get diagonal coefficients, due to the fact that projection and backprojection operators cannot be stored in memory. To tackle this problem, we have taken benefit from the use of a matched pair of projector and backprojector, which was the Separable Footprint (SF) one : based on this pair, we have implemented "squared" projector and backprojector which have enabled us to compute diagonal coefficients on-the-fly. The GPU implementation for these squared operators was the same we used for SF projector and backprojector.

Our tests on simulated data and comparison with joint maximization a posteriori (JMAP) have shown that VBA obtains smoother contours than JMAP and converges faster. Although the memory cost of VBA is higher than the one of JMAP, we have underlined that the algorithm can be applied to estimate the uncertainties in a region-of-interest (ROI). Future works will focus on applications to real and bigger data, as on optimization of GPU implementation of SF pair [15]. Other variational Bayesian algorithms will also be worth to study in order to improve the estimation of uncertainties.

**Funding:** This research was funded by CIFRE Grant 2016/0188 from French Agence Nationale de la Recherche et de la Technologie (ANRT).

## References

1. Feldkamp, L.; Davis, L.; Kress, J. Practical cone-beam algorithm. *JOSA A* **1984**, *1*, 612–619.
2. Fessler, J.A. Statistical image reconstruction methods for transmission tomography. *Hand. Med. Imaging* **2000**, *2*, 1–70.
3. Sidky, E.Y.; Jakob, H.; Pan, X. Convex optimization problem prototyping for image reconstruction in computed tomography with the Chambolle & Pock algorithm. *Phys. Med. Biol.* **2012**, *57*, 3065.
4. McGaffin, M.G.; Fessler, J.A. Alternating dual updates algorithm for X-ray CT reconstruction on the GPU. *IEEE Trans. Comput. Imaging* **2015**, *1*, 186–199.

5. Chapdelaine, C.; Mohammad-Djafari, A.; Gac, N.; Parra, E. A 3D Bayesian Computed Tomography Reconstruction Algorithm with Gauss-Markov-Potts Prior Model and its Application to Real Data. *Fundam. Inform.* **2017**, *155*, 373–405.
6. Xu, Q.; Yu, H.; Mou, X.; Zhang, L.; Hsieh, J.; Wang, G. Low-dose X-ray CT reconstruction via dictionary learning. *IEEE Trans. Med. Imaging* **2012**, *31*, 1682–1697.
7. Vandeghinste, B.; Goossens, B.; Van Holen, R.; Vanhove, C.; Pižurica, A.; Vandenberghe, S.; Staelens, S. Iterative CT reconstruction using shearlet-based regularization. *IEEE Trans. Nucl. Sci.* **2013**, *60*, 3305–3317.
8. Zheng, X.; Ravishankar, S.; Long, Y.; Fessler, J.A. PWLS-ULTRA: An efficient clustering and learning-based approach for low-dose 3D CT image reconstruction. *IEEE Trans. Med. Imaging* **2018**, *37*, 1498–1510.
9. Idier, J. *Bayesian Approach to Inverse Problems*; John Wiley & Sons: Hoboken, NJ, USA, 2008.
10. Fessler, J.A. Mean and Variance of Implicitly Defined Biased Estimators (such as Penalized Maximum Likelihood): Applications to Tomography. *IEEE Trans. Image Process.* **1996**, *5*, 493–506.
11. Pereyra, M. Maximum-A-Posteriori estimation with Bayesian Confidence Regions. *SIAM J. Imaging Sci.* **2017**, *10*, 285–302.
12. Pereyra, M.; Schniter, P.; Chouzenoux, E.; Pesquet, J.C.; Tourneret, J.Y.; Hero, A.O.; McLaughlin, S. A survey of stochastic simulation and optimization methods in signal processing. *IEEE J. Select. Top. Signal Process.* **2016**, *10*, 224–241.
13. Chapdelaine, C. Variational Bayesian Approach and Gauss-Markov-Potts prior model. *arXiv* **2018**, arXiv:1808.09552.
14. Long, Y.; Fessler, J.A.; Balter, J.M. 3D forward and back-projection for X-ray CT using separable footprints. *IEEE Trans. Med. Imaging* **2010**, *29*, 1839–1850.
15. Chapdelaine, C.; Gac, N.; Mohammad-Djafari, A.; Parra, E. New GPU implementation of Separable Footprint Projector and Backprojector: First results. In Proceedings of the 5th International Conference on Image Formation in X-Ray Computed Tomography, 20–23 May 2018.
16. Sauer, K.; Bouman, C. A local update strategy for iterative reconstruction from projections. *IEEE Trans. Signal Process.* **1993**, *41*, 534–548.
17. Besag, J. Spatial interaction and the statistical analysis of lattice systems. *J. R. Stat. Soc. Ser. B (Methodological)* **1974**, *36*, 192–236.
18. Zhao, N.; Basarab, A.; Kouame, D.; Tourneret, J.Y. Joint segmentation and deconvolution of ultrasound images using a hierarchical Bayesian model based on generalized Gaussian priors. *IEEE Trans. Image Process.* **2016**, *25*, 3736–3750.
19. Ayasso, H.; Mohammad-Djafari, A. Joint NDT image restoration and segmentation using Gauss–Markov–Potts prior models and variational bayesian computation. *IEEE Trans. Image Process.* **2010**, *19*, 2265–2277.
20. Ongie, G.; Murthy, N.; Balzano, L.; Fessler, J.A. A Memory-Efficient Algorithm for Large-Scale Sparsity Regularized Image Reconstruction. In Proceedings of the Fifth International Conference on Image Formation in X-Ray Computed Tomography, 2018.
21. Ziegler, A.; Nielsen, T.; Grass, M. Iterative reconstruction of a region of interest for transmission tomography. *Med. Phys.* **2008**, *35*, 1317–1327.

© 2019 by the authors. Licensee MDPI, Basel, Switzerland. This article is an open access article distributed under the terms and conditions of the Creative Commons Attribution (CC BY) license (http://creativecommons.org/licenses/by/4.0/).

*Proceedings*

# Bayesian Approach with Entropy Prior for Open Systems [†]

Natalya Denisova

Institute of Theoretical and Applied Mechanics, 630090 Novosibirsk, Russia; NVDenisova2011@mail.ru

[†] Presented at the 39th International Workshop on Bayesian Inference and Maximum Entropy Methods in Science and Engineering, Garching, Germany, 30 June–5 July 2019.

Published: 12 November 2019

**Abstract:** The Bayesian approach Maximum a Posteriori (MAP) is discussed in the context of solving the image reconstruction problem in nuclear medicine: positron emission tomography (PET) and single photon emission computer tomography (SPECT). Two standard probabilistic forms, Gibbs and entropy prior probabilities, are analyzed. It is shown that both the entropy-based and Gibbs priors in their standard formulations result in global regularization when a single parameter controls the solution. Global regularization leads to over-smoothed images and loss of fine structures. Over-smoothing is undesirable, especially in oncology in diagnosis of cancer lesions of small size and low activity. To overcome the over-smoothing problem and to improve resolution of images, the new approach based on local statistical regularization is developed.

**Keywords:** image reconstruction problem; Bayesian Maximum a Posteriori approach; entropy prior probability; global regularization; local statistical regularization; open systems

## 1. Introduction

Emission tomography techniques, including (PET) and single photon emission computer tomography (SPECT), produce images which are used for clinical diagnosis. Image quality depends substantially on applied reconstruction methods. The deterministic filtered back projection (FBP) method was used for image reconstruction up to the 90-th. Later, statistical iterative methods based on the maximum likelihood principle, entered the nuclear medicine. Statistical algorithms, such as the Maximum Likelihood Expectation Maximization (MLEM) and its accelerated version - Ordered Subsets Expectation Maximization (OSEM)—take into account the stochastic properties of the observed data and provide more accurate images in comparison to FBP method. At the moment, OSEM is the standard algorithm in PET and SPECT systems all over the world. However, OSEM algorithm has fundamental limitations in solving reconstruction problems and unexpected artifacts may arise in the obtained images. These limitations are due to the fact that from the mathematical point of view, image reconstruction belongs to the class of ill-posed inverse problems and regularization is necessary for its correct solution. Regularization is the process of introducing an additional a priori assumption about the solution to obtain a well behaved inverse [1]. Initially, the regularization method was developed by using the deterministic form of prior information. However, the deterministic form of prior information has limitations with the drawback to lead to excessively over-smoothed solutions. The stochastic nature of the observed data requires the probabilistic approach for assigning prior information. V. Turchin suggested to use the Bayesian method of Maximum a Posteriori (MAP) for solving ill-posed problems with stochastic data, naming this approach 'statistical regularization' [2]. The statistical regularization method introduces the prior information in the form of a probability distribution. Two basic probabilistic forms of prior information, widely used in reconstruction tomography, are: Gibbs prior and entropy prior. Both deterministic and statistical regularization methods were developed as 'global regularization', in which a single parameter controls the solution in the whole solution's area. It was expected, that the regularized MAP algorithms should provide more accurate reconstruction of fine structures in comparison to the non-regularized OSEM. However, in [3], rather minor differences between the images obtained by the OSEM algorithm and MAP algorithm with Gibbs prior (MAP-Gibbs) were found. In [4], numerical simulations have shown that by using the entropy-based MAP algorithm (MAP-ENT), the reconstruction errors decrease monotonically with excellent stability, in contrary, the OSEM algorithm behavior was unstable. The OSEM reconstructed image was obtained by stopping the iteration process at the point of minimal error and similarly to the resulting MAP-ENT image. We assume that the cause of the minor difference between OSEM and MAP images is due to the global regularization method, applied in both MAP algorithms. Global regularization smoothes the solution too strong and therefore fine structures may be over-smoothed or lost. Local regularization is needed to improve image quality. The idea of local regularization for statistical MAP approach has not been considered in literature.

The aim of this paper is the theoretical analysis of local regularization method in the frame of statistical Bayesian MAP approach.

## 2. Methods

### 2.1. Bayesian Approach for Solving Tomographic Problems in Nuclear Medicine

In SPECT and PET diagnostic procedures, a patient is administered a radiolabeled pharmaceutical which is distributed with the concentration $n(q)$ in various regions of the body ($q$ is the space coordinate). The function $f(q)$ describes the density of gamma photons produced through radioactive decay. It is assumed that $f(q)$ is a random value which follows the Poisson distribution with mean $\bar{f}$ which is proportional to the radiopharmaceutical concentration $n$. The image reconstruction problem is presented as linear equation:

$$Af = g \tag{1}$$

$A$ is a system matrix which describes data acquisition process and $g$ are registered stochastic data. Gamma photons are emitted by the radiopharmaceutical and are registered by gamma camera. Gamma camera rotates around the patient body and collects gamma photons from different angles. The registered raw data are called projection data. In myocardial perfusion SPECT imaging the data are obtained from 32 or 64 views. An example of clinical raw projection data is shown in Figure 1. The data of 3 from total 64 views are shown.

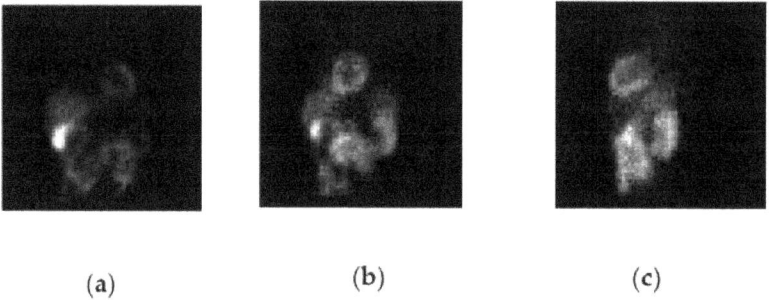

**Figure 1.** Clinical SPECT projection data. The data are demonstrated for three selected views: anterior (**a**), left anterior oblique (**b**) and left lateral (**c**) projections.

The data are obtained in Meshalkin National Medical Research Center (Novosibirsk) by using GE Infinia Hawkeye SPECT/CT system.

The data are assumed to be distributed according to a Poisson law:

$$P(g|\bar{g}) = e^{-\bar{g}} \frac{\bar{g}^g}{g!} \qquad (2)$$

with mean $\bar{g}$:

$$\bar{g} = A\bar{f} \qquad (3)$$

The reconstruction problem (1) is formulated as a classical problem of mathematical statistics: what is the probability density $P(\bar{f}|g)$ for the solution $\bar{f}$ with given data $g$? By using the Bayesian Maximum a Posteriori (MAP) theorem, $P(\bar{f}|g)$ is determined as:

$$P(\bar{f}|g) = \frac{P(\bar{f})P(g|A\bar{f})}{\int P(\bar{f})P(g|A\bar{f})d\bar{f}} \qquad (4)$$

where $P(\bar{f})$ is a prior probability density function and $P(g|A\bar{f})$ is the likelihood distribution of the observed data. The most probable MAP estimation is obtained by maximizing $P(\bar{f}|g)$ in logarithmic form:

$$\tilde{\bar{f}} = \max_{\bar{f}}\{\ln P(\bar{f}) + \ln P(g|A\bar{f})\} \qquad (5)$$

For applications in nuclear medicine, the likelihood distribution in logarithmic form is defined in accordance to (2):

$$\ln P(g|A\bar{f}) = g \ln A\bar{f} - A\bar{f} - \ln g! \qquad (6)$$

### 2.2. Maximum-Likelihood-Based Image Reconstruction Method

When a researcher does not have any prior information the simple way is to assume that all possibilities are equally probable and Bayesian estimation reduces to the Maximum Likelihood (ML) solution:

$$\tilde{\bar{f}} = \arg_{\bar{f}} \max\{\ln P(g|A\bar{f})\} \qquad (7)$$

In practical PET and SPECT applications reconstruction problem is usually discretized. The reconstruction area is assumed to be divided into $J$ voxels in which the unknown source function $\bar{f}$ is distributed with a discrete density $\bar{f}_j$. The projection data are presented in the discrete form as a system of the linear algebraic equations:

$$g_i = \sum_j A_{ij} f_j \qquad (8)$$

where $g_i$ are the measured data in the $i$-th detector cell, matrix $A_{ij}$ is the system matrix describing the possibility that photon emitted in the $j$-th voxel will be detected in $i$-th 'detector'. The ML method (7) provides the following solution:

$$\tilde{\bar{f}}_j^{n+1} = \frac{\tilde{\bar{f}}_j^n}{\sum_i A_{ij}} \sum_i \frac{g_i A_{ij}}{\sum_j A_{ij} \tilde{\bar{f}}_j^n} \qquad (4)$$

$\tilde{f}_j^n$ is an estimation of $\bar{f}_j$ on the $n$-th iteration step. In clinical systems, the OSEM algorithm is used. OSEM is a faster version of ML algorithm. OSEM is not a regularized algorithm, therefore, its behavior in iteration process is unstable and the resulted images are dominated by noisy artifacts. In the first iterations, the error of reconstruction decreases, then it reaches a minimum and begins to increase. At this point a reconstructed image begins to deteriorate. One may think that it is possible to obtain visually good OSEM images by stopping the iteration process at the iteration with minimal error. Nonetheless, image deterioration progresses more rapidly in the zones with lower count statistics. At stopping, a part of the image with low count statistics can become already noisy while another part of the image with high statistics did not reach its optimal resolution yet. This problem is especially important in oncology: a noise in the low count zones imitates virtual 'hot spots' or masks the true 'hot spots'. Currently, stopping of OSEM algorithms is applied in standard SPECT myocardial perfusion imaging protocols. In some cases, the stopping rule may be the cause of artifacts on images.

### 2.3. The Bayesian Image Reconstruction Method with Gibbs Prior

In the literature devoted to medical emission tomography, Bayesian Maximum a Posteriori approach with Gibbs prior is most widely studied. This approach was justified theoretically in [5] and was first applied for SPECT by Geman and McClare in 1984 in [6]. Gibbs prior assumes the Markov Random Field (MRF) distribution for emitted photons. According to the Hammersly-Clifford theorem, the MRF has the distribution which is described by the Gibbs probability:

$$P(\bar{f}) = \frac{1}{Z}\exp(-\beta U(\bar{f})) \qquad (5)$$

where $\beta$ is an unknown parameter, $U(\bar{f})$ is the energy function. Energy function $U(\bar{f})$ is defined as a sum of potentials:

$$U(\bar{f}_j) = \sum_{j \in c} V_c(|\bar{f}_j - \bar{f}_i|) \qquad (11)$$

where $c$ denotes the set of all cliques type, $V_c$ is a potential functions defined on pairwise cliques of neighboring voxels. The Bayesian MAP solution (MAP-Gibbs) of the reconstruction problem (1) is presented in this case as:

$$\bar{f} = \arg_{\bar{f}} \max\{-\beta \sum_i \sum_{j \in c} V_c(|\bar{f}_j - \bar{f}_i|) + g \ln A\bar{f} - A\bar{f} - \ln g!\} \qquad (12)$$

Gibbs prior provides the reconstruction algorithm that specifies local spatial correlations in the source to be reconstructed. A wide range of potential functions $V_c$ have been proposed in the literature. For example, edge-preserving Geman's approximation of the function $V_c$ is defined as follows:

$$V_c = \frac{(\bar{f}_j - \bar{f}_i)^2}{2\delta^2 + (\bar{f}_j - \bar{f}_i)^2} \qquad (13)$$

$\delta$ is edge-preserving parameter. Unfortunately, the physical nature of $V_c$ is not very clear. Huge efforts are aimed at the development of this approach for PET and SPECT applications. However, there are unsolved problems that inhibit the implementation of these algorithms on commercial systems.

## 2.4. Bayesian Image Reconstruction Method with Entropy Prior

Another form of prior, based on the entropy principle, was suggested by Jaynes [7–9]. Maximum-entropy (ME) approach has been successfully applied in the fields of X-ray-, radio- and gamma-astronomy, plasma tomography [10–14], but less in medicine tomography. Applied to radio-astronomical data, the maximum entropy algorithm reveals details not seen by conventional analysis, but which are known to exist. Applied to solve the problem of astronomical image restoring, the Maximum Entropy approach has demonstrated that resolved and unresolved sources can be restored with reliability. These results are of interest in the context of the similarity between astronomical images and 'hot spots' tumor images in nuclear oncology. In the absence of correlations the entropy-based method is superior to the Gibbs' approach in 'hot spots' identifying.

The ME approach goes back to the famous works by Boltzmann and can be formulated as follows: distributions of higher entropy have higher multiplicity and can be realized in more ways. In order to apply the entropy principle to the SPECT and PET imaging problems, the reconstruction area is properly discretized into $J$ boxes. An image is considered as a set of boxes in which a large number of emitting particles are placed. According to the combinatorial theorem, the number of different ways of filling the boxes is given by

$$W = \frac{N_0!}{\prod_j N_j!} \qquad (14)$$

$j$ is a box index, $N_j$ is the number of particles in the $j$-th box and $N_0$ is the total number of particles. One usually uses the Stirling approximation to find the entropy equation in the logarithmic form. The prior probability $P(\bar{f})$ of the unknown function is taken to be proportional to its multiplicity. The image (the number of emitted photons) $\bar{f}_j$ is assumed to be proportional to the density of the emitting particles and, therefore, one obtains

$$\ln P(\bar{f}) = -\beta \sum_{j=1}^{J} \bar{f}_j \ln \bar{f}_j \qquad (15)$$

where $\beta$ is a constant associated with the relation between $N_j$ and $\bar{f}_j$. Maximum *a Posteriori* estimation with entropy prior (MAP-ENT) is written in the discrete form as:

$$\tilde{\bar{f}} = \arg_{\bar{f}} \max\{-\beta \sum_j \bar{f}_j \ln \bar{f}_j + \sum_i [g_i \ln \sum_j A_{ij} \bar{f}_j - \sum_j A_{ij} \bar{f}_j - \ln g_i!]\} \qquad (16)$$

Unlike Gibbs prior, Entropy form does not introduce correlations in the images, beyond those which are required by the data. It is necessary to make the following remark concerning the maximum entropy approach. There are two basic approaches using the entropy prior: the Maximum a Posteriori algorithm with entropy prior and the constrained Maximum Entropy algorithm. A difference between these approaches is as follows: in the MAP algorithm, the data enters as a part of quantity to be maximized, $P(\bar{f}|g)$. With the Maximum Entropy method, the data are added via Lagrange multipliers exterior to the entropy prior $P(\bar{f})$ to be maximized. The Maximum Entropy yields the smoothest solution among all solutions satisfying to given data. A degree of smoothness of the MAP solution is regulated by the regularization parameter.

## 2.5. Bayesian Image Reconstruction Method Based on Open System Theory

Gibbs and entropy forms of prior information result in global regularization method with a single regularization parameter. Global regularization produces too smoothed solution therefore fine structures may be lost on images. Local regularization is needed to improve image quality. The idea of local regularization for statistical MAP approach is considered in this Section. This idea is based on the open systems theory. Open systems are the systems which can be exchanged with environment by energy, matter and information and, from this point of view, emitting objects are open systems. In SPECT and PET diagnostic procedures a radiopharmaceutical is injected into a patient body. Different organs have different metabolism rate for the injected radiopharmaceutical. In the same organ, healthy and ill tissues can also have different metabolism rate. Different radiopharmaceutical accumulation in the organs is associated with a different metabolic rate, which is considered here as a control parameter ' $a$ '. We consider the accumulative model that describes a final steady-state non-equilibrium spatial distribution of radiopharmaceutical in a patient body. Due to radioactive decay, radiopharmaceutical emits gamma photons, so a patient body can be considered as an emitting open system. We assume that processes of nuclear excitation and de-excitation occur at the same rate during all the time of patient examination procedure. So, the distribution of radiopharmaceutical particles in a patient body can be considered as a steady-state open system. The steady-state of the system is defined by a corresponding control parameter. Due to low radiopharmaceutical concentration we can consider an open system of $N$ non-interacting classical particles (ideal gas) in the volume $V$ with the energy $E$. In an equilibrium system, particles are uniformly distributed throughout the volume. Non-uniform spatial distribution of particles can occur in the non-equilibrium steady-state systems. Macroscopic states of an ideal gas can be described through the possible energy distribution of an individual particle. Following [15], in classical case the entropy can be written as:

$$S = -\int gn(q,p,a) \ln n(q,p,a) \frac{dpdq}{\Delta} + \text{const} \tag{17}$$

$q,p$ are spatial and pulse coordinates, $\Delta = (2\pi\hbar)^6$, $g$ is a weight factor associated with different nuclear states of particles. Assuming that energy (pulse) distributions are the same at each point in a discrete physical space, we obtain

$$S = -\sum_j gn(a)_j \ln n(a)_j + \text{const} \tag{18}$$

$n_j(a)$ is radiopharmaceutical concentration in the $j$-th voxel of physical space.

For simplicity, let us focus on one organ, for example, on a liver with the area of healthy tissue and some area of ill tissue. Suppose only two cases with the concentration of radiopharmaceutical particles in the liver $n(a_0)$ and $n(a_1)$. Calculation of entropy can be performed separately for each of these two areas in accordance with the Equation (18). However, definition of total entropy by using the usual Boltzmann formula is not possible because the mean energies of these two states are different. According to Klimontovich S-theorem [16], we can define the total entropy after renormalizing of energies:

$$\langle H(n(a_0)) \rangle = \langle H(n(a_1)) \rangle \tag{19}$$

$H$ is an Hamiltonian function. Integrating (17) by pulses $p$ under condition (19) one obtains the renormalized entropy expressions for the healthy and ill liver areas. Taking into account that mean $\overline{f}$ is proportional to the radiopharmaceutical concentration, we can write the total entropy as:

$$S = -\sum_k \beta_k \sum_{j \in C_k} \overline{f}_j(a_k) \ln \overline{f}_j(a_k) \tag{20}$$

the index $k$ defines the subsystems with different parameter $a$ (different radiopharmaceutical concentration). Maximum *a Posteriori* estimation with entropy prior based on open system theory (MAP-OPEN-ENT) is written in the discrete form as:

$$\widetilde{\overline{f}} = \arg_{\overline{f}} \max \{ -\sum_k \beta_k \sum_{j \in C_k} \overline{f}_j(a_k) \ln \overline{f}_j(a_k) + \sum_i [g_i \ln \sum_j A_{ij} \overline{f}_j - \sum_j A_{ij} \overline{f}_j - \ln g_j !] \tag{21}$$

In solution of SPECT and PET reconstruction problems by using Bayesian approach, the expression (21) leads to local regularization with local regularization parameters $\beta_k$.

## 3. Conclusions

The standard approach for solving the ill-posed inverse problem in image reconstruction is used in the form of 'global regularization', in which a single parameter controls the solution. However, in SPECT and PET imaging, global regularization leads to over-smoothed images and loss of fine structures. Over-smoothing is undesirable, especially in oncology in diagnosis of cancer tumors of small size and low activity. To overcome the over-smoothing problem and to improve resolution of images, the new approach based on local statistical regularization is developed in this work. The theoretical justification of the new method in which the value of the regularization parameter is defined on the base of open system entropy is performed.

**Acknowledgments:** This work was partly supported by Russian Foundation for Basic Research (grant No. 17-52-14004).

**Conflicts of Interest:** The author has no competing interests to disclose.

## References

1. Tikhonov, A.N. Solution of incorrectly formulated problems and the regularization method. *Sov. Math. Doklady* **1963**, *4*, 1035–1038.
2. Turchin, V.F. A solution of Fredholm equation in statistical ensemble of smooth functions. *J. Comput. Math. Math. Phys.* **1967**, *6*, 1270. (In Russian)
3. Nuyts, J.; Fessler, J.A. A penalized-likelihood image reconstruction method for emission tomography, compared to post-smoothed maximum-likelihood with matched spatial resolution. *IEEE Trans. Med. Imaging* **2003**, *22*, 1042–1052.
4. Denisova, N.V.; Terekhov, I.N. A study of myocardial perfusion SPECT imaging with reduced radiation dose using maximum likelihood and entropy-based maximum a posteriori approaches. *Biomed. Phys. Eng. Express* **2016**, *2*, 055015.
5. Geman, S.; Geman, D. Stochastic Relaxation, Gibbs Distributions, and the Bayesian restoration of Images. *IEEE Trans. Pattern Anal. Mach. Intell.* **1984**, *6*, 721–741.
6. Geman, S.; McClure, D. Bayesian image analysis: An application to single photon emission. *Proc. Stat. Comput. Sect.* **1985**, 12–18. Available online: https://discover.libraryhub.jisc.ac.uk/search?q=subject%3A%20Sampling%20%5BStatistics%5D%20Congresses.&rn=11&for=ncl (accessed on 12 November 2019).
7. Jaynes, E.T. Prior Probabilities. *IEEE Trans. Syst. Sci. Cybern.* **1968**, *4*, 227–241.
8. Jaynes, E.T. Information theory and statistical mechanics. *Phys. Rev.* **1957**, *106*, 620–630.
9. Jaynes, E.T. Information theory and statistical mechanics II. *Phys. Rev.* **1958**, *108*, 171–190.
10. Gull, S.F.; Daniell, G.J. Image reconstruction from incomplete and noisy data. *Nature* **1978**, *272*, 686.
11. Skilling, J. The axioms of maximum entropy. In *Maximum Entropy and Bayesian Methods in Science and Engineering*; Ericksonk, G.J., Smith, C.R., Eds.; Springer: Heidelberg, Germany, 1988; Volume 1, pp. 173–188.
12. Frieden, B.R. Restoring with maximum likelihood and maximum entropy. *J. Opt. Soc. Am.* **1972**, *62*, 511–518.
13. Von der Linden, W. Maximum Entropy data analysis. *Appl. Phys. A* **1995**, *60*, 155.
14. Denisova, N. Plasma diagnostics using computed tomography method. *IEEE Trans. Plasma Sci.* **2009**, *37*, 502.
15. Landau, L.D.; Lifshitz, E.M. *Statistical Physics Theoretical Physic*. Pergamon Press: Oxford, UK, 1964. Volume 5.
16. Klimontovich Y.L. S-theorem. *Z. Phys.* **1987**, *66*, 125–127.

© 2019 by the authors. Licensee MDPI, Basel, Switzerland. This article is an open access article distributed under the terms and conditions of the Creative Commons Attribution (CC BY) license (http://creativecommons.org/licenses/by/4.0/).

*Proceedings*

# Effects of Neuronal Noise on Neural Communication [†]

**Deniz Gençağa** [1,*] **and Sevgi Şengül Ayan** [2,*]

[1] Department of Electrical and Electronics Engineering, Antalya Bilim University, Antalya 07190, Turkey
[2] Department of Industrial Engineering, Antalya Bilim University, Antalya 07190, Turkey
* Correspondence: deniz.gencaga@antalya.edu.tr (D.G.); sevgi.sengul@antalya.edu.tr (S.Ş.A.);
  Tel.: +90-242-245-1394 (D.G.); +90-242-245-0321 (S.Ş.A.)
[†] Presented at the 39th International Workshop on Bayesian Inference and Maximum Entropy Methods in Science and Engineering, Garching, Germany, 30 June–5 July 2019.

Published: 19 November 2019

**Abstract:** In this work, we propose an approach to better understand the effects of neuronal noise on neural communication systems. Here, we extend the fundamental Hodgkin-Huxley (HH) model by adding synaptic couplings to represent the statistical dependencies among different neurons under the effect of additional noise. We estimate directional information-theoretic quantities, such as the Transfer Entropy (TE), to infer the couplings between neurons under the effect of different noise levels. Based on our computational simulations, we demonstrate that these nonlinear systems can behave beyond our predictions and TE is an ideal tool to extract such dependencies from data.

**Keywords:** transfer entropy; information theory; Hodgkin-Huxley model

## 1. Introduction

Mathematical models and analysis have been a strong tool to answer many important questions in biology and the work of Hodgkin and Huxley on nerve conduction is one of the best examples of it [1]. In 1952, after many years of theoretical and experimental work of physiologists, a mathematical model was proposed by HH to explain the action potential generation of neurons using conductance models that are defined for different electrically excitable cells [2–4]. Despite the rapid growth in the number of analyses on the communication between neurons, the noise effect has generally been overlooked in the literature. Recently, neuronal noise effects have started to be incorporated into the models, due to a phenomenon, called "Stochastic Resonance" [5]. The communication between neurons is maintained by electrical signals, called "Action Potentials (AP)". If the action potentials, as a response to a stimulant, exceeds a certain threshold value, these signals are referred to as "Spikes". The existence of a spike is determined by the value of a threshold value and additional noise component can easily increase or decrease the value of an AP versus the threshold, thus change the neural spike train code. Therefore, the noise is not merely a nuisance factor and it is capable of changing the meaning of the "neuronal code". For this reason, to better understand how these changes can occur in a very complex system, such as our brain, we must first understand the underlying working principles of neuronal noise, which sets the framework of our investigations.

Here, we utilize information theory to better understand the effects of neuronal noise on the overall communication. Therefore, we generalize the HH model in such a way that the noise can be added to the system beside the coupling among the neurons. In the literature, the effect of coupling among different neurons have been explored by using TE [6], however, to the best of our knowledge, the effects of noise on these interactions have not been fully considered yet.

On the other hand, certain types of models have been suggested to include the noise in the HH model [7] without any coupling between the neurons. Here, we approach the complicated modeling problem by using a simplified version including two neurons, coupling between them, and additional noise terms. We propose utilizing information theory to analyze the relationships in neural communication.

In the literature, information-theoretic quantities, such as Entropy, Mutual Information (MI) and Transfer Entropy (TE) have been successfully utilized to analyze the statistical dependencies and relationships between random variables of highly complex systems [8]. Among these, MI is a symmetric quantity reporting the dependency between two variables, whereas TE, is an asymmetric quantity that can be used to infer the direction of the interaction (as affecting and affected variables) between them [9]. All the above quantities are calculated from observational data by inferring probability distributions. Despite the wide variety of different distribution estimation techniques, the whole procedure still suffers from adverse effects, such as the bias. Most common techniques in probability distribution estimation involve histograms [10], Parzen windows [11] and adaptive methods [12]. In the literature, histogram estimation is widely used due to its computational simplicity. To rely on estimations from data, reporting the statistical significance of each estimate [13] constitutes an important part of the methods.

In this work, we propose utilizing TE to investigate the directional relationships between the coupled neurons of a HH model under noisy conditions. Therefore, we extend the traditional HH model and analyzed the effect of noise on the directional relationships between the coupled neurons. As our first approach to model noisy neuronal interaction, we demonstrate the effect under certain levels of noise power in the simulations. Based on these simulations, we observe that the original interactions are preserved despite many changes in the structure of the neuronal code structure. Our future work will be based on the generalization of this modeling to consider $N$ neurons and the effect of noise on their interactions.

## 2. Materials and Methods

### 2.1. The Hodgkin-Huxley Model

In this study we use Hodgkin-Huxley model which mimics the spiking behavior of the neurons recorded from the squid giant axon. This is the first mathematical model describing the action potential generation and it is one of the major breakthroughs of computational neuroscience [1]. In 1952 two physiologists Hodgkin and Huxley got the Nobel prize after this work and after their work Hodgkin-Huxley type models are defined for many different electrically excitable cells such as cardiomyocytes [2], pancreatic beta cells [3] and hormone secretion [4]. They observed that cell membranes behave much like electrical circuits. The basic circuit elements are the phospholipid bilayer of the cell, which behaves like a capacitor that accumulates ionic charge while the electrical potential across the membrane changes. Moreover, resistors in a circuit are analogue to the ionic permeabilities of the membrane and the electrochemical driving forces are analogous to batteries driving the ionic currents. $Na^+$, $K^+$, $Ca^{2+}$ and $Cl^-$ ions are responsible for almost all the electrical actions in the body. Thus, the electrical behavior of cells is based upon the transfer and storage of ions and Hodgkin and Huxley observed that $K^+$ and $Na^+$ ions are mainly responsible for the HH system.

Mathematical description of the Hodgkin-Huxley model starts with the membrane potential $V$ based on the conservation of electric charge defined as follows

$$C_m \frac{dV}{dt} = I_{ion} + I_{app} \qquad (1)$$

where $C_m$ is the membrane capacitance, $I_{app}$ is the applied current and $I_{ion}$ represents the sum of individual ionic currents and modeled according to Ohm's Law:

$$I_{ion} = -g_K(V - V_K) - g_{Na}(V - V_{Na}) - g_L(V - V_L). \qquad (2)$$

here $g_K$, $g_{Na}$ and $g_L$ are conductances, $V_{Na}$, $V_K$, $V_L$ are the reversal potentials associated with the currents. Hodgkin and Huxley observed that conductances are also voltage dependent. They realize that $g_K$ depends on four activation gates and defined as $g_K = \overline{g_K} n^4$ whereas $g_{Na}$ depends on three activation gates and one inactivation gate and modeled as $g_{Na} = \overline{g_{Na}} m^3 h$. In the HH model, ionic currents are defined as:

$$I_{Na} = \overline{g_{Na}}m^3h(V - V_{Na}) \qquad (3)$$

$$I_K = \overline{g_K}n^4(V - V_K) \qquad (4)$$

$$I_L = \overline{g_L}(V - V_L) \qquad (5)$$

with Na⁺ activation variable $m$ and inactivation variable $h$, and K⁺ activation variable $n$. Here $\overline{(\cdot)}$ denotes maximal conductances. Activation and inactivation dynamics of the channels are changing according to the differential equations below.

$$\frac{dm}{dt} = \frac{m_\infty(V_i) - m}{\tau_m(V_i)} \qquad (6)$$

$$\frac{dh}{dt} = \frac{h_\infty(V_i) - h}{\tau_h(V_i)} \qquad (7)$$

$$\frac{dn}{dt} = \frac{n_\infty(V_i) - n}{\tau_n(V_i)} \qquad (8)$$

The steady state activation and inactivation functions together with time constants are defined as below and the transition rates $\alpha_x$ and $\beta_x$ are given in Table 1.

$$x_\infty(V_i) = \frac{\alpha_x(V_i)}{\alpha_x(V) + \beta_x(V)} \qquad (9)$$

$$\tau_x(V_i) = \frac{1}{\alpha_x(V_i) + \beta_x(V_i)}, \quad x = m, h, n \qquad (10)$$

**Table 1.** Transition rates and parameter values for the HH Model.

|  |  |  |
|---|---|---|
| Transition Rates (ms⁻¹) | $\alpha_m$ | $0.1(40 + V_i)/(1 - \exp(-(55 + V_i)/10)$ |
|  | $\beta_m$ | $4\exp(-(65 + V_i)/18)$ |
|  | $\alpha_h$ | $0.07\exp(-(65 + V_i)/20)$ |
|  | $\beta_h$ | $1/(1 + \exp(-(35 + V_i)))$ |
|  | $\alpha_n$ | $0.01(55 + V_i)/(1 - \exp(-(10 + 55))$ |
|  | $\beta_n$ | $0.125\exp(-(V_i + 65)/80)$ |
| Parameter Values | $C_m$ | $1\ \mu F$ |
|  | $I_{app}$ | $8\ mA$ |
|  | $g_{Na}$ | $120\ \mu S$ |
|  | $g_K$ | $36\ \mu S$ |
|  | $g_L$ | $0.3\ \mu S$ |
|  | $V_{Na}$ | $50\ mV$ |
|  | $V_K$ | $-77\ mV$ |
|  | $V_L$ | $-54.4\ mV$ |

### 2.2. Information Theoretic Quantities

In information theory, Shannon entropy is defined to be the average uncertainty for finding the system at a particular state 'x' out of a possible set of states 'X', where $p(x)$ denotes the probability of that state. Also, it is used to quantify the amount of information needed to describe a dataset. Shannon entropy is given by the following formula

$$H(X) = -\sum_{x \in X} p(x)\log(p(x)) \qquad (11)$$

Mutual information (MI), is another fundamental information-theoretic quantity which is used to quantify the information shared between two datasets. Given two datasets denoted by X and Y, the MI can be written as follows:

$$MI(X,Y) = \sum_{x \in X} \sum_{y \in Y} p(x,y) \log \frac{p(x,y)}{p(x)p(y)} \tag{11}$$

The MI is a symmetric quantity and it can be rewritten as a sum and difference of Shannon entropies by

$$MI(X,Y) = H(X) + H(Y) - H(X,Y) \tag{12}$$

where $H(X,Y)$ is the joint Shannon entropy. If there is a directional dependency between the variables, such as a cause and effect relationship, a symmetric measure cannot unveil the dependency information from data. In the literature, TE was proposed to analyze the directional dependencies between two Markov processes. To quantify the directional effect of a variable X on Y, the TE is defined by the conditional distribution of Y depending on the past samples of both processes versus the conditional distribution of that variable depending only on its own past values [14]. Thus, the asymmetry of TE helps us detect two directions of information flow. The TE definition in both directions (between variables X and Y) are given by the following equations:

$$TE_{XY} = T\left(Y_{i+1} \big| \mathbf{Y}_i^{(k)}, \mathbf{X}_i^{(l)}\right) = \sum_{y_{i+1}, \mathbf{y}_i^{(k)}, \mathbf{x}_i^{(l)}} p\left(y_{i+1}, \mathbf{y}_i^{(k)}, \mathbf{x}_i^{(l)}\right) \log_2 \frac{p\left(y_{i+1} \big| \mathbf{y}_i^{(k)}, \mathbf{x}_i^{(l)}\right)}{p\left(y_{i+1} \big| \mathbf{y}_i^{(k)}\right)} \tag{13}$$

$$TE_{YX} = T\left(X_{i+1} \big| \mathbf{X}_i^{(k)}, \mathbf{Y}_i^{(l)}\right) = \sum_{x_{i+1}, \mathbf{x}_i^{(k)}, \mathbf{y}_i^{(l)}} p\left(x_{i+1}, \mathbf{x}_i^{(k)}, \mathbf{y}_i^{(l)}\right) \log_2 \frac{p\left(x_{i+1} \big| \mathbf{x}_i^{(k)}, \mathbf{y}_i^{(l)}\right)}{p\left(x_{i+1} \big| \mathbf{x}_i^{(k)}\right)} \tag{14}$$

where $\mathbf{x}_i^{(k)} = \{x_i, \ldots, x_{i-k+1}\}$ and $\mathbf{y}_i^{(l)} = \{y_i, \ldots, y_{i-l+1}\}$ are past states, and X and Y are kth and lth order Markov processes, respectively, such that X depends on the k previous values and Y depends on the l previous values. In the literature, k and l are also known as the embedding dimensions.

All the above quantities involve estimation of probability distributions from the observed data. Among many approaches in the literature, we utilize the histogram-based method to estimate the distributions on (14) and (15), due to its computational simplicity. In order to assess the statistical significance of the TE estimations, surrogate data testing is applied, and the p-values are reported.

## 2.3. The Proposed Method

In this paper we focus on the system of two coupled HH neurons with synaptic coupling from neuron 1 to neuron 2. Also, current noise is added with normal distribution for the action potential generation of the squid axons for this two-neuron network. It involves a fast sodium current $I_{i,Na}$, a delayed rectifying potassium current $I_{i,K}$ and a leak current $I_{i,L}$ (measured in $\frac{\mu A}{cm^2}$) for $i = 1,2$. The differential equations for the rate of change of voltage for these neurons are given as follows,

$$C_m \frac{dV_1}{dt} = I_{i,app} - I_{i,Na} - I_{i,K} - I_{i,L} + N(0,\sigma), \tag{15}$$

$$C_m \frac{dV_2}{dt} = I_{i,app} - I_{i,Na} - I_{i,K} - I_{i,L} + k(V_2 - V_1) + N(0,\sigma), \tag{16}$$

where $V_1$ is the membrane voltage for the 1st neuron and $V_2$ is the membrane voltage for the 2nd neuron. Here, $N(0,\sigma)$ shows the noise distribution defined by normal distribution with 0 mean and $\sigma$ standart deviation. Synapting coupling is defined simply $I_{syn} = k(V_2 - V_1)$ with voltage difference and synaptic coupling strength is $k$. When k is between 0 and 0.25, spiking activity occurs with unique stable limit cycle solution. After k = 0.25 system turns back to stable steady state and spiking activity disappears. All other dynamics are same as described in Section 2.1.

First, we propose using TE between $V_2$ and $V_1$, in the case of no noise in (16) and (17). Secondly, we include the noise components in (16) and (17) and utilize TE between $V_2$ and $V_1$, again. This comparison demonstrates the effects of noise on the information flow between the neurons. At a first glance on equations (16) and (17), we can conclude that the direction of the information flow under noiseless case must be from $V_1$ to $V_2$. However, when the noise is added, it is tedius to reach the same conclusion, as the added noise is capable of adding additional spikes and destroying the available ones. The simulation results in the next section demonsrate these findings and provides promising results to generalize our model to more complex neuronal interactions under noise.

The model is implemented in the XPPAUT software [15] using the Euler method (dt = 0.1 ms).

## 3. Results

*Information Flow Changes with Coupling Strength and Noise Level*

Here, we first studied a system of two globally coupled HH model through a synapse by varying the coupling strength $k$, without noise effect. Phase dynamics for our system for two different $k$ coupling strengths are plotted in Figure 1. Here we define two different coupling patterns as shown below. In Figure 1a, neuron 2 fires once after neuron 1 fires twice which we call 2-to-1 coupling with $k = 0.1$. For a larger coupling coefficient ($k = 0.2$) neurons shows different synchronous firing pattern as in Figure 1b. This time, each firing of neuron 2 follows that of neuron 1 which we call 1-to-1 coupling.

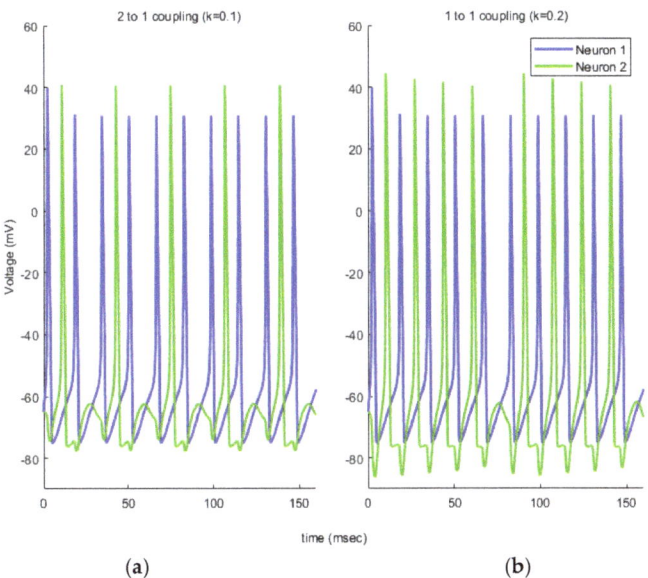

**Figure 1.** Sample spike patterns for two different network configurations: (**a**) 2 to 1 coupling and (**b**) 1 to 1 coupling.

To better understand the effects of noise on our network, we use zero mean Gaussian distributed random variables with standard deviation of $\sigma$. When we incorporate this noise with different variances into our model as illustrated in (16) and (17), we observe a change in the synchronisation of the neurons. Additionally, the obvious patterns disappear totally for larger noise amounts as shown in Figure 2 for each coupled network. Noise can change the synchronization of neurons by inducing or deleting spikes in network. Since the noise plays an important role in changing the dynamics of the network, we need a mechanism to figure out this newly changed patterns under noise effect to explain the behavior of the neuronal network. Therefore, we utilize TE to extract this pattern using the observed voltage data.

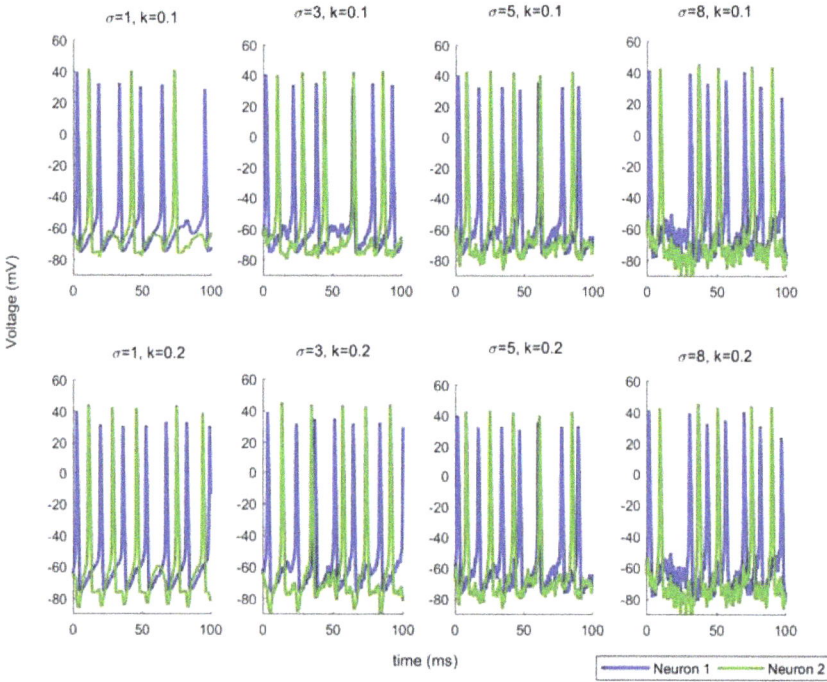

**Figure 2.** Noise changes the synchronization of the network.

To better understand the information flow between the neurons, we estimate TE between the action potentials of two neurons, as described in Section 2.2, using (16) and (17), with an increasing noise intensity. The results are plotted in Figure 3 for both 2-to-1 coupled system and 1-to-1 coupled system. As expected for the network without noise ($\sigma = 0$), transfer entropy value is the highest for both systems. This verifies the changes caused in Neuron 2 by Neuron 1.

To see the effects of noise, we explore the TE between the neurons with increasing $\sigma$ parameter. From Figure 3, we note that with the increasing noise intensity, the values of TE decrease. Although this can be expected, it is of utmost importance to emphasize the case when we do not have any noise. If we do not have noise component in the model, according to Figure 3 and Table 2, we notice that TE value is behaving in opposite way, i.e., the smaller TE, the higher coupling. Another interesting finding is the varying pattern in TE around low $\sigma$ values: The TE increases first and keeps decreasing later. This unexpected result shows that we cannot easily predict the direction of coupling without TE analysis, as the synaptic couplings are nonlinear in nature.

**Table 2.** Transfer entropy values with different noise intensity as $\sigma$ is increasing.

| $\sigma$ | 0 | 0.5 | 1 | 2 | 3 | 5 | 6 | 7 | 8 | 9 | 10 |
|---|---|---|---|---|---|---|---|---|---|---|---|
| K = 0.1 | 0.0665 | 0.017 | 0.0168 | 0.0169 | 0.0129 | 0.0059 | 0.0039 | 0.0014 | 0.0013 | 0.000806 | 0.000512 |
| K = 0.2 | 0.0557 | 0.0264 | 0.0206 | 0.0261 | 0.0242 | 0.0165 | 0.0148 | 0.0095 | 0.0064 | 0.0054 | 0.0033 |

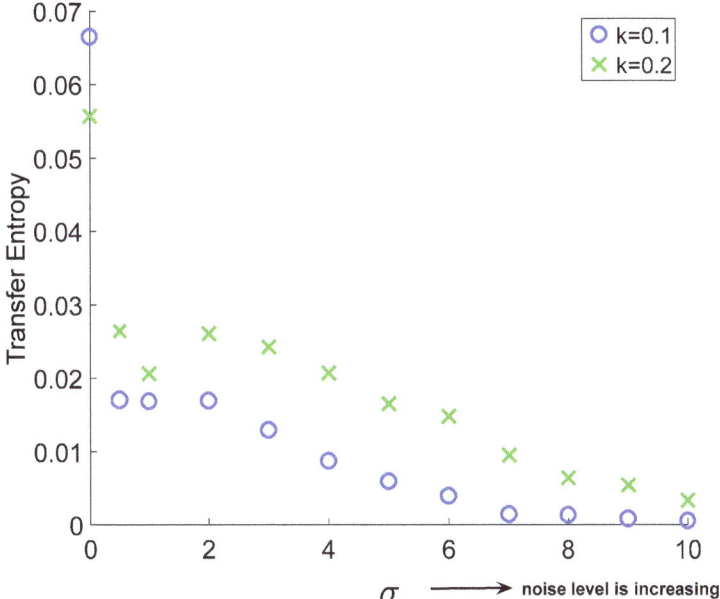

**Figure 3.** Transfer entropy results for 2-to-1 and 1-to-1 coupled HH network.

## 4. Discussion and Conclusions

We study information flow for the coupled network of two neurons under two different coupling states and increasing noise levels, where the neuron models and the synaptic interactions are derived from Hodgkin-Huxley model. Here we propose our model in such a way that we can generate 2-to-1 coupling and 1-to-1 coupling between the neurons. In order to find these relationships from data we propose a TE based approach and analyze the effects of couplings under various noise intensities, successfully. These results help us better understand the interaction between the neurons in real biological systems. This work is of particular importance to explore larger networks with more complex noisy interactions.

**Author Contributions:** D.G. and S.Ş.A. designed the studies; D.G. performed transfer entropy simulations; S.Ş.A. performed mathematical modelling; D.G. and S.Ş.A. analyzed data; D.G. and S.Ş.A. wrote the paper. All authors have seen and approved the final version of the manuscript.

**Funding:** This research is funded by TUBITAK 1001 grant number 118E765.

**Acknowledgments:** We thank graduate student Lemayian Joelponcha for performing some sections of the simulations.

**Conflicts of Interest:** The authors declare no conflict of interest.

## References

1. Hodgkin, A.L.; Huxley, A.F. A quantitative description of membrane current and its application to conduction and excitation in nerve. *J. Physiol.* **1952**, *117*, 500–544.
2. Pandit, S.V.; Clark, R.B.; Giles, W.R.; Demir, S.S. A Mathematical Model of Action Potential Heterogeneity in Adult Rat Left Ventricular Myocytes. *Biophys. J.* **2001**, *81*, 3029–3051.
3. Bertram, R.; Sherman, A. A calcium-based phantom bursting model for pancreatic islets. *Bull. Math. Biol.* **2004**, *66*, 1313–1344.
4. Duncan, P.J.; Şengül, S.; Tabak, J.; Ruth, P.; Bertram, R.; Shipston, M.J. Large conductance $Ca^{2+}$-activated $K^+$ (BK) channels promote secretagogue-induced transition from spiking to bursting in murine anterior pituitary corticotrophs. *J. Physiol.* **2015**, *593*, 1197–1211.

5. Moss, F., Ward, L. M., Sannita, W. G. Stochastic resonance and sensory information processing: a tutorial and review of application. *Clin. Neurophysiol.* **2004**, *115*, 267–281.
6. Li, Z.; Li, X. Estimating temporal causal interaction between spike trains with permutation and transfer entropy. *PLoS ONE* **2013**, *8*, e70894.
7. Goldwyn, J.H.; Shea-Brown, E. The what and where of adding channel noise to the Hodgkin-Huxley equations. *PLoS Comput. Biol.* **2011**, *7*, e1002247.
8. Gencaga, D. *Transfer Entropy (Entropy Special Issue Reprint)*; MDPI: Basel, Switzerland, 2018.
9. Gencaga, D.; Knuth, K.H.; Rossow, W.B. A Recipe for the Estimation of Information Flow in a Dynamical System. *Entropy* **2015**, *17*, 438–470.
10. Knuth, K.H. Optimal data-based binning for histograms. *arXiv* **2006**, arXiv:physics/0605197.
11. Scott, D.W. *Multivariate Density Estimation: Theory, Practice, and Visualization*, 2nd ed.; John Wiley & Sons, Inc.: Hoboken, NJ, USA, 2015.
12. Darbellay, G.A.; Vajda, I. Estimation of the information by an adaptive partitioning of the observation space. *IEEE Trans. Inf. Theory* **1999**, *45*, 1315–1321.
13. Timme, N.M.; Lapish, C.C. A tutorial for information theory in neuroscience. *eNeuro* **2018**, *5*, doi:10.1523/ENEURO.0052-18.2018.
14. Schreiber, T. Measuring information transfer. *Phys. Rev. Lett.* **2000**, *85*, 461–464.
15. XPP-AUT Software. Available online: http://www.math.pitt.edu/~bard/xpp/xpp.html (accessed on 30 June 19).

© 2019 by the authors. Licensee MDPI, Basel, Switzerland. This article is an open access article distributed under the terms and conditions of the Creative Commons Attribution (CC BY) license (http://creativecommons.org/licenses/by/4.0/).

*Proceedings*

# Bayesian Reconstruction through Adaptive Image Notion †

**Fabrizia Guglielmetti [1,*], Eric Villard [2] and Ed Fomalont [2,3]**

1. European Southern Observatory, Karl-Schwarzschild-Str. 2, D-85748 Garching, Germany
2. Joint ALMA Observatory/ESO, Avenida Alonso de Córdova 3107, Vitacura, Santiago 763-0355, Chile; Eric.Villard@alma.cl (E.V.); efomalon@nrao.edu (E.F.)
3. National Radio Astronomy Observatory, 520 Edgemont Road, Charlottesville, VA 22903, USA
* Correspondence: fgugliel@eso.org
† Presented at the 39th International Workshop on Bayesian Inference and Maximum Entropy Methods in Science and Engineering, Garching, Germany, 30 June–5 July 2019.

Published: 5 December 2019

**Abstract:** A stable and unique solution to the ill-posed inverse problem in radio synthesis image analysis is sought employing Bayesian probability theory combined with a probabilistic two-component mixture model. The solution of the ill-posed inverse problem is given by inferring the values of model parameters defined to describe completely the physical system arised by the data. The analysed data are calibrated visibilities, Fourier transformed from the $(u, v)$ to image planes. Adaptive splines are explored to model the cumbersome background model corrupted by the largely varying dirty beam in the image plane. The de-convolution process of the dirty image from the dirty beam is tackled in probability space. Probability maps in source detection at several resolution values quantify the acquired knowledge on the celestial source distribution from a given state of information. The information available are data constrains, prior knowledge and uncertain information. The novel algorithm has the aim to provide an alternative imaging task for the use of the Atacama Large Millimeter/Submillimeter Array (ALMA) in support of the widely used Common Astronomy Software Applications (CASA) enhancing the capabilities in source detection.

**Keywords:** methods: data analysis; methods: statistical; techniques: image processing

---

## 1. Introduction

A software package investigates a method of Bayesian Reconstruction through Adaptive Image Notion (BRAIN) [1] to support and enhance the interferometric (IF) imaging procedure. Extending the works of [2,3], the software kit makes use of Gaussian statistics for a joint source detection and background estimation through a probabilistic mixture model technique. The background is defined as the small scale perturbance of the wanted signal, i.e., the celestial sources. As shown in [3], statistics is rigorously applied throughout the algorithm, so pixels with low intensity can be handled optimally and accurately, without binning and loss of resolution. Following the work of [2], a 2-D adaptive kernel deconvolution method is employed to prevent spurious signal arising from the dirty beam. Continuum, emission and absorption lines detection is foreseen to occur without an explicit subtraction, but propagating the information acquired on the continuum for line detection.

The novel algorithm aims at improving the CASA [4,5] imaging and deconvolution methods currently used, especially for the use of ALMA, but also applicable to next generation instruments as ngVLA [6] and SKA [7]. The developing software is going to be as compatible with CASA as possible by using CASA tasks, data formats, including python scripts. BRAIN has the aim to provide an advancement in current issues as stopping thresholding, continuum subtraction, proper detection of extended emission, separation of point-like sources from diffuse emissions, weak signal, mosaics.

In the following sections, a brief description of the ALMA observatory, the acquired data and radio synthesis imaging is provided. The most used software package for ALMA data analysis, i.e., CASA, is introduced and the complex data transformation for the formation of interferometric images is shown. Last, the technique under development is outlined in the main features as differing from [3].

## 2. ALMA

Located at 5000 m altitude on the Chajnantor plateau (Chile), ALMA is an aperture synthesis telescope operating over a broad range of observing frequencies in the mm and submm regime of the electromagnetic spectrum, covering most of the wavelength range from 3.6 to 0.32 mm (84–950 GHz).

ALMA is characterized by 66 high-precision antennas and a flexible design: Several pairs of antennas (baselines) build a single filled-aperture telescope whose spatial resolution depends on the used configuration and observing frequency. Two main antennas arrangements are identifiable: the 12-m Array and the Atacama Compact Array (ACA). Firstly, with fifty relocatable antennas on 192 stations, the 12-m Array allows for baselines ranging from 15 m to 16 km. Secondly, ACA is composed by twelve closely spaced 7-m antennas (7-m Array) and four 12-m antennas for single dish observations, named Total Power (TP) Array. The 7-m and the TP arrays sample baselines in the (9–30) m and (0–12) m ranges, respectively. The data re-combination from the 7-m and TP arrays overcome the well-known "zero spacing" problem [8] being crucial during the observation of extended sources. The final resolution reached by these arrays is given by the ratio between the observational wavelength and the maximum baseline for a given configuration. Therefore, the 12-m Array is used for sensitive and high resolution imaging, while ACA is preferable for wide-field imaging of extended structures. The 12-m Array and ACA can reach a fine resolution up to 20 mas at 230 GHz with the most extended configuration and 1.44 arcsec at 870 GHz, respectively: These numbers refer to the point spread function (PSF) or Full Width at Half Maximum (FWHM) of the synthesized beam, which is the inverse Fourier transform of a (weighted) $(u,v)$ sampling distribution.

ALMA delivers continuum images and spectral line cubes with frequency characterizing the third axis. The data cubes are provided with up to 7680 frequency channels, with the channel width indicating the spectral resolution (from 3.8 kHz to 15.6 MHz).

### 2.1. ALMA Images

The sky brightness distribution, collected by each antenna, is correlated for given baselines. The measurements obtained from the correlators in an array allow one to reconstruct the complex visibility function of the celestial source. The Van Cittert–Zernike theorem [9] provides the Fourier transform relation between the sky brightness distribution and the array response (or primary beam) [10].

In simplified form, the complex visibility can be stated as follows:

$$\mathcal{V}(u,v) = \iint A(l,m) \cdot I(l,m) e^{2\pi i (ul + vm)} dl dm \approx A e^{i\phi}. \tag{1}$$

In aperture synthesis analysis, the goal is to solve Equation (1) for $I(l,m)$ in the image plane by measuring the complex-valued and calibrated $\mathcal{V}(u,v)$. The coordinates $u,v$, measured in units of wavelength, characterize the spatial frequency domain. The $(u,v)$ plane is effectively composed by the projections of baselines onto a plane perpendicular to the source direction. The calibrated $\mathcal{V}(u,v)$ and $I(l,m)$ are given in units of flux density (1 Jy = $10^{-23}$ ergs cm$^{-2}$ s$^{-1}$ Hz$^{-1}$) and of surface brightness (Jy/beam area), respectively. The antenna reception pattern $A(l,m)$ is relevant for the primary beam correction. The amplitude ($A$) and the phase ($\phi$) inform us about source brightness and position relative to the phase center at spatial frequencies $u$ and $v$.

## 2.2. The Data

A direct application of the inverse Fourier transform of $\mathcal{V}(u,v)$ (Equation (1)) is desirable but suitable only in case of a complete sampling of the spatial frequency domain $B(u,v)$. For a given configuration, the array provides a discrete number of baselines, being the number of baselines provided by $N \cdot (N-1)/2$, where N is the number of antennas. The sampling of the $(u,v)$ plane is consequently limited. Hence, the measured complex visibility function is corrupted by the sampling function: $B(u,v) \cdot \mathcal{V}(u,v)$.

The sampling function $B(u,v)$, composed by sinusoids of various amplitude and phase, is characterized as an ensemble of delta functions accounting for Hermitian symmetry. The inverse Fourier transform of this ensemble of visibilities provides the dirty image $I^D(l,m)$:

$$I^D(l,m) = FT^{-1}\{B(u,v)\mathcal{V}(u,v)\}. \qquad (2)$$

Following the convolution theorem, Equation (2) can be written as follows:

$$I^D(l,m) = b(l,m) \times I(l,m)A(l,m). \qquad (3)$$

The dirty image is the convolution of the sky brightness modified by the antenna primary beam $A(l,m)$ with the dirty beam, given by $b(l,m) = FT^{-1}\{B(u,v)\}$ (FWHM). The true sky brightness reconstruction $I(l,m)$ depends on the image fidelity, since the coverage of the $(u,v)$ plane is by definition incomplete. For this reason the reconstruction of $I(l,m)$ is challenging due to the artefacts introduced by the dirty beam $b(l,m)$, the sparsity in the data and the big data volume.

## 3. Common Astronomy Software Applications

CASA is a suite of tools allowing for calibration and image analyses for radio astronomy data [11]. It is especially designed for the investigation of data observed with ALMA and the VLA [12].

The calibration process determines the net complex correction factors to be applied to each visibility. Corrections are applied to the data due to, e.g., temperature effects, atmospheric effects, antenna gain-elevation dependencies. It includes solvers for basic gain (A,φ), bandpass, polarization effects, antenna-based and baseline-based solutions. Between calibration and imaging procedures, image formation follows in three steps: weighting, convolutional resampling, and a Fourier transform. This is a general procedure employed by most systems over the last 30 years. In order to produce images with improved thermal noise and to customize the resolution, the calibrated visibilities undergo a variety of weighting schemes. The commonly used robust weighting [13] has the property to vary smoothly from natural (provides equal weight to all samples) to uniform (gives equal weight to each measured spatial frequency irrespective of sample density) based on the signal-to-noise ratio of the measurements and a pre-defined noise threshold parameter. The robust weighting scheme has the advantages to make the effective u-v coverage as smooth as possible while allowing for sparse sampling, to modify the resolution and theoretical signal-to-noise of the image. Nonetheless, care has to be taken on the choice of weigthing scheme due to sidelobes effects. The weighted visibilities are resampled onto a regular uv-grid (convolutional resampling) employing a gridding convolution function. A fast Fourier transform is applied on the resampled data and corrections are introduced to transform the image into units of sky brightness. The resulting dirty image $I^D(l,m)$ is deconvolved from the well-known dirty beam $b(l,m)$. A model of the sky brightness distribution is obtained, allowing for the creation of the dirty/residual image, as the result of a convolution of the true sky brightness and the PSF of the instrument. The dirty image arises or from single pointing or from mosaics. Last, CASA offers facilities for simulating observations.

### 3.1. Deconvolution

Most interferometric imaging in CASA is done using the *tclean* task. This task is composed by several operating modes, allowing for the generation of images from visibilities and the reconstruction of a sky model. Continuum images and spectral line cubes are handled. Image reconstruction occurs employing an outer loop of major cycles and an inner loop of minor cycles, following the Cotton-Schwab CLEAN style [14]. The major cycle accounts for the transformation between data and image domains. The minor cycle is designed to operate in the image domain. An iterative weighted $\chi^2$ minimization is implemented to solve for the measurement equations.

Several algorithms for image reconstruction (deconvolvers) in the minor cycle are available, e.g., Hogbom [15], Clark [16], Multi-Scale [17], Multi-Term [18]. Each minor cycle algorithm can be characterized by their own optimization scheme, framework and task interface with the pre-requisite to produce a model image as output. BRAIN has the potentials to become a deconvolution algorithm within the *tclean* task.

*3.2. Simulated ALMA Data*

The CASA simulator takes on input a sky model to create a customized ALMA interferometric or total power observations, including multiple configurations of the 12-m array. The CASA simulator is characterized by two main tasks: *simobserve*, *simanalyze*. The task *simobserve* is used to create a model image (or component list), i.e., a representation of the simulated sky brightness distribution, and the $(u, v)$ data. The dirty image is created with the task *simanalyze*.

In Figure 1, 13 point-like and extended sources are simulated at 97 GHz in the 12-m array for an integration time of 30 s and a total observing time of 1 h. The simulated sources are characterized by a Gaussian flux distribution in the range 0.2–5.0 Jy. Low thermal noise is introduced. The image size is composed by $[1024, 1024]$ pixels, covering 10 arcsec on the side. These and more sofisticated simulated data are planned to be used for the feasibility study.

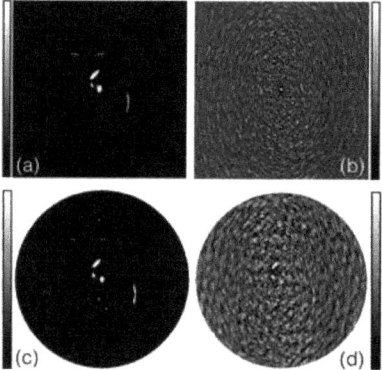

**Figure 1.** Simulated ALMA interferometric data: (**a**) the simulated model image $(2.435–10.685) \times 10^{-5}$ [Jy/pixel], (**b**) the synthesized (dirty) beam $(-0.02–0.04)$ [Brigthness pixel unit], (**c**) the simulated sky $(-0.25–3.76) \times 10^{-3}$ [Jy/beam], (**d**) the dirty image $(-0.05–0.08)$ [Jy/beam].

## 4. BRAIN

An ALMA dirty image (Equation (3)) is characterizable or by continuum or by line detection for a given bandwidth. The detected celestial source intensities can be in absorption or emission with negative or positive values, respectively. We assume that the data went through a robust calibration process. Gaussian statistics describes the data distribution. Nonetheless, efforts are also sought to account for glitches in the calibration process. For those cases, BRAIN is planned to implement an improvement in the data modelling accounting for a Gaussian distribution with a longer tail. The data set in image space is $D = \{d_{ij}\} \in \mathbf{R}$ in Jansky per pixel cell $\{i, j\}$. Following the work of [3], an astronomical image consists of a composition of background and celestial signals. The background is defined as the small scale perturbance of the wanted celestial signal. Two complementary hypotheses are introduced:

$$\begin{cases} B_{ij}: & d_{ij} = b_{ij} + \epsilon_{ij} \\ \overline{B}_{ij}: & d_{ij} = b_{ij} + s_{ij} + \epsilon_{ij}. \end{cases} \quad (4)$$

Hypothesis $B_{ij}$ specifies that the data $d_{ij}$ consists only of background intensity $b_{ij}$ spoiled with noise $\epsilon_{ij}$, i.e., the (statistical) uncertainty associated with the measurement process. Hypothesis $\overline{B}_{ij}$ specifies the case where additional source intensity $s_{ij}$ contributes to the background. Additional assumptions are that negative and positive values for source and background amplitudes are allowed and that the background in average is smoother than the source signal.

The background signal is propagated from the visibilities noise, taking into account that the Fourier Transform is a linear combination of the visibilities with some rotation (phase factor) applied. The noise on the visibilities is mainly introduced, e.g., by cosmic, sky and instrumental (due to receiver, single baseline or antenna, total collecting area, autocorrelations) signals. It is expected that real and imaginary part in the visibility noise is uncorrelated, with the phase factor not affecting the noise. Moreover, the individual visibilities are combined at the phase center and weighted, gridding correction and primary beam correction are increasing noise in the image. The noise in ALMA images is not uniform, with the noise increasing towards the edge. Mosaics of images are particularly cumbersome.

The background amplitude is modelled with a thin-plate-spline, with the support points chosen sparsely in order not to fit the sources [3]. The background model takes into account the dirty beam information, in order to deconvolve the dirty beam from the dirty image. The spline model is under further development to account for a more flexible design. The supporting points are analysed to account for a dynamic setting, where the number and positions of the supporting points are chosen on the basis of the data. This work follows [2] successfully developed in experimental spectra.

Estimates of the hypotheses $B_{ij}$ and $\overline{B}_{ij}$ are the direct effort of this analysis. The likelihood probability for the hypothesis $B_{ij}$ within Gaussian statistics is:

$$p(d_{ij} \mid B_{ij}, b_{ij}) = (2\pi\sigma_{ij}^2)^{-\frac{1}{2}} \exp[-(d_{ij} - b_{ij})^2 / 2\sigma_{ij}^2]. \quad (5)$$

For the alternative hypothesis, a similar equation is applied with included the signal contribution. Similar to what was done in [2,3], the signal is considered as a nuisance parameter. Following the Maximum Entropy distribution, a two-sided exponential function is chosen to describe the mean source intensity in the field ($\lambda_{+/-} = <s_{ij}>$):

$$p(s_{ij}|\lambda_+, \lambda_-) = \begin{cases} \lambda_+^{-1} \exp(-s_{ij}/\lambda_+), & s_{ij} \geq 0 \\ \lambda_-^{-1} \exp(+s_{ij}/\lambda_-), & s_{ij} < 0. \end{cases} \quad (6)$$

The parameters $\lambda_+$ and $\lambda_-$ are positive values, allowing one to introduce two different scales for the signal dependently on its sign. The likelihood for the hypothesis $\overline{B}_{ij}$ is given following the marginalization rule:

$$\begin{aligned} p(d_{ij} \mid \overline{B}_{ij}, b_{ij}, \lambda) &= \int_{-\infty}^{+\infty} ds_{ij} p(d_{ij}|s_{ij}, \overline{B}_{ij}, b_{ij}) p(s_{ij}|\lambda_+, \lambda_-) \\ &= \frac{1}{2|\lambda_+|} \left\{ 1 + \mathrm{erf}\left[ \frac{\lambda_+(d_{ij} - b_{ij}) - \sigma_{ij}^2}{|\lambda_+|\sqrt{2\sigma_{ij}^2}} \right] \right\} \exp\left[ \frac{\lambda_+(d_{ij} - b_{ij}) + \sigma_{ij}^2/2}{\lambda_+^2} \right] + \\ &+ \frac{1}{2|\lambda_-|} \left\{ 1 + \mathrm{erf}\left[ \frac{-\lambda_-(d_{ij} - b_{ij}) - \sigma_{ij}^2}{|\lambda_-|\sqrt{2\sigma_{ij}^2}} \right] \right\} \exp\left[ \frac{-\lambda_-(d_{ij} - b_{ij}) + \sigma_{ij}^2/2}{\lambda_-^2} \right]. \end{aligned} \quad (7)$$

The data are modelled by a two-component mixture distribution in the parameter space within the Bayesian framework. In this way, background and sources with their respective uncertainties are jointly detected. The likelihood for the mixture model combines the probability distribution for the two hypotheses, $B_{ij}$ and $\overline{B}_{ij}$:

$$p(D|\mathbf{b},\mathbf{s},\lambda) = \prod_{ij} \{p(B_{ij}) \cdot p(d_{ij}|B_{ij},b_{ij}) + p(\overline{B}_{ij}) \cdot p(d_{ij}|\overline{B}_{ij},b_{ij},\lambda_+,\lambda_-)\}, \qquad (8)$$

where $\mathbf{b} = \{b_{ij}\}$, $\mathbf{s} = \{s_{ij}\}$, $\lambda = \{\lambda_{+/-}\}$ and $\{ij\}$ corresponds to the pixels of the complete field. The prior pdfs $p(B_{ij})$ and $p(\overline{B}_{ij})$ for the two complementary hypotheses describe the prior knowledge of having background only or additional signal contribution, respectively, in a pixel. These prior pdfs are chosen to be constant, independent of $i$ and $j$: $p(B_{ij}) = \beta$ and $p(\overline{B}_{ij}) = 1 - \beta$. The likelihood for the mixture model allows us to estimate the parameters entering the models from the data. The ultimate goal is to detect sources independently to their shape and intensity and to provide a robust uncertainty quantification. Therefore, Bayes' theorem is used to estimate the probability of the hypothesis $p(\overline{B}_{ij} \mid d_{ij}, b_{ij}, \lambda)$ for detecting celestial sources.

## 5. Concluding Remarks

The work is ongoing for this project. The novel technique, applied to ALMA interferometric data, is designed to create probability maps of source detection, allowing for a joint estimate of background and sources. A 2-D adaptive kernel deconvolution method will strengthen the deconvolution process of the dirty beam from the dirty image, reducing contaminations. BRAIN has the aim to provide an alternative technique in CASA image analysis. It is foreseen that a robust solution as model image is produced. The next effort is applied on the simulated data. Nonetheless, due to the statistical approach employed, BRAIN is also foreseen to be applied on data re-combination. In fact, advanced methods employing Bayesian probability theory have the potentials to address at best the analysis of combined short-spacing single-dish data with those from an interferometer, as provided by ALMA design.

**Acknowledgments:** This work was supported partly by the Joint ALMA Observatory Visitor Program.

## References

1. ALMA Dev. Workshop 2019. Available online: https://zenodo.org/record/3240347#.XVq6PVBS9XQ (accessed on 1 December 2019).
2. Fischer, R.; Hanson, K.M.; Dose, V.; von der Linden, W. Background estimation in experimental spectra. *Phys. Rev.* **2000**, *E61*, 1151–1160.
3. Guglielmetti, F.; Fischer, R.; Dose, V. Background-source separation in astronomical images with Bayesian probability theory—I. The method. *MNRAS* **2009**, *396*, 165–190.
4. CASA Documentation. Available online: https://casa.nrao.edu/casadocs-devel (accessed on 1 December 2019).
5. McMullin, J.P.; Waters, B.; Schiebel, D.; Young, W.; Golap, K. CASA Architecture and Applications. In *Astronomical Data Analysis Software and Systems XVI ASP Conference Series*; Shaw, R.A., Hill, F., Bell, D.J., Eds.; Astronomical Society of the Pacific: San Francisco, CA, USA; Volume 376, pp. 127–130.
6. Next Generaton Very Large Array. Available online: https://ngvla.nrao.edu/ (accessed on 1 December 2019).
7. Square Kilometre Array. Available online: https://www.skatelescope.org/ (accessed on 1 December 2019).
8. Braun, R.; Walterbos, R.A.M. A solution to the short spacing problem in radio interferometry. *Astron. Astrophys.* **1985**, *143*, 307–312.
9. Zernike, F. The concept of degree of coherence and its application to optical problems. *Physica* **1938**, *5*, 785–795.
10. Stanimirovic, S.; Altschuler, D.; Goldsmith, P.; Salter, C. Single-Dish Radio Astronomy: Techniques and Applications. In *ASP Conference Proceedings*; Stanimirovic, S., Altschuler, D., Goldsmith, P., Salter, C., Eds.; Astronomical Society of the Pacific: San Francisco, CA, USA; Volume 278, pp. 375–396.

11. Jaeger, S. The Common Astronomy Software Application (CASA). In *Astronomical Data Analysis Software and Systems ASP Conference Series*; Argyle, R.W., Bunclark, P.S., Lewis, J.R., Eds.; Astronomical Society of the Pacific: San Francisco, CA, USA; Volume 394, pp. 623–626.
12. Very Large Array. Available online: http://www.vla.nrao.edu/ (accessed on 1 December 2019).
13. Briggs, D.S. High Fidelity Interferometric Imaging: Robust Weighting and NNLS Deconvolution. *Bull. Am. Astron. Soc.* **1995**, *27*, 1444.
14. Schwab, F.R. Relaxing the isoplanatism assumption in self-calibration; applications to low-frequency radio interferometry. *Astron. J.* **1984**, *89*, 1076–1081.
15. Hogbom, J.A. Aperture Synthesis with a Non-Regular Distribution of Interferometer Baselines. *Astron. Astrophys. Suppl. Ser.* **1974**, *15*, 417–426.
16. Clark, B.G. An efficient implementation of the algorithm 'CLEAN'. *Astron. Astrophys.* **1980**, *89*, 377–378.
17. Cornwell, T.J. Multiscale CLEAN Deconvolution of Radio Synthesis Images. *IEEE J. Sel. Top. Signal Process.* **2008**, *2*, 793–801.
18. Rau, U.; Cornwell, T.J. A multi-scale multi-frequency deconvolution algorithm for synthesis imaging in radio interferometry. *Astron. Astrophys.* **2011**, *532*, A71.

© 2019 by the authors. Licensee MDPI, Basel, Switzerland. This article is an open access article distributed under the terms and conditions of the Creative Commons Attribution (CC BY) license (http://creativecommons.org/licenses/by/4.0/).

 *proceedings*

*Proceedings*

# TI-Stan: Adaptively Annealed Thermodynamic Integration with HMC [†]

### R. Wesley Henderson *,[‡] and Paul M. Goggans [‡]

Department of Electrical Engineering, University of Mississippi, University, MS 38677, USA; goggans@olemiss.edu
* Correspondence: wesley.henderson11@gmail.com
† Presented at the 39th International Workshop on Bayesian Inference and Maximum Entropy Methods in Science and Engineering, Garching, Germany, 30 June–5 July 2019.
‡ These authors contributed equally to this work.

Published: 22 November 2019

**Abstract:** We present a novel implementation of the adaptively annealed thermodynamic integration technique using Hamiltonian Monte Carlo (HMC). Thermodynamic integration with importance sampling and adaptive annealing is an especially useful method for estimating model evidence for problems that use physics-based mathematical models. Because it is based on importance sampling, this method requires an efficient way to refresh the ensemble of samples. Existing successful implementations use binary slice sampling on the Hilbert curve to accomplish this task. This implementation works well if the model has few parameters or if it can be broken into separate parts with identical parameter priors that can be refreshed separately. However, for models that are not separable and have many parameters, a different method for refreshing the samples is needed. HMC, in the form of the MC-Stan package, is effective for jointly refreshing the ensemble under a high-dimensional model. MC-Stan uses automatic differentiation to compute the gradients of the likelihood that HMC requires in about the same amount of time as it computes the likelihood function itself, easing the programming burden compared to implementations of HMC that require explicitly specified gradient functions. We present a description of the overall TI-Stan procedure and results for representative example problems.

**Keywords:** model comparison; MCMC; thermodynamic integration; HMC

---

## 1. Introduction

Thermodynamic integration (TI) is a numerical technique for evaluating model evidence integrals. The technique was originally developed [2] to estimate the free energy of a fluid. Various improvements and changes have been made over the decades, and the incarnation of the technique that our method is based on is the adaptively-annealed, importance sampling-based method described by Goggans and Chi [3]. Their implementation follows John Skilling's BayeSys [4], and both make use of Binary slice sampling (BSS) and the Hilbert curve to complete the implementation. This article proposes a modification of this method that uses PyStan [5,6] and the No U Turn Sampler (NUTS) [7] instead of BSS and the Hilbert curve. This article is an adaptation of portions of the first author's doctoral dissertation ([1] Chapter 3). A Python 3 implementation of this method by the authors can be found on GitHub (https://github.com/rwhender/ti-stan) [8].

*1.1. Motivation*

The family of adaptively-annealed TI methods are important for solving model comparison problems in engineering, where we frequently need to evaluate complex physics-based mathematical models. TI methods with fixed annealing schedules (e.g., [9,10]) are useful for solving more traditional statistics problems, but tend to fail with the complex models that arise in engineering problems. TI methods that use BSS on the Hilbert curve are useful for a large set of problems; however, these methods see diminishing returns when the number of model parameters grows somewhat large (> 10 or so). These performance issues can be mitigated if the model equation can be decomposed into additive components with identical form and equivalent joint priors on their parameters. However, for problems with many model parameters and with model equations that cannot be decomposed, a different class of methods is required.

*1.2. Background*

From Bayes' theorem, for model vector $M$, data vector $D$, model parameter vector $\Theta$, and prior information $I$, the model evidence is

$$p(D|M, I) = \int p(D|\Theta, M, I) p(\Theta|M, I) \, d\Theta. \tag{1}$$

Here we introduce an inverse temperature parameter, $\beta$, that will control how much the likelihood influences the evidence value,

$$p(D|M, \beta, I) = \int [p(D|\Theta, M, I)]^\beta p(\Theta|M, I) \, d\Theta. \tag{2}$$

The full derivation is omitted here. The result is the thermodynamic integral form of the model evidence,

$$\log p(D|M, \beta, I) = -\int_0^1 \langle E_L(\Theta) \rangle_\beta \, d\beta, \tag{3}$$

where the energy term is defined as the negative log-likelihood:

$$E_L(\Theta) = -\log p(D|\Theta, M, I). \tag{4}$$

The integral in (3) usually cannot be evaluated analytically. For problems with relatively simple models, a fixed temperature ladder can be used, and Markov chain Monte Carlo (MCMC) can be used to estimate the expected energy at each temperature. However, for the class of problems we are concerned with, an approach in which the subsequent temperature is computed based on the conditions observed in the current step is necessary. This process is known as adaptive annealing. The general procedure as described by [3] is as follows:

1. Start at $\beta = 0$ where $p(\Theta|M, D, \beta, I) = p(\Theta|M, I)$, and draw $C$ samples from this distribution (the prior).
2. Compute the Monte Carlo estimator for the expected energy at the current $\beta$,

$$\langle E_L(\Theta) \rangle_\beta \approx \frac{1}{N} \sum_{t=1}^{C} E_L(\Theta_t), \tag{5}$$

where $\Theta_t$ is the current position of the $t$-th Markov chain.

3. Increment $\beta$ by $\Delta \beta_i$, where

$$\Delta \beta_i = \frac{\log \frac{\max w_j}{\min w_j}}{\max E_L(\Theta_i) - \min E_L(\Theta_i)}, \tag{6}$$

$j$ is the index on the chains, $w_j$ is the weight associated with chain $j$, and

$$w_j = \exp[-\Delta \beta_i E_L(\Theta_j)]. \tag{7}$$

4. Re-sample the population of samples using importance sampling.
5. Use MCMC to refresh the current population of samples. This yields a more accurate sampling of the distribution at the current temperature. This step can be easily parallelized, as each sample's position can be shifted independently of the others.
6. Return to step 2 and continue until $\beta_i$ reaches 1.
7. Estimate (3) using quadrature and the expected energy estimates built up using (5).

In this procedure, steps 3 and 4 are closely connected. In order to refresh the sample population most effectively, the importance sampling step should discard and replace at most 1 sample per temperature. New temperatures are chosen in a way that encourages this behavior. The term $\log \frac{\max w_j}{\min w_j}$ is a method parameter that can be set to make the adaptive annealing process more or less aggressive. Values of this parameter only slightly greater than one encourage a slow annealing, while higher values encourage a faster process.

## 2. Materials and Methods

The main innovation of this article relates to the implementation of step 5. As of Summer 2018, a survey of the available modern implementations of MCMC methods indicated that MC Stan (or simply Stan) [5], was the gold standard for general purpose MCMC. Stan uses NUTS [7] as the basis for its sampling functions. NUTS is based on Hamiltonian Monte Carlo (HMC) [11], which uses the gradient of the log-likelihood function to more efficiently explore the posterior distribution. NUTS improves upon HMC by automatically choosing optimal values for HMC's tunable method parameters. NUTS has been shown to sample complex distributions effectively. We sought to build an improved thermodynamic integration implementation by using Stan instead of binary slice sampling and leapfrog sampling to refresh the sample population at each temperature within TI. The result, Thermodynamic integration with Stan (TI-Stan), is described in this section.

The TI-Stan algorithm is shown in Algorithm 1.

Our implementation is in Python, so we made use of the PyStan interface to Stan [6]. Stan defines its own language for defining statistical models, which allows it to efficiently compute the derivatives needed for HMC via automatic differentiation. For a particular problem, it is therefore necessary to write a Stan file that contains the Stan-formatted specification of the model, in addition to the pure-Python energy functions necessary for TI with BSS. Once one is familiar with the simple Stan language, this additional programming cost becomes trivial compared to the time savings achieved by using this method instead of BSS.

**Algorithm 1** Thermodynamic integration with Stan
---
1: **procedure** TI(P, S, C, W, data)
2:     **Inputs**: P–Number of parameters, S–Number of Stan iterations per temperature, C–Number of chains, W–Ratio to control adaptive annealing, data–Data
3:     **for** $m \leftarrow 1, C$ **do**
4:         **for** $j \leftarrow 1, P$ **do**
5:             $\alpha_j^m \leftarrow \text{RAND}(0,1)$
6:         **end for**
7:         $E_m^* \leftarrow \text{ENERGY}(\alpha^m, data)$
8:     **end for**
9:     $i \leftarrow 1$
10:     Compute $\langle E^* \rangle_i$
11:     $\beta_1 \leftarrow \min\{\log(W)/[\max(E^*) - \min(E^*)], 1\}$
12:     $w \leftarrow \exp(-\beta_1 E^*)$
13:     IMPORTANCESAMPLING($w, \alpha, E^*, C$)
14:     **while** $\beta_i > 0$ and $\beta_i < 1$ **do**
15:         **for** $m \leftarrow 1, C$ **do**
16:             STANSAMPLING($\alpha^m, E_m^*, C, P, S, \beta_i, data$)
17:         **end for**
18:         $i \leftarrow i + 1$
19:         $\Delta\beta \leftarrow \log(W)/[\max(E^*) - \min(E^*)]$
20:         $\beta_i \leftarrow \min(\beta_{i-1} + \Delta\beta, 1)$
21:         **if** $\beta_{i-1} + \Delta\beta > 1$ **then**
22:             $\Delta\beta \leftarrow 1 - \beta_{i-1}$
23:         **end if**
24:         $w \leftarrow \exp(-\Delta\beta E^*)$
25:         IMPORTANCESAMPLING($w, \alpha, E^*, C$)
26:     **end while**
27:     Estimate (3) using trapezoid rule and $\{\beta_i\}$ and $\{\langle E^* \rangle_i\}$
28: **end procedure**

*2.1. Tests*

We use two test problems to demonstrate TI-Stan in practice. These test problems are described below.

2.1.1. Twin Gaussian Shells

The first example is the twin Gaussian shell problem from [12]. In [12], the authors present results for this problem in up to 30 dimensions. Handley, et al. [13] also use this problem in 100 dimensions to test their algorithm. This problem presents a few interesting challenges. Because the likelihood takes the form of a thin, curved density whose mass centers on a hyper-spherical shell, MCMC moves are difficult to make efficiently. The bimodal distribution is also challenging to sample effectively. Finally, the examples we explore are high-dimensional to the point that standard numerical integration techniques would be useless.

The likelihood function in the twin Gaussian shells problem takes the form,

$$\mathcal{L}(\Theta) = \frac{1}{\sqrt{2\pi}w_1} \exp\left[-\frac{(|\Theta - \mathbf{c}_1| - r_1)^2}{2w_1^2}\right] + \frac{1}{\sqrt{2\pi}w_2} \exp\left[-\frac{(|\Theta - \mathbf{c}_2| - r_2)^2}{2w_2^2}\right]. \tag{8}$$

Following [12], we set the parameters as follows: $w_1 = w_2 = 0.1$, $r_1 = r_2 = 2$, $\mathbf{c}_1 = [-3.5, 0, \cdots, 0]^T$, and $\mathbf{c}_2 = [3.5, 0, \cdots, 0]^T$. We use a uniform prior over the hypercube that spans $[-6, 6]$ in each dimension.

## 2.1.2. Detection of Multiple Stationary Frequencies

For the second test, we estimate the number of stationary frequencies present in a signal. This problem is similar to the problem of multiple stationary frequency estimation in [14, Chapter 6], with the additional task of determining the number of stationary frequencies present. Differences among log-evidence values for models containing either the most probable number of frequencies or more tend to be small, meaning that a precise estimate of these log-evidence values is essential to the task of determining the most probable model.

Each stationary frequency (*j*) in the model is determined by three parameters: the in-phase amplitude ($A_j$), the quadrature amplitude ($B_j$), and the frequency ($f_j$). Given *J* stationary frequencies, the model at time step $t_i$ takes the following form:

$$g[t_i; \Theta] = \sum_{j=1}^{J} A_j \cos(2\pi f_j t_i) + B_j \sin(2\pi f_j t_i), \tag{9}$$

where $\Theta$ is the parameter vector

$$\Theta = \begin{bmatrix} A_1 \ B_1 \ f_1 \ \cdots \ A_J \ B_J \ f_J \end{bmatrix}^T.$$

For the purposes of this test the noise variance used to generate the simulated data is known, hence we use a Gaussian likelihood function,

$$\mathcal{L}(\Theta) = \prod_{i=1}^{K} \exp\left\{ -\frac{[g(t_i; \Theta) - d_i]^2}{2\sigma^2} \right\}, \tag{10}$$

for *K* simulated data $d_i$ and noise variance $\sigma^2$. The log-likelihood function is then

$$\log \mathcal{L}(\Theta) = -\sum_{i=1}^{K} \frac{[g(t_i; \Theta) - d_i]^2}{2\sigma^2}. \tag{11}$$

Each model parameter is assigned a uniform prior distribution with limits as shown in Table 1.

**Table 1.** Prior bounds for multiple stationary frequency model parameters.

|       | Lower Bound | Upper Bound |
|-------|-------------|-------------|
| $A_j$ | −2          | 2           |
| $B_j$ | −2          | 2           |
| $f_j$ | 0 Hz        | 6.4 Hz      |

Our test signal is a sum of two sinusoidal components, and zero-mean Gaussian noise with variance $\sigma^2 = 0.01$. This signal is sampled at randomly-spaced instants of time, in order to demonstrate that this time-domain method does not require uniform sampling to perform spectrum estimation. Bretthorst [15] demonstrates that the Nyquist critical frequency in the case of nonuniform sampling is $1/2\Delta T'$, where $\Delta T'$ is the dwell time. The dwell time is not defined for arbitrary-precision time values as used in this example, so we must choose another limiting value. A more conservative limit is given by $1/10\Delta T_{avg}$, where $\Delta T_{avg}$ is the average spacing between time steps, 1/64 s. This formulation yields a prior maximum limit of 6.4 Hz, as shown in Table 1. The parameters used to generate the simulated data are shown in Table 2.

**Table 2.** Parameters used to generate simulated signal.

| j | $A_j$ | $B_j$ | $f_j$ (Hz) |
|---|---|---|---|
| 1 | 1.0 | 0.0 | 3.1 |
| 2 | 1.0 | 0.0 | 5.9 |

## 3. Results

For these tests, performance is compared among the Thermodynamic integration with binary slice sampling (TI-BSS) method and the TI-Stan method. The settings used for TI-BSS are shown in Table 3, while the settings used for TI-Stan are shown in Table 4. For each example, the user-defined annealing control constant W was set to both 1.5 and 2.0. For the box-plots in this section, the middle line represents the median value, the box is bounded by the upper and lower quartiles, and the whiskers extend to the range of the data that lies within 1.5 times the inter-quartile range. Any data points past this threshold are plotted as circles.

**Table 3.** Parameters for TI-BSS examples.

| Parameter | Value | Definition |
|---|---|---|
| S | 200 | Number of binary slice sampling steps |
| M | 2 | Number of combined binary slice sampling and leapfrog steps |
| C | 256 | Number of chains |
| B | 32 | Number of bits per parameter in SFC |

**Table 4.** Parameters for TI-Stan examples.

| Parameter | Value | Definition |
|---|---|---|
| S | 200 | Number of steps allowed in Stan |
| C | 256 | Number of chains |

These results were generated on a Google Cloud instance with 32 virtual Intel Broadwell CPUs and 28.8 GB of RAM.

First, we present results for the twin Gaussian shells distribution with 10 dimensions. A box-plot summarizing the log-evidence estimates over 20 runs each for TI-Stan and Thermodynamic integration with binary slice sampling and the Hilbert curve (TI-BSS-H) and for each value of W is shown in Figure 1a. A box-plot summarizing the run times over 20 runs each for the TI methods is shown in Figure 1b.

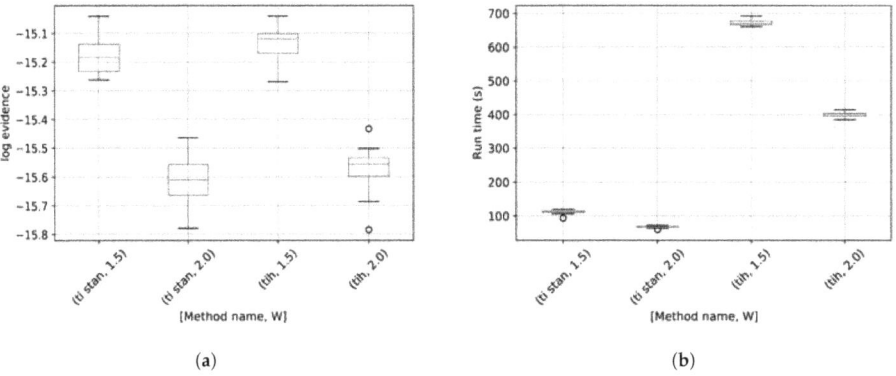

**Figure 1.** Twin Gaussian shell test results. (**a**) Box-plot of log-evidence for the 10-D twin Gaussian shell problem for TI-Stan and TI-BSS-H; (**b**) Box-plot of run time in seconds for the 10-D twin Gaussian shell problem for TI-Stan and TI-BSS-H.

Second, we present results for the detection of multiple stationary frequencies problem. Box-plots of log-evidence values for a model assuming one, two, and three frequencies present are shown in Figures 2a, 3a, and 4a. For the models with one and three frequencies present, results are shown for TI-Stan, TI-BSS-H, and Thermodynamic integration with binary slice sampling and the Z-order curve (TI-BSS-Z) [16]. For the model with two frequencies present (the model also used to generate the test signal), results for TI-BSS-Z are not shown. For this model, TI-BSS-Z ended early here and did not arrive at a reasonable result. Box-plots of the run time for models assuming one, two, and three frequencies present are shown in Figures 2b, 3b, and 4b.

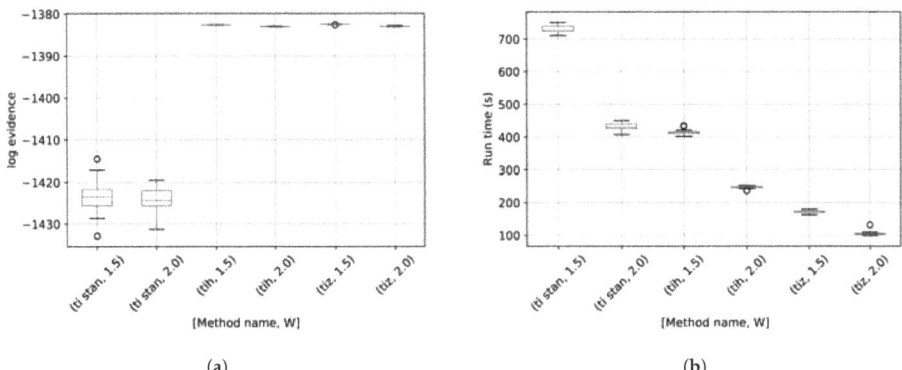

**Figure 2.** MSF model with $J = 1$ results. (a) Box-plot of log-evidence for the one stationary frequency model for TI-Stan, TI-BSS-H, and TI-BSS-Z, for two values of $W$; (b) Box-plot of run time for the one stationary frequency model for TI-Stan, TI-BSS-H, and TI-BSS-Z, for two values of $W$.

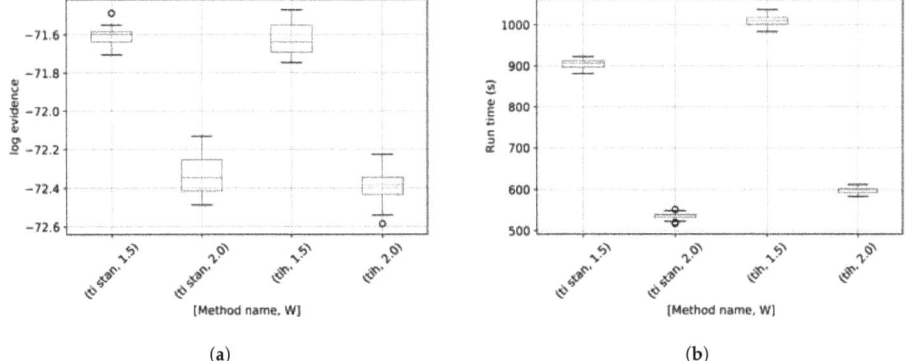

**Figure 3.** MSF model with $J = 2$ results. (a) Box-plot of log-evidence for the two stationary frequency model for TI-Stan and TI-BSS-H, for two values of $W$; (b) Box-plot of run time for the two stationary frequency model for TI-Stan and TI-BSS-H, for two values of $W$.

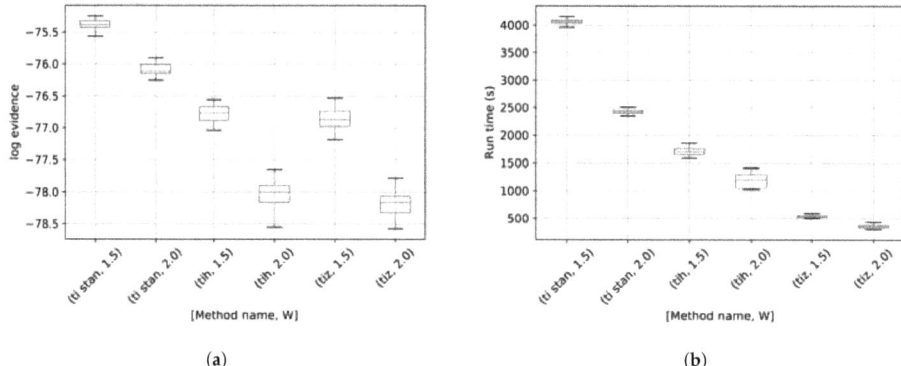

**Figure 4.** MSF model with $J = 1$ results. (**a**) Box-plot of log-evidence for the three stationary frequency model for TI-Stan, TI-BSS-H, and TI-BSS-Z, for two values of $W$; (**b**) Box-plot of run time for the three stationary frequency model for TI-Stan, TI-BSS-H, and TI-BSS-Z, for two values of $W$.

## 4. Discussion

Regarding the twin Gaussian shells test, the analytical log-evidence for this distribution [12] is $-14.59$. None of the configurations tested actually reached that value (Figure 1a, but the runs using $W = 1.5$ got closest, suggesting that a value of $W$ closer to 1 would perhaps approach the correct value more closely. Figure 1b shows that the run time drastically increases as $W$ approaches 1. It also shows that TI-BSS-H takes about 6 times longer, on average, than TI-Stan to compute its estimate of the log-evidence. According to Figure 1a, the two methods have comparable accuracy and precision, so this difference in run time illustrates the difficulty the Hilbert curve-based method has with distributions of high dimension.

Regarding the detection of multiple stationary frequencies test, there are no analytical log-evidence values available. We argue that a method is successful if the model used to generate the data clearly has the highest log-evidence, with a good margin between it and the log-evidence for the other models. Figures 2a and 4a show some significant disagreement among the various methods for the "wrong" models (those with one and three frequencies), but Figure 3a shows that the methods are in much closer agreement for the two frequency model. For TI-Stan and TI-BSS-H and for both values of $W$, the two frequency model is clearly the maximum-log-evidence choice. Even with the variations in the runs, the results do not overlap at any point from model to model, and the closest model-to-model margins are all greater than 2.3, which corresponds to an odds of 10.

In Figure 2b, TI-Stan has the greatest run time for both values of $W$, suggesting that its adaptive sampling process had trouble efficiently sampling distributions based on this high-error model. TI-BSS-H was much faster, and TI-BSS-Z was faster still. In Figure 3b, the run times of TI-Stan and TI-BSS-H are comparable. This suggests that TI-Stan was able to more effectively sample the distribution based on the lower-error model. Figure 4b shows a similar pattern in the run times to Figure 2b. The fact that this model is able to fit the noise in the data (yielding especially sharp distributions) and the fact that the distribution is increasingly multi-modal as the number of frequencies increases may explain why TI-Stan took a long time to compute a result here.

These preliminary results indicate that TI-Stan is a promising method for computing model evidence for problems with complex physics-based mathematical models. Results for further problems, including the twin Gaussian shell problem with up to 100 dimensions, can be found in ([1] Chapter 3). Future work could further evaluate this method's usefulness by solving real complex model comparison problems in engineering.

**Funding:** This research received no external funding.

**Conflicts of Interest:** The authors declare no conflict of interest.

## Abbreviations

The following abbreviations are used in this manuscript:

TI  Thermodynamic integration
BSS  Binary slice sampling
TI-Stan  Thermodynamic integration with Stan
TI-BSS  Thermodynamic integration with binary slice sampling
TI-BSS-H  Thermodynamic integration with binary slice sampling and the Hilbert curve
TI-BSS-Z  Thermodynamic integration with binary slice sampling and the Z-order curve
HMC  Hamiltonian Monte Carlo
NUTS  No U Turn Sampler
MCMC  Markov chain Monte Carlo

## References

1. Henderson, R.W. Design and analysis of efficient parallel bayesian model comparison algorithms. Doctoral Dissertation, University of Mississippi, Oxford, MS, USA, 2019.
2. Kirkwood, J.G. Statistical Mechanics of Fluid Mixtures. *J. Chem. Phys.* **1935**, *3*, 300–313.
3. Goggans, P.M.; Chi, Y. Using thermodynamic integration to calculate the posterior probability in Bayesian model selection problems. *AIP Conf. Proc.* **2004**, *707*, 59–66. doi:10.1063/1.1751356.
4. Skilling, J. *BayeSys and MassInf*; Maximum Entropy Data Consultants Ltd.: London, UK, 2004.
5. Carpenter, B.; Gelman, A.; Hoffman, M.D.; Lee, D.; Goodrich, B.; Betancourt, M.; Brubaker, M.; Guo, J.; Li, P.; Riddell, A. Stan : A Probabilistic Programming Language. *J. Stat. Softw.* **2017**, *76*. doi:10.18637/jss.v076.i01.
6. Stan Development Team. *PyStan: The Python Interface to Stan*; 2018. Available online:http://mc-stan.org (accessed on 21 Novemver 2019)
7. Hoffman, M.D.; Gelman, A. The No-U-Turn Sampler: Adaptively Setting Path Lengths in Hamiltonian Monte Carlo. *J. Mach. Learn. Res.* **2014**, *15*, 1593–1623.
8. Henderson, R.W. TI-Stan. 2019. original-date: 2019-07-04T09:56:19Z. Available online: https://github.com/rwhender/ti-stan. (accessed on 21 Novemver 2019)
9. Gelman, A.; Meng, X.L. Simulating normalizing constants: From importance sampling to bridge sampling to path sampling. *Stat. Sci.* **1998**, *13*, 163–185.
10. Oates, C.J.; Papamarkou, T.; Girolami, M. The Controlled Thermodynamic Integral for Bayesian Model Evidence Evaluation. *J. Am. Stat. Assoc.* **2016**, *111*, 634–645. doi:10.1080/01621459.2015.1021006.
11. Neal, R.M. MCMC using Hamiltonian dynamics. In *Handbook of Markov Chain Monte Carlo*; Brooks, S., Gelman, A., Jones, G., Meng, X.L., Eds.; Chapman & Hall / CRC Press: New York, NY, USA, 2011.
12. Feroz, F.; Hobson, M.P.; Bridges, M. MultiNest: an efficient and robust Bayesian inference tool for cosmology and particle physics. *Mon. Not. R. Astron. Soc.* **2009**, *398*, 1601–1614. doi:10.1111/j.1365-2966.2009.14548.x.
13. Handley, W.; Hobson, M.; Lasenby, A. PolyChord: Next-generation nested sampling. *Mon. Not. R. Astron. Soc.* **2015**, *453*, 4384–4398. doi:10.1093/mnras/stv1911.
14. Bretthorst, G.L. *Bayesian Spectrum Analysis and Parameter Estimation*; Springer: Berlin/Heidelberg, Germany, 1988.
15. Bretthorst, G.L. Nonuniform sampling: Bandwidth and aliasing. *AIP Conf. Proc.* **2001**, *567*, 1–28. doi:10.1063/1.1381847.
16. Henderson, R.W.; Goggans, P.M. Using the Z-order curve for Bayesian model comparison. In *Bayesian Inference and Maximum Entropy Methods in Science and Engineering. MaxEnt 2017*; Springer Proceedings in Mathematics & Statistics; Polpo, A., Stern, J., Louzada, F., Izbicki, R., Takada, H., Eds.; Springer: Berlin/Heidelberg, Germany, 2018; Volume 239, pp. 295–304.

© 2019 by the authors. Licensee MDPI, Basel, Switzerland. This article is an open access article distributed under the terms and conditions of the Creative Commons Attribution (CC BY) license (http://creativecommons.org/licenses/by/4.0/).

*Proceedings*

# Carpets Color and Pattern Detection Based on Their Images [†]

Sayedeh Marjaneh Hosseini [1,‡], Ali Mohhamad-Djafari [2,‡], Adel Mohammadpour [3,*,‡,§], Sobhan Mohammadpour [4,5,‡] and Mohammad Nadi [6,‡]

1. Department of Computer Science, Amirkabir University of Technology, Tehran, Iran; marjanehosseini@aut.ac.ir
2. Laboratoire des Signaux & Systèmes (Supélec, CNRS, Université Paris Sud), 91190 Gif-sur-Yvette, France; ali.mohammad-djafari@l2s.centralesupelec.fr
3. Department of Statistics, Amirkabir University of Technology (Tehran Polytechnic), Tehran, Iran
4. Department of Computer Engineering, Sharif University of Technology, Tehran, Iran; sobhan.mohammadpour@umontreal.ca
5. Department of Computer Science and Operation Research, Université de Montréal, Montreal, QC H3T 1J4, Canada
6. E-CarpetGallery, Montreal, QC H4E 4N7, Canada; mohammad.nadi@gmail.com
* Correspondence: adel@aut.ac.ir
† Presented at the 39th International Workshop on Bayesian Inference and Maximum Entropy Methods in Science and Engineering, Garching, Germany, 30 June–5 July 2019.
‡ Authors' names are alphabetically arranged, the 4th author contributed 40% to this work.
§ Present address: Department of Mathematics and Statistics, McGill University, Montreal, QC H3A 0B9, Canada.

Published: 24 December 2019

**Abstract:** In these days of fast-paced business, accurate automatic color or pattern detection is a necessity for carpet retailers. Many well-known color detection algorithms have many shortcomings. Apart from the color itself, neighboring colors, style, and pattern also affects how humans perceive color. Most if not all, color detection algorithms do not take this into account. Furthermore, the algorithm needed should be invariant to changes in brightness, size, and contrast of the image. In a previous experiment, the accuracy of the algorithm was half of the human counterpart. Therefore, we propose a supervised approach to reduce detection errors. We used more than 37,000 images from a retailer's database as the learning set to train a Convolutional Neural Network (CNN, or ConvNet) architecture.

**Keywords:** color detection; pattern detection; carpet images; convolutional neural network

## 1. Introduction

The wave of digitalization is taking over many business areas; carpet selling is no exception. With its recent exponential growth, electronic carpet retailers are selling more carpets than ever. Like most businesses, these retailers need to automate their process, which typically involves getting a container of carpets, tagging them and adding them to an online catalog. While the first and the last phase have already been automated, due to the lack of a common and extensive standard between major carpet exporters from major hand-turfed carpet hubs like Iran, India, and Afghanistan, many retailers have to employ carpet experts to generate the metadata visually. Current literature does not seem to provide any acceptable computational model that requires a minimal change in the production pipeline. We took e-Carpet Gallery (www.ecarpetgallery.com) (e-CG) work-flow as an example and tried to automate its second step.

E-CG's work-flow consists of three steps. First, a container full of carpet arrives at the warehouse, the employees unload the container, take pictures of the rugs and fill out the empty fields of features of carpets for each carpet one by one then move them to storage. Due to the volume of operation, employees have a limited time to tag the carpets. The repetitive nature of the task mixed with its complexity makes either the process slow(multiple checks) or very error-prone. The markets have started automating the more deterministic and less opinioned parts of the process like the measurement of weight, width, length, and shape of carpets. However, there is no method to classify nominal features like material and quality of the carpets. Accurate determination of qualitative or hybrid features such as color, pattern, and style, are essential for shoppers but is very difficult. While this process has matured over the years, its accuracy much to desire. In this work, we apply a few selected image processing algorithms to e-CG's database. This e-CG is a production-level database with a reasonable error rate and contains fields like top-down image, pattern, and color (main color). We focus on the automatic labeling of two qualitative features: color and pattern. Well-known and intuitive color detection method such as the pixel count of clustered image's pixels fail due to the difference between a visual human color assessment of a carpet(or any other abstract image) and, e.g., counting the number of high-frequency pixels. This is due to the association between color and pattern in a carpet. A reoccurring example is that the color of a carpet with a big red flower in its center is usually red based on a human assessment, even when most of the carpet is a lighter color like beige. To the authors' knowledge, CNNs are not yet applied as a classification tool for patterns and colors of carpets. However, CNN applies in color detection [1], for example, in the vehicle color recognition [2]. In the next section, we introduce two features of carpets and their levels (or labels). In Section 3, we point out a few technical issues. We apply a few classification methods as well as CNN for pattern and color.

## 2. Database and Features

E-CG has a database with more than one million carpet images. We only consider a sample of 37,000 carpets from the last two years to speed up the classification process. We introduce three features: color, pattern, and style via frequency bar charts. The level names of each feature are listed alphabetically in the corresponding graph. Figure 1 shows color levels and a sample of images to demonstrate color levels in a real database. One can find that the red color has a high-frequency level and is one of the unbalancing element of the database. Levels' frequency chart of patterns is plotted in Figure 2. The number of pattern levels is fixed based on the dataset. Figure 3 illustrates sample image patterns and their corresponding edges to clarify the different patterns.

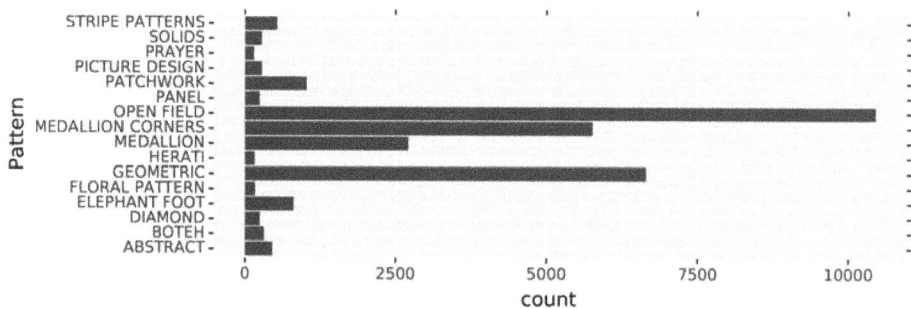

**Figure 1.** Frequency bar chart of pattern levels. Level names are sorted alphabetically.

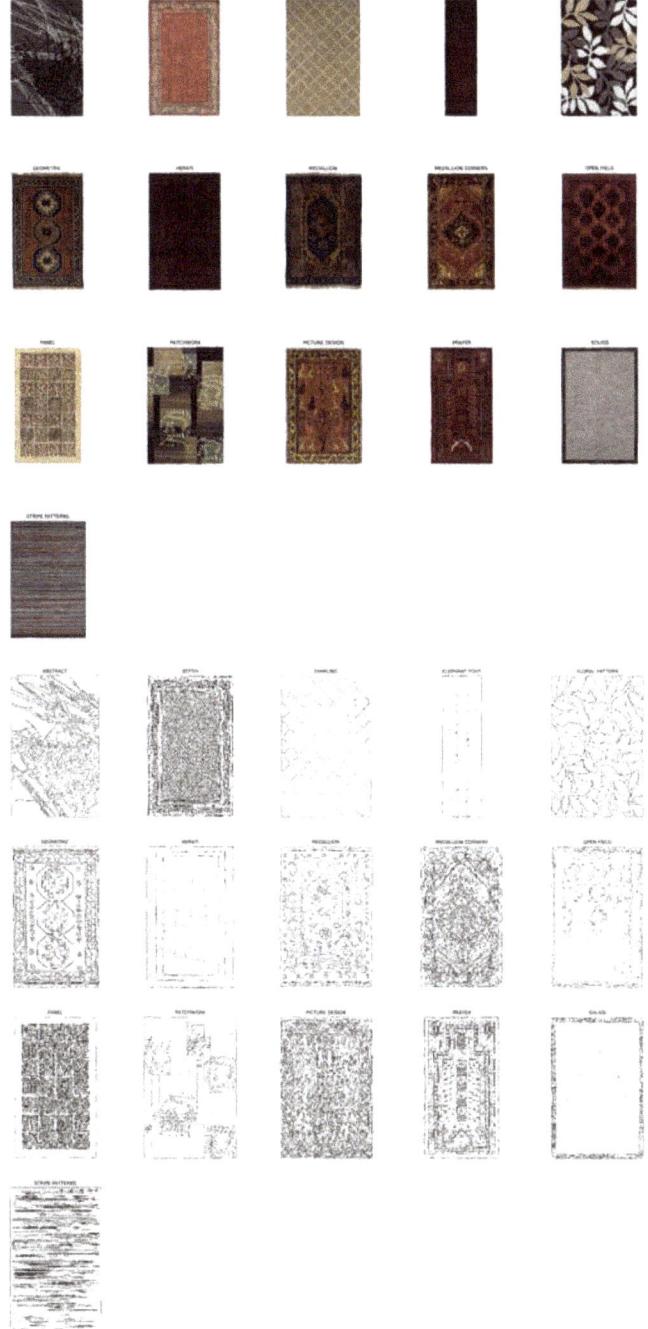

**Figure 2.** (**Top**): Sample carpet with their pattern levels. (**Bottom**): Corresponding edges of top images.

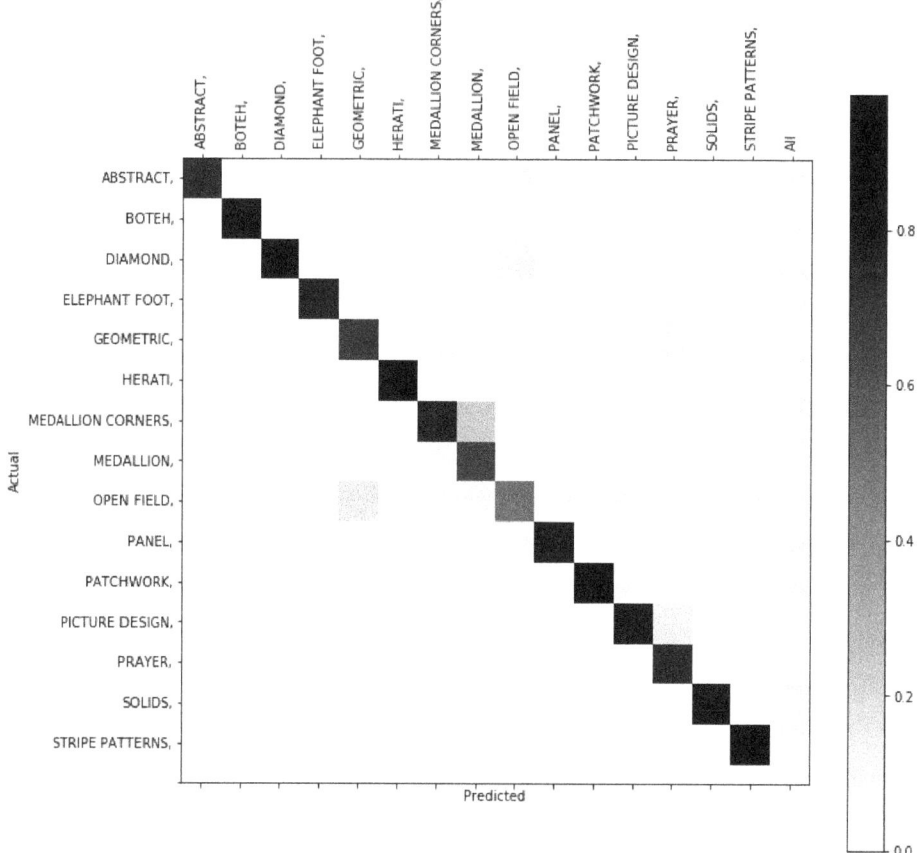

**Figure 3.** Pattern classification confusion matrix.

We consider a sample of recorded data that each feature level had at least 100 occurrences in our selected time frame. While 100 may have arbitrarily selected, this ensures that we do not have to deal with features and categories that do not often appear in production, thus reducing the model complexity and the error rate introduced by using unbalanced data.

## 3. Classification

Before turning to the methodology of classification algorithms, we point out to essential technical issue:

- Color accuracy has always been an issue for professional photographers, designers, and printers to deal with it on a daily basis. The images we got were not very uniformly photographed. They were not very uniformly lighted, with the top of the image being the brightest and bottom being darkest.
- The size of carpets are very irregular; this makes the use of CNN rather hard because runner carpets will end up with a lot of white space.

*3.1. Pattern*

For pattern, we decided that it would be reasonable to start with CNN. Since we did not have the computational power to train a CNN from scratch, we used the XCeption architecture [3], with pre-trained weights on ImageNet, a model with a high top 1 accuracy on ImageNet, which seems good at understanding patterns. XCeption architecture has fewer parameters and more accuracy with respect to a few tested architectures. However, one may find a better CNN architecture with high-level hardware. We removed the top layers and added two dense layers, one with twice as many nodes as the number of patterns with an ELU (Exponential Linear Unit) activation and a softmax layer to classify. We trained the network with an Adam optimizer [4], for 60 epoch with randomly rotated, sheared, and flipped images. Due to the unbalancedness of the database, we decided to under-sample everything. It's worth noting that patterns that are similar or not well defined like "open field" and "diamond" tend to get misclassified often by humans and machines.

In Figure 4, we report the percentage of classified test dataset that was 20% of database images; this figure present pattern classification confusion matrix. The actual and predicted values are row and column of the confusion matrix, respectively. We make a column-wise normalization for a better interpretation. The last column shows the sum of misclassified that can be considered as a bias indicator. One can observe that the diagonal of the matrix is bolded that shows the performance of the proposed algorithm is more than 80%.

*3.2. Color*

We Used *K*-means clustering algorithm to find a value representing each image, then run a classification algorithm, AdaBoost [5], on the result. This two-step classification method is used to speed up AdaBoost algorithm. The accuracy was around 45%. Three contributing factors where observed:

1. Colors are tightly stacked, and very similar colors like ivory and beige or dark copper and red get misclassified often,
2. The image was not calibrated.
3. The majority does not always mean most dominant as having a small area of a dominant color like red or black will result in a red or black carpet. Therefore, we decided to reuse the network from Pattern Recognition.

With all of this in mind were-trained the CNN we used for pattern recognition and got a similar result to the two steps classification. As we had expected, the network often confused similar colors like the previous experiment. It's possible that unless the pictures are very well calibrated nothing can give an accurate result. The results were acceptable despite the low accuracy because color ranges are not really well defined and the network mostly confused resemblant colors like beige and ivory.

The confusion matrix of CNN for color, similar to the pattern feature, is plotted in Figure 5. We can observe that the performance is similar to the above two steps *k*-means-AdaBoost algorithm is less than 50%.

Proceedings 2019, 33, 28

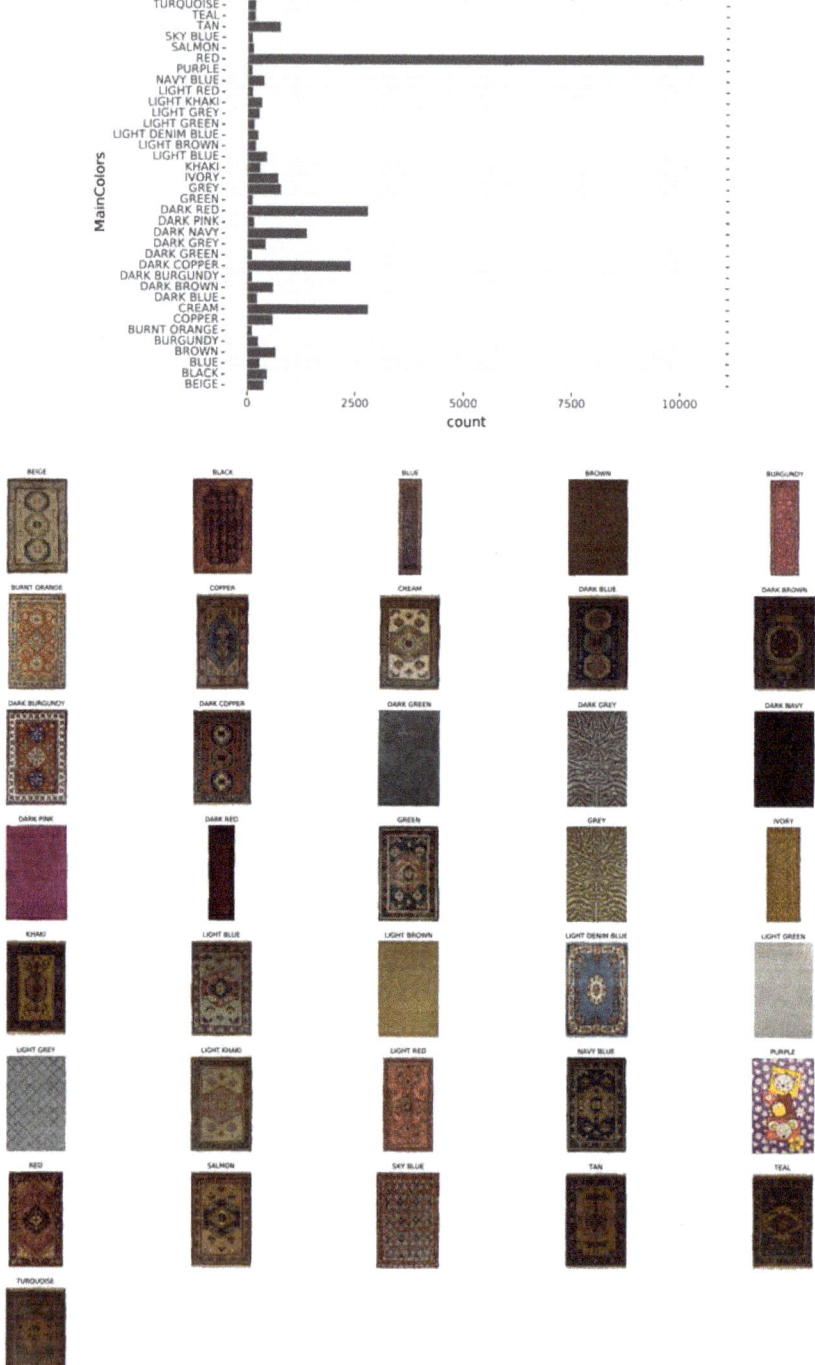

**Figure 4.** Frequency bar chart and a sample of color level labels. Label names are sorted alphabetically.

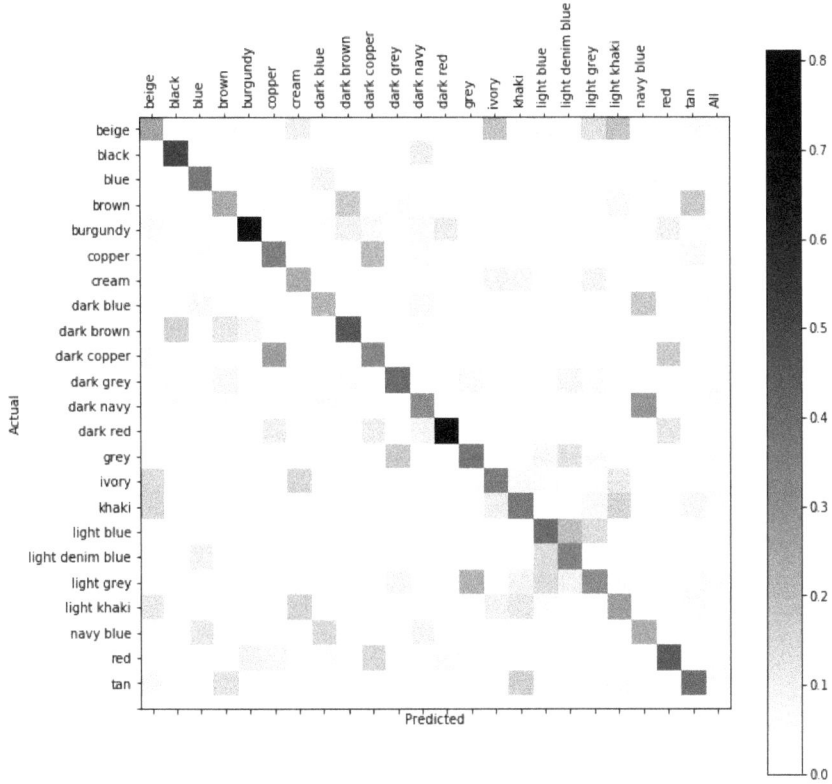

**Figure 5.** Color classification confusion matrix.

**Acknowledgments:** The authors would like to thank two anonymous referees for their helpful comments and for careful reading that greatly improved the article. We thank E-CarpetGallery for access to their database and support of three last authors during their research.

## References

1. Luong, T.X.; Kim, B.; Lee, S. Color image processing based on nonnegative matrix factorization with convolutional neural network. In Proceedings of the International Joint Conference on Neural Networks (IJCNN), Beijing, China, 6–11 July 2014; pp. 2130–2135.
2. Zhang, Q.; Zhuo, L.; Li, J.; Zhang, J.; Zhang, H.; Li, X. Vehicle color recognition using multiple-layer feature representations of lightweight convolutional neural network. *Signal Process.* **2018**, *147*, 146–153.
3. Chollet, F. Xception: Deep learning with depthwise separable convolutions. In Proceedings of the IEEE Conference on Computer Vision and Pattern Recognition (CVPR), Honolulu, HI, USA, 21–26 July 2017; pp. 1800–1807.
4. Kingma, D.P.; Ba. J.L. Adam: A Method for Stochastic Optimization. In Proceedings of the International Conference on Learning Representation (ICLR 2015), San Diego, CA, USA, 7–9 May 2015.
5. Schapire, R.; Freund Y. *Boosting: Foundations and Algorithms*; MIT: Cambridge, MA, USA.

 © 2019 by the authors. Licensee MDPI, Basel, Switzerland. This article is an open access article distributed under the terms and conditions of the Creative Commons Attribution (CC BY) license (http://creativecommons.org/licenses/by/4.0/).

*Proceedings*

# An Entropic Dynamics Approach to Geometrodynamics [†]

**Selman Ipek * and Ariel Caticha**

Physics Department, University at Albany-SUNY, Albany, NY 12222, USA; acaticha@albany.edu
* Correspondence: sipek@albany.edu
† Presented at the 39th International Workshop on Bayesian Inference and Maximum Entropy Methods in Science and Engineering, Garching, Germany, 30 June–5 July 2019.

Published: 27 November 2019

**Abstract:** In the Entropic Dynamics (ED) framework quantum theory is derived as an application of entropic methods of inference. The physics is introduced through appropriate choices of variables and of constraints that codify the relevant physical information. In previous work, a manifestly covariant ED of quantum scalar fields in a fixed background spacetime was developed. Manifest relativistic covariance was achieved by imposing constraints in the form of Poisson brackets and of intial conditions to be satisfied by a set of local Hamiltonian generators. Our approach succeeded in extending to the quantum domain the classical framework that originated with Dirac and was later developed by Teitelboim and Kuchar. In the present work the ED of quantum fields is extended further by allowing the geometry of spacetime to fully partake in the dynamics. The result is a first-principles ED model that in one limit reproduces quantum mechanics and in another limit reproduces classical general relativity. Our model shares some formal features with the so-called "semi-classical" approach to gravity.

**Keywords:** quantum gravity; quantum field theory; entropy; geometry; Hamiltonian dynamics; quantum foundations

---

## 1. Introduction

Efforts to develop a theory of quantum gravity (QG) have been dedicated, principally, to two main candidates—string theory (ST) and loop quantum gravity (LQG). (For a review see e.g., [1]) But despite the general sentiment that these programs are markedly different, they share some fundamental commonalities not usually discussed. Namely, these programs share a rather strict view of QG in which gravitation is itself a quantum force, akin to the electroweak and strong forces of the standard model; such approaches may aptly be referred to as *quantized gravity* theories. But QG, broadly construed, is not simply a program of quantized gravity. Indeed, although interest in QG is driven by a variety of reasons (see e.g., [2]), a very basic motivation is that of *consistency*: if matter, which is well-described by quantum theory (QT), is a source of energy, momentum, and so on, and if gravity couples to matter, then there must be some theory, or shared framework that brings the two together. We conjecture that this common thread is, in fact, *entropy*.

We propose here a QG candidate formulated entirely within the Entropic Dynamics (ED) framework. ED is a scheme for generating dynamical theories that are consistent with the entropic and Bayesian rules for processing information (For a review of entropic and Bayesian methods, see e.g., [3,4]). Among the successes of the ED framework are principled derivations of several aspects of the quantum formalism (For a review of current work, see e.g., [5]) that are also sensitive to, and indeed, help to clarify many conceptual issues that plague QT [6]. The standout feature of ED that makes this possible is that a clear delineation is maintained between the *ontological*, or physical, aspects of the theory and those that are of *epistemological* significance. The distinction is important: the ontic, or physical, variables are the subject of our inferences, and it is their values we wish to predict. The ability to make such predictions, however, depends on the available information, which in ED is codified in the constraints. In short, ED is a dynamics driven by constraints.

In previous work [7], ED was utilized to derive the standard quantum field theory on curved space-time (QFTCS) by making a particular choice of constraints, appropriate for a scalar field $\chi_x$ propagating on a *fixed* background space-time. One constraint involved the introduction of a drift potential $\phi$ that guides the flow of the probability $\rho$, while another was the employment of a canonical formalism, i.e., *Hamiltonians*, for driving their joint dynamics. But in the context of a manifestly relativistic theory, further constraints are needed. In [7] these were supplied by the adoption of the covariant canonical methods of Dirac, Hojman, Kuchař, and Teitelboim (DHKT). We briefly review their work. Within the DHKT scheme, time evolution unfolds as an accumulation of local deformations in three-dimensional space, constrained by the requirement that the evolution of three-dimensional space be such that it sweeps a four-dimensional space-time. The criterion for accomplishing this, called *path independence* by Kuchař and Teitelboim, results in an "algebra" to be satisfied by the generators of deformations (The quotes in "algebra" are meant as a reminder that local deformations do not form a true group; while two deformations can be composed to form another, the composition depends on the original surface). An interesting aspect of the DHKT approach is that whether one deals with an externally prescribed space-time, or whether the background geometry is itself truly dynamical, the "algebra" to be satisfied remains the *same* [8] (see also [9]). The true distinction manifests in the choice of variables to describe the evolving geometry.

Our goal here is to pursue an ED in which the background is itself a full partner in the dynamics. But we also wish to proceed conservatively; our aim is not, after all, to simply discard information which has already been proven valuable in ED. Thus we proceed in a minimalist fashion by taking the constraints of [7] and altering them to account for the additional information; which amounts here to a different choice of *variables* to describe the geometry. Following the seminal work of Hojman, Kuchař, and Teitelboim in [10], we make a choice in which the metric of three-dimensional space is itself a canonical variable. The ED that emerges from this choice bears great formal resemblance to the so-called "semi-classical" Einstein equations, but the conceptual differences are enormous: (a) in direct contradiction to the standard (Copenhagen) interpretation of QT, the scalar matter field is ontic and thus has definite values at all times. (b) There are no quantum probabilities. The ED approach is in strict adherence to the Bayesian and entropic principles of inference. (c) There are none of the paradoxes associated to the quantum measurement problem—which is the main source of objection to semi-classical gravity (see [11] and references therein). (d) Our geometrodynamics is not a forced marriage between classical gravity and quantum field theory. It is a first principles framework that not only derives the correct coupling of matter to gravity, but also reconstructs the theory of quantum mechanics itself.

The paper is outlined as follows. In Section 2, we give a quick review of the ED of infinitesimal steps, while we review some basic space-time kinematics in Section 3. In Section 4 we introduce the notion of a relativistic notion of entropic time. Many of the new results are in Sections 5 and 6, where we construct a geometrodynamics driven by entropic matter. We discuss our results in Section 7.

## 2. Reviewing the Entropic Dynamics of Infinitesimal Steps

We adopt the notations and conventions of [7] throughout. We study a single scalar field $\chi(x) \equiv \chi_x$ whose values are posited to be definite, but unknown. An entire field configuration, denoted $\chi$, lives on a 3-dimensional curved space $\sigma$, the points of which are labeled by coordinates $x^i$ ($i = 1, 2, 3$). The space $\sigma$ is a three-dimensional curved space equipped with a metric $g_{ij}$ that is currently fixed, but that will later become dynamical. A single field configuration $\chi$ is a point in an $\infty$-dimensional configuration space $\mathcal{C}$. Our uncertainty in the values of the field is then given by a probability distribution $\rho[\chi]$ over $\mathcal{C}$, so that the proper probability that the field attains a value $\hat{\chi}$ in an infinitesimal region of $\mathcal{C}$ is $\text{Prob}[\chi < \hat{\chi} < \chi + \delta\chi] = \rho[\chi] D\chi$, where $D\chi$ is an integration measure over $\mathcal{C}$.

**Maximum Entropy**—The goal is to predict the evolution of the scalar field $\chi$. To this end we make one major assumption: in ED, the fields follow continuous trajectories such that finite changes can be analyzed as an accumulation of many infinitesimally small ones. Thus we are interested in obtaining the probability $P[\chi'|\chi]$ of a transition from an initial configuration $\chi$ to a neighboring $\chi' = \chi + \Delta\chi$. This is accomplished via the Maximum Entropy (ME) method by maximizing the entropy functional,

$$S[P,Q] = -\int D\chi' P[\chi'|\chi] \log \frac{P[\chi'|\chi]}{Q[\chi'|\chi]}, \tag{1}$$

relative to a prior $Q[\chi'|\chi]$ and subject to appropriate constraints.

The prior $Q[\chi'|\chi]$ that incorporates the information that the fields change by infinitesimally small amounts, but which is otherwise maximally uninformative is a product of Gaussians (This prior can itself be derived from the ME method, and in that case, the $\alpha_x$ appear as Lagrange multipliers),

$$Q[\chi'|\chi] \propto \exp -\frac{1}{2}\int dx\, g_x^{1/2} \alpha_x (\Delta \chi_x)^2 \tag{2}$$

where $g_x^{1/2} = \det|g_{ij}|^{1/2}$ is a scalar density of weight one. (For notational simplicity we write $dx'$ instead of $d^3x'$). Continuity is enforced in the limit that the scalar-valued parameters $\alpha_x \to \infty$. As argued in [7], a single additional constraint is required to develop a richer quantum dynamics (Note that since $\chi_x$ and $\Delta\chi_x$ are scalars, in order that (3) be invariant under coordinate transformations of the surface the derivative $\delta/\delta\chi_x$ must transform as a scalar density),

$$\langle \Delta\phi \rangle = \int_{\mathcal{C}} D\chi' P[\chi'|\chi] \int dx\, \Delta\chi_x \frac{\delta\phi[\chi]}{\delta\chi_x} = \kappa', \tag{3}$$

which involves the introduction of a "drift" potential $\phi[\chi]$ whose complete justification is still a subject of future investigation (There is strong reason to believe more compelling answers will come from the ED of Fermions). Maximizing (1) subject to (3) and normalization, we obtain a Gaussian transition probability distribution,

$$P[\chi'|\chi] = \frac{1}{Z[\alpha_x, g_x]} \exp -\frac{1}{2} \int dx\, g_x^{1/2} \alpha_x \left( \Delta\chi_x - \frac{1}{g_x^{1/2} \alpha_x} \frac{\delta\phi[\chi]}{\delta\chi_x} \right)^2, \tag{4}$$

where $Z[\alpha_x, g_x]$ is the normalization constant. The Gaussian form of (4) allows us to present a generic change,

$$\Delta\chi_x = \langle \Delta\chi_x \rangle + \Delta w_x, \tag{5}$$

as resulting from an expected drift $\langle \Delta\chi_x \rangle$ plus fluctuations $\Delta w_x$. While $\langle \Delta w_x \rangle = 0$, because the distribution is Gaussian, the square fluctuations and expected short steps do not. That is,

$$\langle \Delta w_x \Delta w_{x'} \rangle = \frac{1}{g_x^{1/2} \alpha_x} \delta_{xx'} \quad \text{and} \quad \langle \Delta\chi_x \rangle = \frac{1}{g_x^{1/2} \alpha_x} \frac{\delta\phi[\chi]}{\delta\chi_x} \equiv \Delta\bar{\chi}_x. \tag{6}$$

Therefore, as in [7], the fluctuations dominate the trajectory leading to a Brownian motion.

## 3. Some Space-Time Kinematics

In geometrodynamics, the primary object of interest is the three-dimensional metric of space $g_{ij}$, whose time evolution is required to sweep a four-dimensional space-time. Thus coordinates $X^\mu$ ($\mu, \nu, \ldots$

$= 0, 1, 2, 3$) can still be assigned to space-time, even if its geometry *a priori* unknown. By construction, space-time can be foliated by a sequence of space-like surfaces $\{\sigma\}$; that is, its topology is globally hyperbolic. The embedding of a surface in space-time is defined by four embedding functions $X^\mu = X^\mu(x^i)$.

An infinitesimal deformation of the surface $\sigma$ to a neighboring surface $\sigma'$ is determined by the deformation vector

$$\delta \xi^\mu = \delta \xi^\perp n^\mu + \delta \xi^i X_i^\mu, \tag{7}$$

where $n^\mu$ is the unit normal to the surface ($n_\mu n^\mu = -1$, $n_\mu X_i^\mu = 0$) and where $X_i^\mu(x) = \partial_{ix} X^\mu(x)$ are the space-time components of vectors tangent to the surface. The normal and tangential components of $\delta \xi^\mu$, also known as the lapse and shift, are collectively denoted $(\delta \xi^\perp, \delta \xi^i) = \delta \xi^A$ and are given by

$$\delta \xi_x^\perp = -n_{\mu x} \delta \xi_x^\mu \quad \text{and} \quad \delta \xi_x^i = X^i_{\mu x} \delta \xi_x^\mu, \tag{8}$$

where $X^i_{\mu x} = g^{ij} g_{\mu\nu} X^\nu_{jx}$.

## 4. Entropic Time

In ED, entropic time is introduced as a tool for keeping track of the accumulation of many short steps. (For additional details on entropic time, see e.g., [3]). Here we introduce a manifestly covariant notion of entropic time, along the lines of that in [7].

**Ordered instants**—Central to our formulation of entropic time is the notion of an instant of time. In a properly relativistic theory in curved space-time, such a notion is provided by an *arbitrary* space-like surface, denoted $\sigma$ (see e.g., [9]). This allows us to define the epistemic state at the instant $\sigma$, characterized by the probability $\rho_\sigma[\chi]$, the drift potential $\phi_\sigma[\chi]$, etc.

Having established the notion of an instant, including a assignment of the instantaneous probability $\rho_\sigma[\chi]$, our task turns to updating from one instant to the next. Such dynamical information is encoded in the short-step transition probability from Equation (4), or better yet, the joint probability $P[\chi', \chi] = P[\chi'|\chi] \rho_\sigma[\chi]$. A straightforward applications of the "sum rule" of probability theory suggests that the probability at the next instant is given by

$$\rho_{\sigma'}[\chi'] = \int D\chi \, P[\chi'|\chi] \rho_\sigma[\chi]. \tag{9}$$

This is the basic dynamical equation for the evolution of probability.

**Duration**—To complete our construction of time we must specify the duration between instants. In ED time is defined so that motion looks simple. Since for short steps the dynamics is dominated by fluctuations, Equation (6), the specification of the time interval is achieved through an appropriate choice of the multipliers $\alpha_x$. Moreover, following [7], since we deal here with the duration between curved spaces, this notion of separation should be local, and it is natural to define duration in terms of the local proper time $\delta_x^\perp$. More specifically, let

$$\alpha_x = \frac{1}{\delta \xi_x^\perp} \quad \text{so that} \quad \langle \Delta w_x \Delta w_{x'} \rangle = \frac{\delta \xi_x^\perp}{g_x^{1/2}} \delta_{xx'}. \tag{10}$$

**The local-time diffusion equations**—The dynamics expressed in integral form by (9) and (10) can be rewritten in differential form as an infinite set of local equations, one for each spatial point,

$$\frac{\delta \rho_\sigma}{\delta \xi_x^\perp} = -g_x^{-1/2} \frac{\delta}{\delta \chi_x} \left( \rho_\sigma \frac{\delta \Phi_\sigma}{\delta \chi_x} \right) \quad \text{with} \quad \Phi_\sigma[\chi] = \phi_\sigma[\chi] - \log \rho_\sigma^{1/2}[\chi]. \tag{11}$$

As shown in [7], this set of equations describes the flow of the probability $\rho_\sigma[\chi]$ in the configuration space $\mathcal{C}$ in response to the geometry and the functional $\Phi_\sigma[\chi]$, which will eventually be identified as the Hamilton-Jacobi functional, or the phase of the wave functional in the quantum theory. Moreover, when a particular foliation is chosen, these equations collectively form a Fokker-Planck equation, thus justifying the name, the Local-Time Fokker-Planck (LTFP) equations for Equation (11).

## 5. Geometrodynamics Driven by Entropic Matter

In an *entropic* dynamics, evolution is driven by information codified into constraints. An entropic geometrodynamics, it follows, consists of dynamics driven by a specific choice of constraints, which we discuss here. In [7], QFTCS was derived under the assumption that the background remains fixed. But such assumptions, we know, should break down when one considers states describing a non-negligible concentration of energy and momentum. Thus we must revise our constraints appropriately. A natural way to proceed is thus to allow the geometry itself to take part in the dynamical process: the geometry affects $\rho_\sigma[\chi]$ and $\phi_\sigma[\chi]$, they then act back on the geometry, and so forth. Our goal here is to make this interplay concrete.

**Path independence**—In a relativistic theory there are many ways to evolve from an initial instant to a final one, and each way must agree. This is the basic insight by DHKT in their development of manifestly covariant dynamical theories. The implementation of this idea, through the principle of *path independence*, leads to a set of Poisson brackets (see e.g., [8])

$$\{H_{\perp x}, H_{\perp x'}\} = (g_x^{ij} H_{jx} + g_{x'}^{ij} H_{jx'}) \partial_{ix} \delta(x, x'), \tag{12}$$

$$\{H_{ix}, H_{\perp x'}\} = H_{\perp x} \partial_{ix} \delta(x, x'), \tag{13}$$

$$\{H_{ix}, H_{jx'}\} = H_{ix'} \partial_{jx} \delta(x, x') + H_{jx} \partial_{ix} \delta(x, x'), \tag{14}$$

supplemented by the constraints

$$H_{\perp x} \approx 0 \quad \text{and} \quad H_{ix} \approx 0, \tag{15}$$

where "$\approx$" is understood as a *weak* equality [12]. (From a practical viewpoint, the Poisson bracket relations are essentially constraints on the allowed functional form of the generators $H_{Ax}$ for arbitrary choices of the dynamical variables. On the other hand, the weak constraints $H_{Ax} \approx 0$ are meant to restrict the allowed choices of initial conditions for a given form of the generators $H_{Ax}$.)

**The phase space**—The Equations (12)–(15) of path independence are *universal*. That is, if the dynamics is to be relativistic, these equations must hold. Whatever the choice of canonical variables, or whether the background is fixed or dynamical, the same "algebra" must hold.

As one might expect, the ED formulated here with a dynamical background shares some similarities with the theory developed in [7], in the context of a fixed background. Most obvious is that the variables $\rho$ and $\phi$, or equivalently, $\rho$ and $\Phi$ remain canonically conjugate, forming the so-called *ensemble phase space*, or e-phase space for short. However a crucial difference emerges with respect to the treatment of the geometry. Here, deviating from [7], we instead follow Hojman, Kuchař, and Teitelboim (HKT) by describing the dynamics of the geometry by taking the metric $g_{ij}(x)$ as a central dynamical variable. Of course, for the scheme to be canonical, we must also introduce a set of auxiliary variables $\pi^{ij}$, which must have the character of tensor densities. At this juncture the sole interpretation of the $\pi^{ij}$ are as the momenta conjugate to $g_{ij}$, defined by the canonical Poisson bracket relations

$$\{g_{ij}(x), g_{kl}(x')\} = \{\pi^{ij}(x), \pi^{kl}(x')\} = 0, \quad \{g_{ij}(x), \pi^{kl}(x')\} = \frac{1}{2}\left(\delta_i^k \delta_j^l + \delta_j^k \delta_i^l\right) \delta(x, x'). \tag{16}$$

Here we have introduced the notion of a Poisson bracket, which is an anti-symmetric bi-linear product that allows for the notion of an algebra. Written in local coordinates, the Poisson brackets take the form

$$\{F,G\} = \int dx \left( \frac{\delta F}{\delta g_{ij}(x)} \frac{\delta G}{\delta \pi^{ij}(x)} - \frac{\delta G}{\delta g_{ij}(x)} \frac{\delta F}{\delta \pi^{ij}(x)} \right) + \int D\chi \left( \frac{\tilde{\delta} F}{\tilde{\delta}\rho[\chi]} \frac{\tilde{\delta} G}{\tilde{\delta}\Phi[\chi]} - \frac{\tilde{\delta} G}{\tilde{\delta}\rho[\chi]} \frac{\tilde{\delta} F}{\tilde{\delta}\Phi[\chi]} \right), \quad (17)$$

for arbitrary functionals $F$ and $G$ of the phase space variables $(\rho, \Phi; g_{ij}, \pi^{ij})$.

**The super-momentum**—We now turn our attention to the local Hamiltonian generators $H_{Ax}$, and more specifically, we look to provide explicit expressions for these generators in terms of the canonical variables by solving the Poisson brackets Equations (12)–(14). We begin with the tangential generator $H_{ix}$, which generates changes in the canonical variables by dragging them parallel to the space $\sigma$. As shown in [8,10], the tangential generator can be shown to split

$$H_{Ax}[\rho, \Phi; g_{ij}, \pi^{ij}] = H^G_{Ax}[g_{ij}, \pi^{ij}] + \tilde{H}_{Ax}[\rho, \Phi], \quad (18)$$

into components we identify as an ensemble super-momentum $\tilde{H}_{ix}$ and a gravitational super-momentum $H^G_{ix}$. This then leads to Equation (14) similarly decomposing into ensemble and gravitational pieces so that each can be solved independently of the other. The appropriate super-momentum for the ensemble sector was given in [7], with the result that

$$\tilde{H}_{ix} = -\int D\chi \rho[\chi] \frac{\delta \Phi[\chi]}{\delta \chi_x} \partial_{ix} \chi_x, \quad \text{while} \quad H^G_{ix} = -2\partial_j \left( \pi^{jk} g_{ik} \right) + \pi^{jk} \partial_i g_{jk} \quad (19)$$

is the gravitational contribution, determined by Hojman et al. in [10]. The so-called super-momentum constraint from Equation (15) is then just

$$H_{ix} = -2\partial_j \left( \pi^{jk} g_{ik} \right) + \pi^{jk} \partial_i g_{jk} - \int D\chi \rho[\chi] \frac{\delta \Phi[\chi]}{\delta \chi_x} \partial_{ix} \chi_x \approx 0. \quad (20)$$

**The super-Hamiltonian**—As pertains to the super-Hamiltonian, a similar decomposition does not occur. But following Teitelboim [8] let us suggestively rewrite $H_{\perp x}$ as

$$H_{\perp x} = H^G_{\perp x}[g_{ij}, \pi^{ij}] + \tilde{H}_{\perp x}[\rho, \Phi; g_{ij}, \pi^{ij}].$$

Note we make no assumptions in writing $H_{\perp x}$ in this way, as this simply *defines* the contribution of "matter". The assumption comes, instead, the requirement that the ensemble super-Hamiltonian $\tilde{H}_{\perp x}$ to be *independent* of the momentum variable $\pi^{ij}$ (Although many interesting models remain after this assumption, some models of physical interest are, indeed, excluded by this simplification. We leave it to future work to relax this requirement). (This simplification is called the assumption of "non-derivative" coupling, since it can be *proven* [8] that this implies $\tilde{H}_{\perp x}$ is just a local function (not functional) of $g_{ij}$, not its derivatives). Another consequence of this assumption is that Equation (12)—the most difficult Poisson bracket—decomposes completely into gravitational and matter sectors. Thus each can be approached independently. That is, the gravitational side can proceed *as if* there were no sources, while the matter side can proceed along lines similar to [7].

The solution to the gravitational piece, which is quite involved, was first given by HKT [10] and takes the form (Note that the solution given in [10] relies on the assumption that geometrodynamics is time-reversible. An alternative derivation in [9] obtains the same result without this assumption, but uses a Lagrangian instead)

$$H^G_{\perp x} = \kappa G_{ijkl} \pi^{ij} \pi^{kl} - \frac{g^{1/2}}{2\kappa} R, \quad \text{where} \quad G_{ijkl} = g^{-1/2} \left( g_{ik} g_{jl} + g_{il} g_{jk} - g_{ij} g_{kl} \right) \quad (21)$$

is the DeWitt super metric, $R$ is the Ricci scalar for three-dimensional space, and $\kappa$ is a constant coefficient. (We have set the cosmological constant $\lambda = 0$). The determination of the ensemble super-Hamiltonian is subject not only to satisfying the Poisson bracket Equation (12), but also to the requirement that the evolution generated by $\tilde{H}_{\perp x}$ reproduces the LTFP Equation (11). In [7] it was shown that an acceptable (but not exhaustive) family of ensemble super-Hamiltonians is given by

$$\tilde{H}_{\perp x}[\rho, \Phi] = \tilde{H}^0_{\perp x}[\rho, \Phi; g_{ij}] + F_x[\rho; g_{ij}] \tag{22}$$

with

$$\tilde{H}^0_{\perp x}[\rho, \Phi] = \frac{1}{2} \int D\chi \rho \left( \frac{1}{g_x^{1/2}} \left( \frac{\delta \Phi}{\delta \chi_x} \right)^2 + g_x^{1/2} g^{ij}(x) \partial_{ix} \chi_x \partial_{jx} \chi_x \right), \tag{23}$$

where the functional $F_x[\rho; g_{ij}]$ is restricted to the simple form

$$F_x[\rho; g_{ij}] = \int D\chi \rho \left( g_x^{1/2} V_x(\chi_x) + \frac{\beta}{g_x^{1/2}} \left( \frac{\delta \log \rho}{\delta \chi_x} \right)^2 \right), \tag{24}$$

where the potential $V_x(\chi)$ is a function only of the field and $\beta$ is a coupling constant. For future convenience we set to $\beta = 1/8$.

From Equations (21) and (22) the super-Hamiltonian constraint is then just

$$H_{\perp x} = H^G_{\perp x} + \tilde{H}_{\perp x} \approx 0, \tag{25}$$

where the gravitational and ensemble pieces are those given in Equations (21) and (22). Note that a solution of this constraint requires fixing a set of variables in terms of another set—i.e., the gravitational variables are not necessarily independent of the probability $\rho$ and phase $\Phi$!

## 6. The Dynamical Equations

In the previous section we have identified a representation of the relations Equations (12)–(15) in terms of the canonical variables $(\rho, \Phi; g_{ij}, \pi^{ij})$. The resulting evolution, generated by these $H_{Ax}$, leads to a fully covariant geometrodynamics driven by entropic matter. But to do this, we first pick a foliation of space-time with parameter $t$, specified by a particular choice of *lapse* and *shift* functions, which are, respectively, given by

$$N(x,t) = \frac{\delta \tilde{\xi}_x^\perp}{dt} \quad \text{and} \quad N^i(x,t) = \frac{\delta \tilde{\xi}_x^i}{dt}. \tag{26}$$

**The Schrödinger equation**—We are interested in the dynamical evolution of the ensemble variables $\rho$ and $\Phi$, however, this very same dynamics can be expressed equivalently by the introduction of complex variables $\Psi_t = \rho_t^{1/2} e^{i\Phi_t}$ and $\Psi_t^* = \rho_t^{1/2} e^{-i\Phi_t}$ (we use units in which $\hbar = 1$). The reason these variable turn out to be useful, is that the dynamical equations turn out to take a familiar form. In particular, we have

$$i\partial_t \Psi_t[\chi] = i \int dx \left[ \{\Psi_t[\chi], H_{\perp x}\} N(x,t) + \{\Psi_t[\chi], H_{ix}\} N^i(x,t) \right]$$
$$= \int dx \left[ N(x,t) \hat{H}_{\perp x} + N^i(x,t) \hat{H}_{ix} \right] \Psi_t[\chi], \tag{27}$$

where $\hat{H}_{ix}$ and $\hat{H}_{\perp x}$ are given by

$$\hat{H}_{ix} = i \partial_i \chi_x \frac{\delta}{\delta \chi_x} \quad \text{and} \quad \hat{H}_{\perp x} = -\frac{1}{2 g^{1/2}} \frac{\delta^2}{\delta \chi_x^2} + \frac{g^{1/2}}{2} g^{ij} \partial_i \chi_x \partial_j \chi_x + g^{1/2} V_x(\chi_x; g_{ij}), \tag{28}$$

respectively. Notice that although Equation (27) is ostensibly a linear equation for the functional $\Psi_t$, which may suggest calling this a "Schrödinger equation", closer inspection reveals this to be misleading. Indeed, owing to the constraint Equations (20) and (25), the metric $g_{ij}$ that appears in this equation itself depends on the variables $\Psi$ and $\Psi^*$, leading instead to a *non-linear* equation!

**Geometrodynamics**—To complete the description of the dynamics we will determine the evolution of the geometrical variables $(g_{ij}, \pi^{ij})$. Beginning with the metric, its time evolution, generated by the super-Hamiltonians $H_{Ax}$ given above, after a straightforward computation yields

$$\partial_t g_{ij} = 2N(x,t) g_x^{-1/2} \kappa \left( 2\pi_{ij}(x) - \pi(x) g_{ij}(x) \right) + \nabla_i N_j(x,t) + \nabla_j N_i(x,t), \tag{29}$$

where $\nabla_i$ is the metric compatible covariant derivative (Note that Equation (29) relates the conjugate momentum $\pi^{ij}$ to the extrinsic curvature of the surface $K_{ij}$). The equation for the conjugate momentum $\pi^{ij}$ is more interesting

$$\begin{aligned}\partial_t \pi^{ij} &= -\frac{N g^{1/2}}{2\kappa} \left( R^{ij} - \frac{1}{2} g^{ij} R \right) + \frac{N\kappa}{g^{1/2}} g^{ij} \left( \pi^{kl} \pi_{kl} - \frac{1}{2} \pi^2 \right) - 4\frac{N\kappa}{g^{1/2}} \left( \pi^{ik} \pi_k^j - \frac{1}{2} \pi \pi^{ij} \right) \\ &+ \frac{g^{1/2}}{2\kappa} \left( \nabla^i \nabla^j N - g^{ij} \nabla^k \nabla_k N \right) + \nabla_k \left( \pi^{ij} N^k \right) - \pi^{ik} \nabla_k N^j - \pi^{kj} \nabla_k N^i - N \frac{\partial \tilde{H}_{\perp x}}{\partial g_{ij}},\end{aligned} \tag{30}$$

where we have introduced $R^{ij}$, the Ricci tensor (We have used the fact that the non-derivative coupling assumption implies that $\tilde{H}_{\perp x}$ is local in $g_{ij}$). Note that the evolution of $\pi^{ij}$ is driven by $\partial \tilde{H}_{\perp x} / \partial g_{ij}$, which is a functional of the epistemic state.

## 7. Conclusions and Discussion

The ED developed here has several interesting features. Although written in the relatively less common language of geometrodynamics, the evolution Equations (29) and (30), together with the constraints (20) and (25) are mathematically equivalent to the so-called "semi-classical" Einstein equations (SCEE), which are typically written as [2]

$$G_{\mu\nu} = 8\pi\kappa \langle \hat{T}_{\mu\nu} \rangle, \tag{31}$$

where $G_{\mu\nu}$ is the Einstein tensor and $\langle \hat{T}_{\mu\nu} \rangle$ is the expected value of energy-momentum operator of a quantum scalar field. Such a theory of gravity has long been seen as a desirable step intermediate to a full theory of QG, in part because it contains well-established physics—QFTCS and classical general relativity—in the limiting cases where they are valid. But there has been much debate (see e.g., [11]), on the other hand, as to the status of semi-classical theories as true QG candidate; with many harboring a negative view.

Here we do not propose a definitive rebuke of all these critics, but note that the ED formulation of SCEE has certain features that allow it to evade the most cogent criticisms. For one, a problem that is often raised against the SCEE is that it is proposed in a rather *ad hoc* manner, based on heuristic arguments. But in ED these equations are *derived* on the basis of well-defined assumptions and constraints. Period. Another argument commonly raised against the SCEE is that the left hand side, featuring the gravitational field, is a real "physical" field, while the right hand side contains the quantum state $\Psi$, which is epistemic. In the ED approach, however, the physical variables are the field $\chi_x$; the geometrical variables are more properly viewed as constraints. In other words, these variables are not measured directly in an experiment, but rather, their values are *inferred* from an ensemble of measurements, i.e., the geometry *is* epistemic! While this may seem a controversial viewpoint, recent work [14] suggests that the geometry of space may, indeed, be of entropic origin as well.

Finally, the Schrödinger equation we obtain here is quite unorthodox. As mentioned above, the dynamics of $\Psi$ follows a *non-linear* equation. This non-linearity is not, however, an artifact of a bad approximation, but the prediction of a theory derived from first principles. This loss of linearity, while highly problematic for many standard approaches to QT, follows naturally in ED. This begs the question, is linearity just a misguided prejudice? Will the superposition principle become the first casualty of quantum gravity?

## References

1. Carlip, S. Quantum gravity: A progress report. *Rep. Prog. Phys* **2001**, *64*, 885, arXiv:0108040; Rovelli, C. Loop quantum gravity: The first 25 years. *Class. Quantum Gravity* **2011**, *28*, 153002, arXiv:1012.4707; Mukhi, S. String theory: A perspective over the last 25 years. *Class. Quantum Gravity* **2011**, *28*, 153001, arXiv:1110.2569.
2. Butterfield, J.; Isham, C. Space-time and the philosophical challenge of quantum gravity. *arXiv* **1999**, arXiv:9903072.
3. Caticha, A. Entropic Inference and the Foundations of Physics.
4. Rosenkrantz, R.D. (Ed.) *E. T. Jaynes: Papers on Probability, Statistics and Statistical Physics*; Reidel: Dordrecht, The Netherlands, 1983.
5. Ipek, S.; Caticha, A. An Entropic Dynamics Approach to Geometrodynamics. *Ann. Phys.* **2019**, *531*, 1700408, arXiv:1711.02538.
6. Johnson, D.T.; Caticha, A. *AIP Conference Proceedings: Vol. 1443, No. 1*; AIP: 2012; Vanslette, K.; Caticha, A. *AIP Conference Proceedings: Vol. 1853. No. 1*; AIP Publishing: 2017.
7. Ipek, S.; Abedi, M.; Caticha, A. *AIP Conference Proceedings: Vol. 1853. No. 1*; AIP Publishing: 2017; Ipek, S.; Abedi, M.; Caticha, A. Entropic dynamics: Reconstructing quantum field theory in curved space-time. *arXiv* **2018**, arXiv:1803.07493.
8. Teitelboim, C. How commutators of constraints reflect the spacetime structure. *Ann. Phys.* **1973**, *79*, 542–557; Teitelboim, C. The Hamiltonian Structure of Spacetime. Ph.D. Thesis, Princeton University, Princeton, NJ, USA, 1973.
9. Kuchař, K. A Bubble-Time Canonical Formalism for Geometrodynamics. *J. Math. Phys.* **1972**, *13*, 768–781; Auhtor name. Title. *Math. Phys.* **1974**, *15*, 708–715; Kuchař, K. Geometry of hyperspace. I. *J. Math. Phys.* **1976**, *17*, 777–791.
10. Hojman, S.A.; Kuchař, K.; Teitelboim, C. Title. *Ann. Phys.* **1976**, *96*, 88–135.
11. Arguments for the quantization of gravity; Eppley, K.; Hannah, E. Title. *Found. Phys.* **1977**, *7*, 51–68; Unruh, W.G. *Quantum Theory of Gravity. Essays in Honor of the 60th Birthday of Bryce S. DeWitt*; 1984; Page, D.N.; Geilker, C.D. Indirect evidence for quantum gravity. *Phys. Rev. Lett.* **1981**, *47*, 979. Arguments denying the necessity to quantize gravity by Huggett Nick; Callender, C. Why quantize gravity (or any other field for that matter)? *Philos. Sci.* **2001**, *68*, S382–S394; Wüthrich, C. To quantize or not to quantize: Fact and folklore in quantum gravity. *Philos. Sci.* **2005**, *72*, 777–788; Albers, M.; Kiefer, C.; Reginatto, M. Measurement analysis and quantum gravity. *Phys. Rev. D* **2008**, *78*, 064051, arXiv:0802.1978; Kibble, T.W.B. *Is a Semi-Classical Theory of Gravity Viable?* Quantum Gravity II; Clarendon Press: Oxford, UK, 1981.
12. Bergmann, P.G. Non-linear field theories. *Phys. Rev.* **1949**, *75*, 680; Dirac, P.A.M. Title; Courier Corporation: 2013.
13. Arnowitt, R.; Deser, S.; Misner, C.W. Canonical variables for general relativity. *Phys. Rev.* **1960**, *117*, 1595; Arnowitt, R.; Deser, S.; Misner, C.W. Republication of: The dynamics of general relativity. *Gen. Relativ. Gravit.* **2008**, *40*, 1997–2027, arXiv:0405109.
14. Caticha, A. *AIP Conference Proceedings: Vol. 1757. No. 1*; AIP Publishing: 2016, arXiv:1512.09076; Caticha, A. *In These Proceedings*; 2019.

© 2019 by the authors. Licensee MDPI, Basel, Switzerland. This article is an open access article distributed under the terms and conditions of the Creative Commons Attribution (CC BY) license (http://creativecommons.org/licenses/by/4.0/).

*Proceedings*

# Learning Model Discrepancy of an Electric Motor with Bayesian Inference [†]

David N. John [1,2,*], Michael Schick [2] and Vincent Heuveline [1,3]

1. Engineering Mathematics and Computing Lab (EMCL), Interdisciplinary Center for Scientific Computing (IWR), Heidelberg University, 69120 Heidelberg, Germany; vincent.heuveline@uni-heidelberg.de
2. Corporate Research, Robert Bosch GmbH, 71272 Renningen, Germany; michael.schick3@de.bosch.com
3. Data Mining and Uncertainty Quantification Group (DMQ), Heidelberg Institute for Theoretical Studies (HITS) GmbH, 69118 Heidelberg, Germany
* Correspondence: david.john@de.bosch.com
† Presented at the 39th International Workshop on Bayesian Inference and Maximum Entropy Methods in Science and Engineering, Garching, Germany, 30 June–5 July 2019.

Published: 25 November 2019

**Abstract:** Uncertainty Quantification (UQ) is highly requested in computational modeling and simulation, especially in an industrial context. With the continuous evolution of modern complex systems demands on quality and reliability of simulation models increase. A main challenge is related to the fact that the considered computational models are rarely able to represent the true physics perfectly and demonstrate a discrepancy compared to measurement data. Further, an accurate knowledge of considered model parameters is usually not available. e.g., fluctuations in manufacturing processes of hardware components or noise in sensors introduce uncertainties which must be quantified in an appropriate way. Mathematically, such UQ tasks are posed as inverse problems, requiring efficient methods to solve. This work investigates the influence of model discrepancies onto the calibration of physical model parameters and further considers a Bayesian inference framework including an attempt to correct for model discrepancy. A polynomial expansion is used to approximate and learn model discrepancy. This work extends by discussion and specification of a guideline on how to define the model discrepancy term complexity, based on the available data. Application to an electric motor model with synthetic measurements illustrates the importance and promising perspective of the method.

**Keywords:** inverse problem; calibration; bayesian inference; model discrepancy

## 1. Introduction

The increasing complexity of technical systems yields high demands on model quality and numerical accuracy. Quantification of these models under uncertainty is desired to make statements about reliability and accuracy. Consequently, there is a need for efficient solvers and new additional Uncertainty Quantification (UQ) methods, which are able to cope with the soaring complexity of models and that take into account all sources of uncertainty. Model calibration requires the solution of an inverse problem with uncertainties, such as parametric, observation, structural model and solution method uncertainty. Let $\mathcal{G} : \mathcal{X} \to \mathcal{Y}$, for $\mathcal{X} \subseteq \mathbb{R}^d, \mathcal{Y} \subseteq \mathbb{R}^n$, represent a simulation model that maps some input $x$ to an output $y = \mathcal{G}(x)$. An inverse problem is the task of finding an $x \in \mathcal{X}$ for a given measurement $y \in \mathbb{R}^n$ such that $y = \mathcal{G}(x)$. Generally, this equality does not hold as the measurements $y$ are usually corrupted. Hence, one considers $y = \mathcal{G}(x) + \eta$, where $\eta \in \mathbb{R}^n$ represents the observation uncertainty due to measurement noise or other errors. Simply inverting $\mathcal{G}$ is not possible as $\eta$ is unknown and in general $\mathcal{G}^{-1}$ does not exist. A classical approach to solve this problem is by minimizing the data misfit, i.e., $\min_{x \in \mathcal{X}} \frac{1}{2}\|y - \mathcal{G}(x)\|_{\mathcal{Y}}^2$. However, this problem is typically ill-posed in the sense of Hadamard, i.e., multiple solutions might exist and stability might be a problem. To obtain a well-posed problem regularization is necessary. One approach is Tikhonov regularization, also known as ridge regression in statistics [1]. However, the particular choice of regularization is somewhat arbitrary [2]. With the Bayesian approach one is interested in finding a probability measure $\mu^y(x)$ on $\mathcal{X}$ with probability density $\pi(x|y)$ that expresses how likely $\mathcal{G}(x)$ for a certain $x \in \mathcal{X}$ describes $y$ under consideration of noise. Then the problem is well-posed under slight assumptions and leads to a natural way of regularization due to the definition of prior distributions $\pi_0(x)$ for unknown parameters, see [1–3]. Generally, the posterior distribution $\pi(x|y)$ is intractable and one can not sample from it directly, hence approximative methods, such as filtering, variational and sampling methods, are required [2,4,5]. Metropolis-Hastings Markov-Chain Monte-Carlo (MH-MCMC) sampling is used in this work [4,6,7]. Note that methods to improve sampler efficiency [8–10] and surrogate modeling of $\mathcal{G}$, e.g. by Polynomial Chaos expansions [11,12] or GPs [13,14], could leverage overall simulation time and speed up the inference, if required.

A difficulty in solving the inverse problem is to capture and separate all sources of uncertainty, often called "identification problem". A main challenge is related to the fact that the considered computational models $\mathcal{G}$ are rarely able to represent the true physics perfectly and demonstrate a discrepancy compared to measurement data. To be more specific the term model discrepancy in this work denotes the difference between simulation model for the true physical parameters and the true system, hence structural model uncertainty (e.g., lack of knowledge, missing physics), but also implementation and numerical errors. However, this model discrepancy is usually unknown.

The Kennedy and O'Hagan (KO) framework [13] is one of the first attempts to model and explicitly take account of all the uncertainty sources that arise in the calibration of computer models. Model discrepancy is considered by an additional term and modeled by an Gaussian process (GP). Following [13], Arendt et al. [15,16] suggest a modular Bayesian approach and discuss the identification problem. Examples illustrate that sometimes this separation is possible under mild assumptions, e.g., smoothness of the model discrepancy, but also that it is not possible in other cases. In the companion work [16] they show an approach how to improve identifiability by using multiple responses and representing correlation between responses. Another work using multiple responses is [17]. Brynjarsdóttir and O'Hagan [18] state that with the KO framework in order to infer physical parameters and model discrepancy simultaneously sufficient prior distributions for at least one of those must be given. However, Tuo and Wu [19,20] showed that the choice of the model discrepancy prior has a permanent influence onto the parameter posterior distribution even in the large data limit. Plumlee [21] presented an approach to improve identifiability by defining a prior distribution of the model discrepancy that is orthogonal to the gradient of the model. However, computational costs are high. Nagel et al. [22] modeled the model discrepancy term by an polynomial expansion, assuming smoothness for the true underlying model discrepancy.

Following all these works we consider a Bayesian model with a term for measurement noise and a model discrepancy. We adapt the idea of representing model discrepancy by a polynomial expansion [22]. The major contribution of this work is to provide answers on how to select a polynomial degree by keeping the complexity of the model low while still providing high accuracy in discrepancy modeling. This is shown in a practical guideline, which recommends how to select a sufficient maximum polynomial degree of the truncated polynomial expansion, based on the available data and the estimation of measurement noise. Furthermore, critical points conditioning the choice of the model discrepancy term are discussed in detail. The framework is applied to the calibration of a direct current (DC) electric motor model in a synthetic setting. i.e., synthetic measurements are created and a modified electric motor model, containing an artificial model error are used to infer model parameters. Due to the synthetic setup, available references allow a quantitative evaluation of the considered methods performance and accuracy.

This paper is structured in the following way: Section 2 formulates the considered example setup. Section 3 specifies a Bayesian inference model, followed by a discussion of the model discrepancy term complexity. Numerical results are presented in Section 4. This work concludes in Section 5.

## 2. Electric Motor Model

A DC electric motor is a rotating electrical machine converting electrical energy into mechanical energy. Let $I$ [$A$] denote the electric current and $\omega$ [$rad/s$] the angular velocity. For $t > 0$, the ODEs

$$L\dot{I}(t) = -RI(t) - c_m \omega(t) + U, \quad (1)$$

$$J\dot{\omega}(t) = c_g I(t) - d\omega(t) + T_L, \quad (2)$$

with initial conditions $I(0) = 0$, $\omega(0) = 0$, describe the electro-mechanical behavior of a DC electric motor [23]. The real valued parameters are: cable harness resistance $R$ [$\Omega$], motor constants $c_m$ [$Vs/rad$], $c_g$ [$Nm/A$], voltage $U$ [$V$], friction $d$ [$kg\,m/s$], inductivity $L$ [$H$], inertia $J$ [$kg\,m^2$] and constant torque $T_L$ [$Nm$] required by the load. Here, only $R$ is considered uncertain and all other parameters are fixed to $c_m = 0.01$, $c_g = 0.01$, $U = 24$, $d = 10^{-6}$, $L = 10^{-4}$, $J = 5 \times 10^{-5}$, $T_L = 0$.

Let $\hat{I}(t_i)$ and $\hat{\omega}(t_i)$ denote numerical approximations of $I$ and $\omega$ at equidistant time points $\{t_i\}_{i=1,\ldots,M} \in D := [0, T]$ for $M = 501$ and $T = 0.5[s]$. For this work the explicit Runge-Kutta method dopri5 (Dormand and Prince) is used, see [24]. For notational convenience we define

$$\mathcal{G}(R) := [\hat{I}(t_1), \ldots, \hat{I}(t_M), \hat{\omega}(t_1), \ldots, \hat{\omega}(t_M)] \in \mathbb{R}^{2M}, \quad (3)$$

depending only on $R \in \mathcal{X}$. Synthetic measurements for a fixed reference value $R_0 = 0.1$ are created by $y := \mathcal{G}(R_0) + \varepsilon$, where $\varepsilon := [\varepsilon_I, \varepsilon_\omega]$ is a realization of $\mathcal{N}(0, \Sigma)$, with covariance $\Sigma \in \mathbb{R}^{2M \times 2M}$. For independent and identically distributed (iid) measurement noise for current and angular velocity at each time step let $\Sigma := diag(\sigma_I^2 I_M, \sigma_\omega^2 I_M)$, with $\sigma_{I,\omega} > 0$. Figure 1 displays a measurement of $I$ and $\omega$ with $\sigma_{I,0} = 2$, $\sigma_{\omega,0} = 10$.

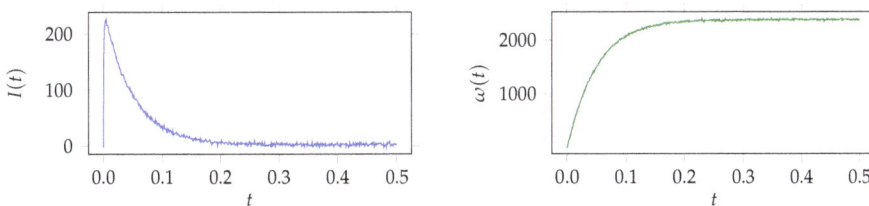

**Figure 1.** Noisy synthetic measurements of electric current $I(t)$ and angular velocity $\omega(t)$.

As the goal of this work is to learn about model discrepancy an artificial model error $\phi \in \mathbb{R}$ is introduced into $J\dot{\omega}(t) = \phi c_g I(t) - d\omega(t) + T_L$. Let $\mathcal{G}_\phi(R)$ be the numerical approximation of the model for a given $\phi$ and denote it as simulation model. Obviously, $\mathcal{G} = \mathcal{G}_\phi \iff \phi = 1$. $\phi \neq 1$ can be interpreted as missing physics, i.e. by damping or amplifying a part of the equation, or as a misspecified model. Define the discrepancy $d(R) = [d_I, d_\omega](R) := \mathcal{G}(R_0) - \mathcal{G}_\phi(R) \in \mathbb{R}^{2M}$ and the noisy discrepancy $d^\varepsilon(R) = [d_I^\varepsilon, d_\omega^\varepsilon](R) := y - \mathcal{G}_\phi(R) \in \mathbb{R}^{2M}$. Denote for $\phi = 0.9$ the reference discrepancies by $d_0 := d(R_0)$ and $d_0^\varepsilon := d^\varepsilon(R_0)$.

## 3. Bayesian Inference Solution Process

Consider the inverse problem of finding an $R \in \mathcal{X}$ such that $y = \mathcal{G}_\phi(R)$, where $y \in \mathbb{R}^{2M}$ are noisy measurements and $\mathcal{G}_\phi$ is the simulation model containing an artificial model error $\phi$. The goal here is not just to infer an optimal $R \in \mathcal{X}$, but also one that is as close as possible to the reference $R_0$, i.e., the true physical value, which was used to create the measurements $y$. Following [22] we introduce two Bayesian models that differ in their complexity. The simpler model only considers measurement noise, whereas the more complex model considers model discrepancy additionally.

### 3.1. Bayesian Model 1 (BM1): Measurement Noise

As the measurements $y$ are noisy, additive noise $\varepsilon = [\varepsilon_I, \varepsilon_\omega] \in \mathbb{R}^{2M}$ is added to the simulation model. Here, $\varepsilon_I, \varepsilon_\omega$ are considered as zero mean Gaussian $\mathcal{N}(0, \sigma_I^2 I_M)$ and $\mathcal{N}(0, \sigma_\omega^2 I_M)$ with unknown standard deviations $\sigma_I, \sigma_\omega > 0$, respectively. The Bayesian model 1 (BM1) is

$$y = \mathcal{G}_\phi(R) + \varepsilon(\sigma_I, \sigma_\omega). \tag{4}$$

Hence, the likelihood is $y|R, \sigma_I, \sigma_\omega \sim \mathcal{N}(\mathcal{G}_\phi(R), \Sigma(\sigma_I, \sigma_\omega))$, with $\Sigma(\sigma_I, \sigma_\omega) = diag(\sigma_I^2 I_M, \sigma_\omega^2 I_M)$. By expressing a-priori knowledge in terms of a prior probability density function $\pi(R, \sigma_I, \sigma_\omega) = \pi(R)\pi(\sigma_I)\pi(\sigma_\omega)$, Bayes' formula yields for the posterior probability density $\pi(R, \sigma_I, \sigma_\omega|y) \sim \pi(y|R, \sigma_I, \sigma_\omega)\pi(R, \sigma_I, \sigma_\omega)$. The $\pi(\sigma_I), \pi(\sigma_\omega)$ are defined as uninformative Inverse Gamma distributions $InvGamma(1,1)$. This is a common choice for conjugate priors of scale parameters in Bayesian statistics [4], in particular for a Gaussian likelihood with given mean.

### 3.2. Bayesian Model 2 (BM2): Measurement Noise and Model Discrepancy

BM2 extends BM1 by considering model discrepancy with an additive term $\delta = [\delta_I, \delta_\omega] \in \mathbb{R}^{2M}$, additionally to iid measurement noise. Note that this is already the discretized version and $\delta_I = [\delta_I(t_0), \ldots, \delta_I(t_M)]$, $\delta_\omega = [\delta_\omega(t_1), \ldots, \delta_\omega(t_M)]$ are vectors of the, yet to define, model discrepancy terms $\delta_I, \delta_\omega : D \to \mathbb{R}$ evaluated at $t_i, i = 1, \ldots, M$. Omitting the superscripts for a moment, we now assume that $\delta \in L^2(D)$. Let $\{p_j\}_{j \in \mathbb{N}} \subseteq L^2(D)$ be a basis of functions $p_j : D \to \mathbb{R}$ dense in $L^2(D)$. Then $\delta$ can be represented by the expansion $\delta(t) = \sum_{j=0}^\infty a_j p_j(t)$. For practicability reasons the expansion is truncated after a $K \in \mathbb{N}$. Such an truncated expansion was also used in [22]. With this let, for the truncation parameter $K \in \mathbb{N}$, the truncated functional expansions

$$\delta_I^K(t) = \sum_{j=0}^K a_j^{(I)} p_j(t), \quad \delta_\omega^K(t) = \sum_{j=0}^K a_j^{(\omega)} q_j(t), \tag{5}$$

be approximative models for the model discrepancy terms $\delta_I, \delta_\omega$. The bases $\{p_j\}_{j=0,\ldots,K}$ and $\{q_j\}_{j=0,\ldots,K}$ are not necessarily identical. Note that each expansion could be truncated with own truncation parameters $K_I, K_\omega$, but for notational convenience and due to later usage we stick to $K = K_I = K_\omega$. Let $a = [a^{(I)}, a^{(\omega)}] = [a_0^{(I)}, \ldots, a_K^{(I)}, a_0^{(\omega)}, \ldots, a_K^{(\omega)}] \in \mathbb{R}^{2K+2}$ be the coefficient vector and $\delta_I^K(a) = [\delta_I^K(t_0), \ldots, \delta_I^K(t_M)]$, $\delta_\omega^K(a) = [\delta_\omega^K(t_1), \ldots, \delta_\omega^K(t_M)]$. Hence $\delta^K(a) = [\delta_I^K, \delta_\omega^K](a) \in \mathbb{R}^{2M}$ denotes the approximation of the true underlying model discrepancy $\delta$. The Bayesian model 2 (BM2) is

$$y = \mathcal{G}_\phi(R) + \delta^K(a) + \varepsilon(\sigma_I, \sigma_\omega). \tag{6}$$

The number of unknown parameters is $3 + 2K + 2$. With the additional unknown coefficients $a$, the prior probability density function is defined as $\pi(R, \sigma_I, \sigma_\omega, a) = \pi(R)\pi(\sigma_I)\pi(\sigma_\omega)\pi(a)$, where $\pi(a) = \prod_{j=0}^K \pi(a_j^{(I)})\pi(a_j^{(\omega)})$. The prior distributions for $\pi(\sigma_I)$ and $\pi(\sigma_\omega)$ are specified as above. Now, with the likelihood $y|R, \sigma_I, \sigma_\omega, a \sim \mathcal{N}(\mathcal{G}_\phi(R) + \delta^K(a), \Sigma(\sigma_I, \sigma_\omega))$, the posterior is given by $\pi^K(R, \sigma_I, \sigma_\omega, a|y) \sim \pi(y|R, \sigma_I, \sigma_\omega, a)\pi(R, \sigma_I, \sigma_\omega, a)$, where $\pi^K$ denotes the dependence on $K$.

Choosing basis functions and priors for the coefficients: If knowledge about the discrepancy is available, this should be modeled accordingly by defining an appropriate prior distribution for $\delta$, i.e., in the case of $\delta^K(a)$ appropriate choices for $a, K$ and $\{p_j\}_{j=0,\ldots,K} \subset L^2(D)$. However, in general this knowledge is not available and some modeling assumptions need to be made. With the assumption that $\delta$ is rather smooth, orthonormal polynomials $p_j : D \to \mathbb{R}$ with $deg(p_j) = j$ are a reasonable choice. As [22], we also opt for Legendre polynomials that are orthogonal to a constant weight. In [22] they argue about the prior for the coefficients $\pi(a_j^{(I)})$, $\pi(a_j^{(\omega)})$ and decide for zero mean Laplace distribution $Laplace(0, b)$ with the arbitrary choice of $b = 10$. It assigns the highest probability around zero and decays exponentially towards the tails, see [25].

Choosing the truncation parameter K: With the choice of Legendre polynomials $K$ corresponds to the maximum polynomial degree. This $K$ determines the complexity of the model discrepancy term $\delta^K(a)$. Important factors in order to specify the model discrepancy term complexity are: accuracy, computational costs, bias-variance tradeoff and non-identifiability. An optimal $K$ should be large enough such that $\delta^K(a)$ is accurate enough to approximate the underlying discrepancy correct, but at the same time it should be as small as possible since a large $K$ increases the number of unknown parameters and consequently computational costs. Furthermore, a large $K$ might yield non-identifiability of all unknowns, as the prior of $\delta^K(a)$ gets non-informative. In [18] they state that in order to infer model discrepancy and model parameters at the same time, at least for one of those an informative prior must be given. The term bias-variance tradeoff describes the fact that with increasing model complexity $K$ the bias of an estimator decreases, but at the same time the variance increases with the model complexity [26]. Consequently, the overall error as sum of squared bias and variance is minimal for an optimal model complexity.

Taking all these factors into account the approach in this work on how to find an optimal $K$ is following: Start with an initial $K = K_0 \in \mathbb{N}$ and increase $K$ iteratively until the marginal posterior distribution $\pi^K(\sigma_I, \sigma_\omega|y)$ of the noise standard deviations stabilizes. i.e., that some distance $D(\pi^K(\sigma_I, \sigma_\omega|y), \pi^{K+L}(\sigma_I, \sigma_\omega|y)) < tol$ for a given tolerance $tol > 0$ and $L \in \mathbb{N}$. Then select the $K$ such that this condition holds. Why is this sufficient? If a model discrepancy is present in BM1, then the noise term is the only instance to capture it. As the noise term is modeled with zero mean, the standard deviation might be overestimated consequently. By adding the model discrepancy term in BM2 captures a fraction of the model discrepancy. As a consequence, the noise term $\varepsilon$ needs to represent only the remaining discrepancy and is estimated by a smaller value. If the estimated standard deviation of the noise does not change anymore (within the tolerance) from $K$ to $K + L$ for $L \in \mathbb{N}$ the smallest sufficient $K$ is found. For this $K$ the model discrepancy term $\delta^K(a)$ should represent the underlying model discrepancy best within its current specification and a separation of model uncertainty and measurement uncertainty is achieved.

## 4. Numerical Results

The numerical results for BM1 and BM2 applied to the electric motor model are presented and compared to the reference. The prior distribution for $R$ is chosen as $\mathcal{N}(R_0, 0.2R_0)$. Variations of the prior mean by $+/-15\%$ did not influence the upcoming results much, thus simply $R_0$ was chosen as mean. An MH-MCMC implementation of the Python package PyMC3 [7] is used to approximate the posterior distribution probability density function. The marginal posterior moments are empirically approximated using Monte Carlo integration with a certain number of the obtained MH-MCMC samples. In order to compare and evaluate the solutions we define error measures.

**Definition 1.** *Define the relative error for a parameter estimate $\hat{x}$ as $\epsilon_{rel}(x) = |\hat{x} - x_0|/|x_0|$, where $x_0$ is the reference value. $\hat{x}$ might be the empirical approximation of the marginal posterior mean $\mathbb{E}[x|y]$.*

**Definition 2.** *Let $x$ with $\pi(x|y)$ be an estimation for $x_0 \in \mathbb{R}$ and given $y$. The mean square error (MSE) of $x$ is $MSE(x) := \mathbb{E}[(x - x_0)^2|y] = Bias(x)^2 + V[x|y]$, with $Bias(x) = (\mathbb{E}[x|y] - x_0)$. The mean is with respect to $\pi(x|y)$. In case of $\pi^K(x|y)$ depending on $K$ write $MSE_K(x) = \mathbb{E}_K[(x - x_0)^2|y] = Bias_K(x)^2 + V_K[x|y]$.*

### 4.1. Results Bayesian Model 1

In Figure 2 the index BM1 denotes marginal posterior moments for parameters $R, \sigma_I$ and $\sigma_\omega$ obtained with BM1 in comparison to the reference and BM2. With respect to a burn in phase, only the last 700 samples of three independently sampled Markov chains, each of total length 1500, are used to obtain the results. The marginal posterior distribution for $\sigma_\omega$ is close to the reference, but for $R$ and $\sigma_I$ the marginal posteriors concentrate at values different to the references and do not even assign a significant probability to the references. The marginal posterior mean values of $R, \sigma_I$ and $\sigma_\omega$ and their respective relative error with respect to the reference are displayed in Table 1.

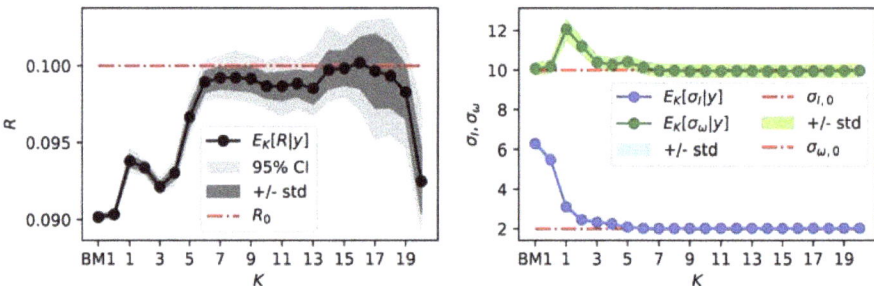

**Figure 2.** Results obtained with BM1 (noted by index BM1) and BM2 for $K = 0, \ldots, 20$. Moments of the marginal posterior distributions of $R$ are displayed on the left and moments of $\sigma_I, \sigma_\omega$ on the right.

**Table 1.** Marginal posterior mean and relative error of parameters $R, \sigma_I$ and $\sigma_\omega$ for BM1 and BM2 with $K = 9$. The relative errors are with respect to the reference values $R_0 = 0.1, \sigma_{I,0} = 2$, and $\sigma_{\omega,0} = 10$.

|  | Marginal Posterior Mean | | | Relative Error | | |
| --- | --- | --- | --- | --- | --- | --- |
|  | $\hat{R}$ | $\hat{\sigma}_I$ | $\hat{\sigma}_\omega$ | $\epsilon_{rel}(\hat{R})$ | $\epsilon_{rel}(\hat{\sigma}_I)$ | $\epsilon_{rel}(\hat{\sigma}_\omega)$ |
| BM1 | $9.03 \times 10^{-2}$ | $6.29 \times 10^0$ | $1.03 \times 10^1$ | $9.66 \times 10^{-2}$ | $2.14 \times 10^0$ | $2.79 \times 10^{-2}$ |
| BM2 ($K = 9$) | $9.92 \times 10^{-2}$ | $2.02 \times 10^0$ | $9.95 \times 10^0$ | $8.48 \times 10^{-3}$ | $1.03 \times 10^{-2}$ | $4.89 \times 10^{-3}$ |

The noisy discrepancy $d^\varepsilon(\hat{R})$ for $\hat{R}$ is displayed in Figure 3. (Remark: For simplicity the mean $\mathbb{E}[d^\varepsilon(R)|y] = y - \mathbb{E}[\mathcal{G}_\phi(R)|y] \approx y - \mathcal{G}_\phi(\mathbb{E}[R|y])$ is approximated by assuming linearity for $\mathcal{G}_\phi$ in a neighborhood of $\hat{R}$.) For $\omega$ this is basically the measurement noise, since $\mathbb{E}[d^\varepsilon_\omega] \approx 0$ and $V[d^\varepsilon_\omega]^{1/2} = 10.18 \approx \sigma_{\omega,0} \approx \hat{\sigma}_\omega$. But for $d^\varepsilon_I$ there is some discrepancy, at least in the first half of the time interval, different to measurement noise. Calculating the empirical standard deviation for a fixed zero mean $\sigma_I^\dagger := \frac{1}{M-1} \sum_{i=1}^M (d^\varepsilon_I(\hat{R})_i)^2$, where $d^\varepsilon_I(\hat{R})_i$ denotes the $i$-th component of $d^\varepsilon_I(\hat{R})$, one obtains $\sigma_I^\dagger = 6.23 \approx \hat{\sigma}_I$. The obtained value corresponds to the overestimated marginal posterior distribution of $\sigma_I$ that centers around a similar value, see Figure 2 and Table 1. Obviously, BM1 leads to biased and overconfident parameter estimates.

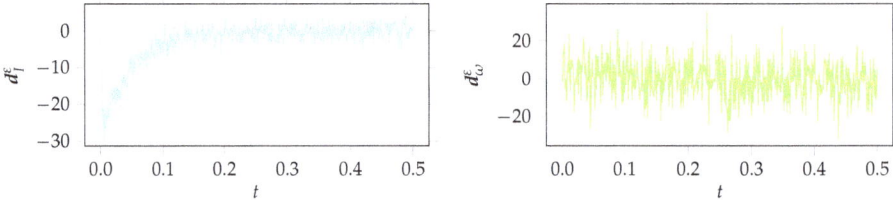

**Figure 3.** Remaining discrepancy $d^\varepsilon(\hat{R}) = [d^\varepsilon_I, d^\varepsilon_\omega](\hat{R}) = y - \mathcal{G}_\phi(\hat{R})$, for $\hat{R}$ obtained with BM1.

## 4.2. Results Bayesian Model 2

Figure 2 also displays moments of the marginal posterior distributions of $R, \sigma_I$ and $\sigma_\omega$, obtained via BM2 for $K = 0, \ldots, 20$. For each $K$ the last $0.4 \times 10^5$ samples of three independently sampled Markov chains of length $10^5$ are used to approximate the moments. In contrast to BM1 a larger number of MH-MCMC samples is required for convergence, due to an increased number of unknowns in BM2. For comparison, the index BM1 in the following figures denotes results without $\delta$. Following the guideline specified in Section 3 the marginal posterior distributions of the noise standard deviations of $\sigma_I$ and $\sigma_\omega$ in Figure 2 are considered to pick an appropriate K. Both marginal posterior distributions $\pi^K(\sigma_I|y)$ and $\pi^K(\sigma_\omega|y)$ stabilize for $K \geq 7$ and are almost identical for $K \geq 9$, considering mean and standard deviation of the marginal posterior distribution. This indicates that $K = 9$ is a sufficient polynomial degree and increasing K further does not improve the results with the current specification of the model discrepancy term. Figure 4 displays for $K = 9$ the concentration of the posterior distribution of $\delta^K(a) = [\delta_I^K, \delta_\omega^K](a)$ around the reference discrepancy $d_0$.

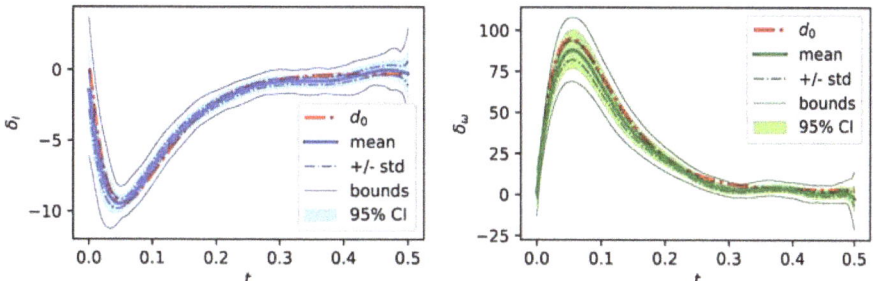

**Figure 4.** Posterior moments of model discrepancy $\delta^K(a) = [\delta_I^K, \delta_\omega^K](a)$ for $K = 9$ in comparison to the reference discrepancy $d_0 = [d_{I,0}, d_{\omega,0}]$.

The relative error $nBias_K(\delta_*^K - d_{0,*})/\|d_{0,*}\|$ is $1.0 \times 10^{-1}$ for I and $9.2 \times 10^{-2}$ for $\omega$. Also the marginal posterior distribution of R concentrates close to the reference value and the relative error of the posterior mean reduces around one order of magnitude compared to the result of BM1, see Table 1. Also, both noise standard deviations concentrate around the reference and significantly improve in relative errors compared to BM1. Overall, the posterior distribution of $\mathcal{G}_\phi(R) + \delta^K(a) + \varepsilon$ yields a good approximation of the measurements $y$ with only a small variance band. Figure 2 shows that for $K = 6, \ldots, 19$ the marginal posterior mean of R roughly stabilizes in some kind of a plateau close to the reference $R_0$, but the posterior standard deviation increases with K. For $K = 0, \ldots, 5$ the model discrepancy term $\delta^K(a)$ is not flexible enough to approximate the underlying discrepancy appropriate enough. As a consequence, the standard deviations of the measurement noise are overestimated and the estimation of the parameter R is still biased for those values of K.

Adding on this, Figure 5 displays $Bias_K$, variance $V_K$ and $MSE_K$ of R. The MSE of R is minimal around $K = 7, 8, 9$ and $K = 15$, indicating those values as an optimal model complexity and backing up the decision of $K = 9$ as sufficient polynomial degree. Figure 5 largely corresponds to the bias-variance tradeoff, since the variance increases with K and the bias decreases with K, at least until $K = 15$. For $K \geq 15$ the bias increases as non-identifiability occurs additionally. Especially for $K = 20$ overfitting and non-identifiability can be observed as the posterior distribution of R is spread out and biased, see Figure 2. For complimentary numerical results (e.g., posterior moments of $\delta^K(a)$ for $K = 20$) see the accompanying preprint of this work [27].

## 5. Conclusions

In this work a method to infer model parameters and model discrepancy is considered and applied to synthetic measurement data of an electric motor. The suggested Bayesian model 2 (BM2) considers measurement noise and model discrepancy, in order to separate these two sources of uncertainty and improve physical parameter estimation. The model discrepancy term is modeled as a truncated polynomial expansion $\delta^K(a)$ with unknown coefficients $a$ and an maximum polynomial degree K.

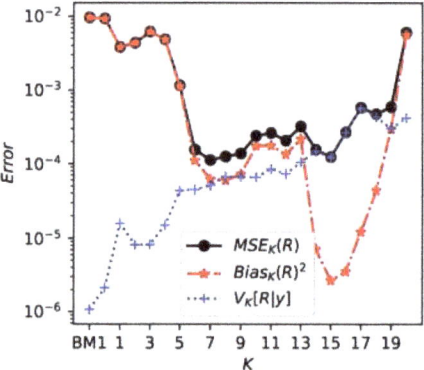

**Figure 5.** Bias-variance tradeoff: Bias $Bias_K(R)$, variance $V_K[R|y]$ and mean square error $MSE_K(R)$ of $R$ with respect to $R_0$, depending on $K$, for $K = 0, \ldots, 20$ from BM2. The index BM1 denotes the previous results without $\delta$, obtained via Bayesian model 1.

A discussion and a guideline on how to define appropriate model discrepancy term complexity, i.e., here the maximum polynomial degree $K$, based on the marginal posterior distribution of measurement noise standard deviation is presented. The framework applied to the electric motor showed promising perspectives by an improved estimation of the model parameters. Furthermore a good approximation of the a-priori unknown model discrepancy is learned with BM2 for a sufficient maximum polynomial degree $K = 9$. An appropriate choice of $K$ is crucial. For $K$ too small the accuracy of $\delta^K(a)$ is not sufficient and estimation is just slightly improved. If $K$ is too large the prior contains to less information about the reference discrepancy and consequently the posterior distribution does not converge anymore. Consequently, in order to identify both, the underlying parameter value and the reference discrepancy of the test scenario, it is important to find an optimal $K$ and with this formulate at least for one of the unknowns a prior containing some information about the reference.

Next steps are to apply this approach to higher dimensional problems and further to real field data to test its capabilities. Then, as real data implies a more complex simulator, surrogate modeling will be mandatory to leverage computational expenses. Additionally, evidence approximation could be an another criteria to select a Bayesian model, which will be considered in future work.

**Author Contributions:** D.N.J. implemented and conceived the numerical results. D.N.J. and M.S. analyzed the data. D.N.J., M.S. and V.H. wrote the paper.

**Acknowledgments:** We thank Robert Bosch GmbH for funding this work.

**Conflicts of Interest:** The authors declare no conflict of interest. The funders had no role in the design of the study; in the collection, analyses, or interpretation of data; in the writing of the manuscript, or in the decision to publish the results.

## References

1. Kaipio, J.; Somersalo, E. *Statistical and Computational Inverse Problems*; Applied Mathematical Sciences; Springer Science+Business Media, Inc.: New York, NY, USA, 2005.
2. Stuart, A.M. Inverse problems: A Bayesian perspective. *Acta Numer.* **2010**, *19*, 451–559.
3. Dashti, M.; Stuart, A.M. The Bayesian Approach to Inverse Problems. In *Handbook of Uncertainty Quantification*; Ghanem, R., Higdon, D., Owhadi, H., Eds.; Springer: Cham, Switzerland, 2017; pp. 311–428.
4. Gelman, A.; Carlin, J.B.; Stern, H.S.; Dunson, D.B.; Vehtari, A.; Rubin, D.B.; Stern, H.S. *Bayesian Data Analysis*, 3rd ed.; Texts in Statistical Science Series; CHAPMAN & HALL/CRC and CRC Press: Boca Raton, FL, USA, 2013.

5. Sullivan, T.J. *Introduction to Uncertainty Quantification*; Texts in Applied Mathematics 0939-2475; Springer: Berlin, Germany, 2015; Volume 63.
6. Hastings, W.K. Monte Carlo Sampling Methods Using Markov Chains and Their Applications. *Biometrika* **1970**, *57*, 97.
7. Salvatier, J.; Wiecki, T.V.; Fonnesbeck, C. Probabilistic programming in Python using PyMC3. *PeerJ Comput. Sci.* **2016**, *2*, e55.
8. Schillings, C.; Schwab, C. Scaling limits in computational Bayesian inversion. *ESAIM Math. Model. Numer. Anal.* **2016**, *50*, 1825–1856.
9. Sprungk, B. *Numerical Methods for Bayesian Inference in Hilbert Spaces*, 1st ed.; Universitätsverlag der TU Chemnitz: Chemnitz, Germany, 2018.
10. Hoffman, M.D.; Gelman, A. The No-U-turn sampler: adaptively setting path lengths in Hamiltonian Monte Carlo. *J. Mach. Learn. Res.* **2014**, *15*, 1593–1623.
11. Xiu, D.; Karniadakis, G.E. The Wiener-Askey polynomial chaos for stochastic differential equations. *SIAM J. Sci. Comput.* **2002**, *24*, 619–644.
12. Glaser, P.; Schick, M.; Petridis, K.; Heuveline, V. Comparison between a Polynomial Chaos surrogate model and Markov Chain Monte Carlo for inverse Uncertainty Quantification based on an electric drive test bench. In Proceedings of the ECCOMAS Congress 2016, Crete Island, Greece, 5–10 June 2016.
13. Kennedy, M.C.; O'Hagan, A. Bayesian calibration of computer models. *J. R. Stat. Soc. Ser. B Stat. Methodol.* **2001**, *63*, 425–464.
14. Rasmussen, C.E.; Williams, C.K. *Gaussian Process for Machine Learning*; MIT Press: Cambridge, MA, USA, 2006.
15. Arendt, P.D.; Apley, D.W.; Chen, W. Quantification of Model Uncertainty: Calibration, Model Discrepancy, and Identifiability. *J. Mech. Des.* **2012**, *134*, 100908.
16. Arendt, P.D.; Apley, D.W.; Chen, W.; Lamb, D.; Gorsich, D. Improving Identifiability in Model Calibration Using Multiple Responses. *J. Mech. Des.* **2012**, *134*, 100909.
17. Paulo, R.; García-Donato, G.; Palomo, J. Calibration of computer models with multivariate output. *Comput. Stat. Data Anal.* **2012**, *56*, 3959–3974.
18. Brynjarsdóttir, J.; O'Hagan, A. Learning about physical parameters: The importance of model discrepancy. *Inverse Probl.* **2014**, *30*, 114007.
19. Tuo, R.; Wu, C.F.J. Efficient calibration for imperfect computer models. *Ann. Stat.* **2015**, *43*, 2331–2352.
20. Tuo, R.; Jeff Wu, C.F. A Theoretical Framework for Calibration in Computer Models: Parametrization, Estimation and Convergence Properties. *SIAM/ASA J. Uncertain. Quantif.* **2016**, *4*, 767–795.
21. Plumlee, M. Bayesian Calibration of Inexact Computer Models. *J. Am. Stat. Assoc.* **2016**, *112*, 1274–1285.
22. Nagel, J.B.; Rieckermann, J.; Sudret, B. Uncertainty Quantification in Urban Drainage Simulation: Fast Surrogates for Sensitivity Analysis and Model Calibration. 2017. Available online: http://arxiv.org/abs/1709.03283 (accessed on 23 March 2018).
23. Toliyat, H.A. (Ed.) *Handbook of Electric Motors*, 2nd ed.; Electrical and Computer Engineering; Dekker: New York, NY, USA; Basel, Switzerland, 2004; Volume 120.
24. Wanner, G.; Hairer, E. *Solving Ordinary Differential Equations I*; Springer: Berlin, Germany, 1991; Volume 1.
25. Kotz, S.; Kozubowski, T.; Podgorski, K. *The Laplace Distribution and Generalizations: A Revisit with Applications to Communications, Economics, Engineering, and Finance*; Springer Science & Business Media: Berlin, Germany, 2012.
26. Hastie, T.; Tibshirani, R.; Friedman, J.H. *The Elements of Statistical Learning: Data Mining, Inference, and Prediction*, 2nd ed.; Springer Series in Statistics; Springer: New York, NY, USA, 2009.
27. John, D.; Schick, M.; Heuveline, V. Learning model discrepancy of an electric motor with Bayesian inference. *Eng. Math. Comput. Lab* **2018**, doi:10.11588/emclpp.2018.1.51320.

© 2019 by the authors. Licensee MDPI, Basel, Switzerland. This article is an open access article distributed under the terms and conditions of the Creative Commons Attribution (CC BY) license (http://creativecommons.org/licenses/by/4.0/).

*Proceedings*

# Intracellular Background Estimation for Quantitative Fluorescence Microscopy [†]

**Yannis Kalaidzidis [1,2,*], Hernán Morales-Navarrete [1], Inna Kalaidzidis [1] and Marino Zerial [1]**

[1] Max Planck Institute of Molecular Cell Biology and Genetics. Pfotenhauerstr. 108, 01307 Dresden, Germany; moralesn@mpi-cbg.de (H.M.-N.); ikalaidz@gmail.com (I.K.); zerial@mpi-cbg.de (M.Z.)
[2] Faculty of Bioengineering and Bioinformatics, Moscow State University, 119991 Moscow, Russia
\* Correspondence: kalaidzi@mpi-cbg.de
[†] Presented at the 39th International Workshop on Bayesian Inference and Maximum Entropy Methods in Science and Engineering, Garching, Germany, 30 June–5 July 2019.

Published: 6 December 2019

**Abstract:** Fluorescently targeted proteins are widely used for studies of intracellular organelles dynamic. Peripheral proteins are transiently associated with organelles and a significant fraction of them are located at the cytosol. Image analysis of peripheral proteins poses a problem on properly discriminating membrane-associated signal from the cytosolic one. In most cases, signals from organelles are compact in comparison with diffuse signal from cytosol. Commonly used methods for background estimation depend on the assumption that background and foreground signals are separable by spatial frequency filters. However, large non-stained organelles (e.g., nuclei) result in abrupt changes in the cytosol intensity and lead to errors in the background estimation. Such mistakes result in artifacts in the reconstructed foreground signal. We developed a new algorithm that estimates background intensity in fluorescence microscopy images and does not produce artifacts on the borders of nuclei.

**Keywords:** Bayesian image analysis; fluorescence microscopy; background estimation

## 1. Introduction

The development of technologies for creating genetically encoded chimeric conjugates of proteins of interest with fluorescent proteins opened a new era in study of intracellular processes by means of quantitative fluorescence microscopy [1], and it is widely used in studies of spatio-temporal dynamics of intracellular organelles in live and fixed cells [2–4]. Several approaches for the quantification of cytosolic and membrane-bound proteins have been developed. However, most of them fail when applied to cases where the existence of large non-stained organelles (e.g., nuclei) produce sudden changes in the cytosol fluorescent intensity. Many peripheral membrane proteins dynamically switch between cytosolic and membrane-bound state. Whereas, they generate compact fluorescent images of intracellular organelles when they are in membrane-bound state, fuzzy fluorescent background is generated when they are in cytosolic state. As an example of such peripheral membrane proteins, we analyzed images of the small GTPase Rab5 conjugated with Green Fluorescent Protein (GFP), which dynamics orchestrates intracellular endocytic transport [5]. Quantitative analysis of endosome-associated proteins requires discrimination of fluorescent endosomes from fluorescent cytosolic background (Figure 1a). The problem of discriminating high spatial frequency (bright compact) structures, i.e., endosomes, from low-frequency (cytoplasmic) background has been extensively studied and many solutions (using heuristic as well as Bayesian approaches) have been developed. However, they mostly rely on the assumption that background corresponds to low-frequency signal, which is not the case of peripheral proteins. Usually, images of peripheral proteins (Figure 1a) show large dark areas with sharp boundaries, which are imprints of nuclei in the fluorescent cytoplasm. Multiple unlabeled organelles are observed as dark areas in the cytoplasm with spatial frequencies similar to those of endosomes. This spatial frequency similarity is exemplified by the intensity profile along yellow line on Figure 1a, which is presented on Figure 1b (black line). In a recent study [6], the problem of non-smooth background was addressed for time series (live cell imaging) by using conditional random fields to estimate the background as well as to segment the motile organelles. However, this algorithm is not applicable for single images. Unfortunately, state-of-the-arts algorithms for background estimation on single fluorescence microscopy images [7–9] explicitly rely on the smoothness of background signal and in this respect are not better than textbook "rolling ball" algorithm [10]. All of them produce artefacts (false-positive foreground rim) on border of cell nucleus (see Figure 1b,c). In present work we proposed new algorithm (TBL) based on probability distribution for two background levels: one in cytosol and another in possibly presented "dark" nucleus, each background is smooth, but transition between them could be abrupt.

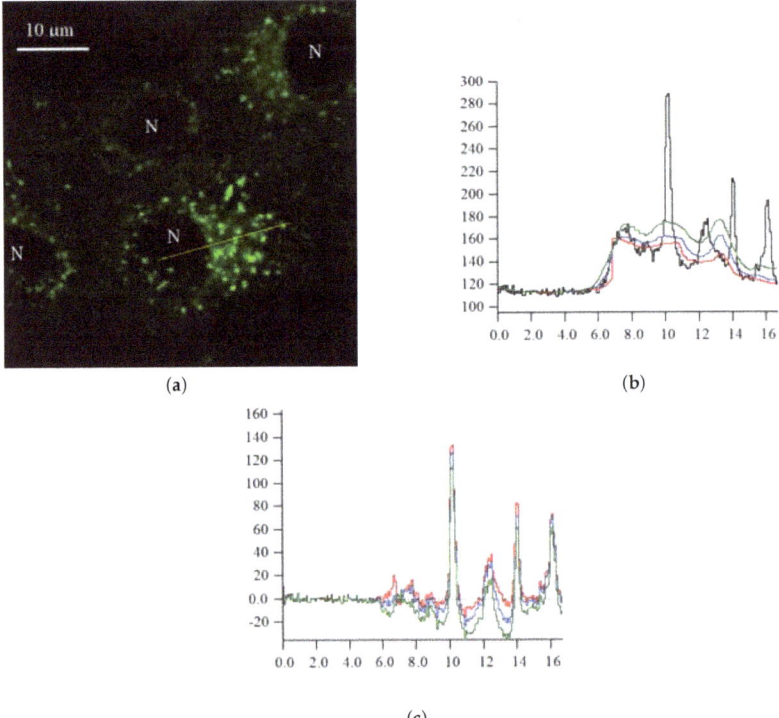

**Figure 1.** (a) A431 cells with GFP-tagged Rab5a. The cytosol is labelled by soluble fraction of GFP-Rab5a. Bright structures of different size and shape are endosomes, which are labelled by membrane-bound GFP-Rab5a. Images were obtained by spinning disk microscope (Andor-Olympus-IX71 inverted stand microscope; scan head CSU-X1 Yokogawa, objective Olympus UPlanSApo 63x 1.35oil, Optovar 1.6). Letters N denote nuclei. Yellow line marks the intensity profile presented on panel b. (b) Intensity profiles along the yellow line (Figure 1). Black curve is intensity of original image. Red, green, blue and cyan curves are background estimation by "rolling ball" [10], Gaussian, median and FMOR [7] filters respectively. (c) Difference between original intensities and estimated background. Green, blue and red curves correspond to background estimation by Gaussian, median and FMOR filters respectively. Filter window was $4\mu m$ for all filters.

## 2. Two Background Level Estimation (TBL) Algorithm

### 2.1. Probabilistic Model of Intensity

Images of fluorescence microscopy are dominated by Poisson noise of photo-electron flux in photomultiplier tube (PMT) or CMOS/CCD camera. Probability to detect $n$ photo-electrons is $P(n) = \frac{\lambda^n e^{-\lambda}}{\Gamma(n+1)}$. Assuming that intensity $I$ linearly depends on number photo-electrons, we got $I = \alpha \cdot n + I_0$, where offset $I_0$ can be as positive as negative, dependent on microscope settings. Therefore, variance of intensity $\sigma^2 = \alpha \cdot I + \zeta$, where $\zeta = \varepsilon^2 - \alpha \cdot I_0$ and $\varepsilon^2$ is variance of zero-mean Gaussian noise of electronic circuits.

First, we found parameters $\alpha$ and $\zeta$ for a single image as it was described in [11]. In most practical cases $\varepsilon^2$ is small, therefore we approximated $I_0 = -\frac{\zeta}{\alpha}$ and subtracted it form the image: $J_i = \max\left(0, I_i + \frac{\zeta}{\alpha}\right)$. Second, we calculated estimation of variance $\sigma_i^2$ for each pixel. Third, we approximated Poisson noise distribution by truncated Gaussian distribution (since it allows get integrals analytically):

Therefore, probability of intensity in absence of foreground, given background intensity $B_i$, was approximated as:

$$P(J_i|B_i, \sigma_i) = \sqrt{\frac{2}{\pi}} \frac{e^{-\frac{1}{2}\frac{(J_i-B_i)^2}{\sigma_i^2}}}{\sigma_i \left(-\frac{1}{\sqrt{2}}\frac{B_i}{\sigma_i}\right)}, B_i \geq 0 \tag{1}$$

In presence of foreground signal $F_i$ it was approximated as:

$$P(J_i, \sigma_i|B_i) = \int_0^\infty \sqrt{\frac{2}{\pi}} \frac{e^{-\frac{1}{2}\frac{(J_i-F_i-B_i)^2}{\sigma_i^2}}}{\sigma_i erfc\left(-\frac{1}{\sqrt{2}}\frac{B_i}{\sigma_i}\right)} P(F_i) \, dF_i, \tag{2}$$

where $P(F_i)$ is prior of foreground. By maximum entropy principle, we chose prior distribution for foreground

$$P(F_i|\mu_i) = \frac{1}{\mu} e^{-\frac{F_i}{\mu}} \tag{3}$$

The parameter of the prior (mean expected amplitude of foreground $\mu$) was found outside of the analyzed pixel (see Appendix A).

After substitution (3) to (2) we got:

$$P(J_i, \sigma_i|B_i) = \frac{1}{\mu} \frac{e^{-\frac{1}{2}\frac{(J_i-B_i)^2}{\sigma_i^2}}}{\Psi\left(\frac{\sigma_i}{\mu} - \frac{J_i-B_i}{\sigma_i}\right)} \frac{e^{\frac{1}{2}\left(\frac{B_i}{\sigma}\right)^2} \Psi\left(\frac{\sigma_i}{\mu} + \frac{B_i}{\sigma_i}\right) \Psi\left(-\frac{B_i}{\sigma_i}\right)}{\Psi\left(\frac{\sigma_i}{\mu} + \frac{B_i}{\sigma_i}\right) + \Psi\left(-\frac{B_i}{\sigma_i}\right)}, \tag{4}$$

where $\Psi(x) = \sqrt{\frac{2}{\pi}} \frac{e^{-\frac{1}{2}x^2}}{erfc\left(\frac{1}{\sqrt{2}}x\right)}$

### 2.2. Probabilistic Model of Two Background Levels

Assuming that background is slow varying signal, we approximate it by constant in the vicinity of the pixel $i$. The vicinity window is defined by characteristic scale discriminating background and foreground. We introduced latent variables $z_{i,k}$, $k = 1, 2, 3, 4$ to define 4 possible states of pixel intensity: background $B_1$, background $B_2$, background $B_1$ with foreground and background $B_2$ with foreground. Then probability of intensities in the vicinity window $\Omega$ In the pixel having two background levels in the vicinity window, the probability of intensities we got:

$$P(\{J_i, z_i\} | B_1, B_2, \{\sigma_i\}) =$$

$$= \prod_{i \in \Omega} e^{-\frac{1}{2}\frac{J_i^2}{\sigma_i^2}} \left( \begin{array}{l} e^{\frac{J_i B_{i,1}}{\sigma_i^2}} \Psi\left(-\frac{B_{i,1}}{\sigma_i}\right) \left( \frac{\delta_{z_i,1}}{\sigma_i}(1-\beta_0) + \frac{\delta_{z_i,2}}{\mu} \frac{\beta_0}{\Psi\left(\frac{\sigma_i}{\mu} - \frac{J_i-B_{i,1}}{\sigma_i}\right)} \frac{\Psi\left(\frac{\sigma_i}{\mu} + \frac{B_{i,1}}{\sigma_i}\right)}{\Psi\left(\frac{\sigma_i}{\mu} + \frac{B_{i,1}}{\sigma_i}\right) + \Psi\left(-\frac{B_{i,1}}{\sigma_i}\right)} \right) + \\ + e^{\frac{J_i B_{i,2}}{\sigma_i^2}} \Psi\left(-\frac{B_{i,2}}{\sigma_i}\right) \left( \frac{\delta_{z_i,3}}{\sigma_i}(1-\beta_0) + \frac{\delta_{z_i,4}}{\mu} \frac{\beta_0}{\Psi\left(\frac{\sigma_i}{\mu} - \frac{J_i-B_{i,2}}{\sigma_i}\right)} \frac{\Psi\left(\frac{\sigma_i}{\mu} + \frac{B_{i,2}}{\sigma_i}\right)}{\Psi\left(\frac{\sigma_i}{\mu} + \frac{B_{i,2}}{\sigma_i}\right) + \Psi\left(-\frac{B_{i,2}}{\sigma_i}\right)} \right) \end{array} \right) \quad (5)$$

where $\beta_0$ is prior probability of presence of foreground in the pixel (see Appendix A). If $\frac{B}{\sigma} > 1$, then the expression can be simplified as:

$$P(\{J_i, z_i\} | B_1, B_2, \{\sigma_i\}) =$$

$$= \prod_{i \in \Omega} e^{-\frac{1}{2}\frac{J_i^2}{\sigma_i^2}} \left( \begin{array}{l} e^{\frac{J_i B_{i,1}}{\sigma_i^2}} \Psi\left(-\frac{B_{i,1}}{\sigma_i}\right) \left( \frac{\delta_{z_i,1}}{\sigma_i}(1-\beta_0) + \frac{\delta_{z_i,2}}{\mu} \beta_0 \frac{1}{\Psi\left(\frac{\sigma_i}{\mu} - \frac{J_i-B_{i,1}}{\sigma_i}\right)} \right) + \\ + e^{\frac{J_i B_{i,2}}{\sigma_i^2}} \Psi\left(-\frac{B_{i,2}}{\sigma_i}\right) \left( \frac{\delta_{z_i,3}}{\sigma_i}(1-\beta_0) + \frac{\delta_{z_i,4}}{\mu} \beta_0 \frac{1}{\Psi\left(\frac{\sigma_i}{\mu} - \frac{J_i-B_{i,2}}{\sigma_i}\right)} \right) \end{array} \right) \quad (6)$$

We used EM algorithm to maximize likelihood $L(B_1, B_2 | \{J_i, z_i\}, \{\sigma_i\})$ over backgrounds $B_1, B_2$. Resulting backgrounds are presented on Figure 2a.

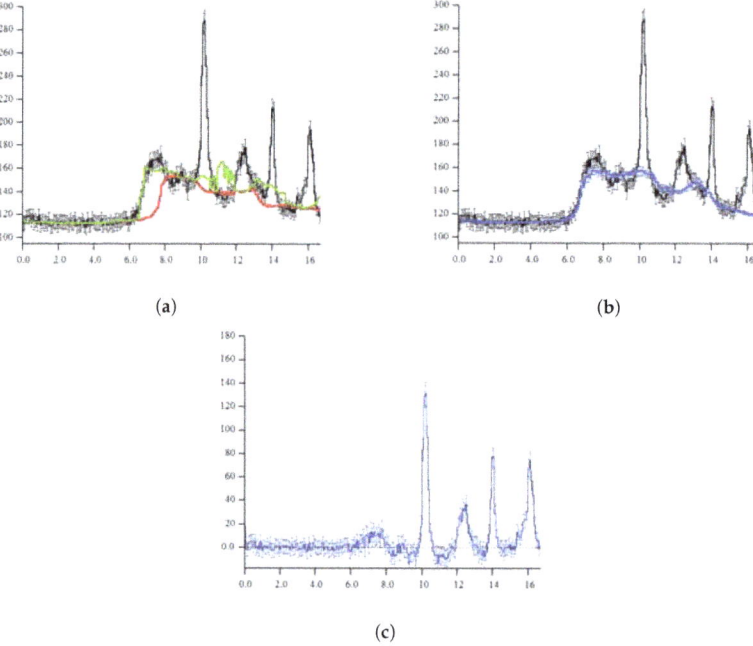

**Figure 2.** (a) Original intensity (as on Figure 1) curve (black) and two estimated background levels (red and green curves). (b) Original intensity curve (black) and background estimation by TBL (blue) curve. (c) Difference between original intensities and TBL background.

Then we calculated probability of presence of foreground in the pixel $i$ (assuming $B_2 > B_1$ without loss of generality):

$$p_i^{signal} = \begin{cases} \frac{p_{i,z_i=4}}{p_{i,z_i=4}+p_{i,z_i=3}}, & I_i > B_2 \\ \frac{p_{i,2}}{p_{i,z_i=2}+p_{i,z_i=1}} \end{cases} \tag{7}$$

where

$$p_{i,z_i=1} = \frac{(1-\beta_0)}{\sigma_i} e^{\frac{I_i B_{i,1}}{\sigma_i^2}} \Psi\left(-\frac{B_{i,1}}{\sigma_i}\right); p_{i,z_i=3} = \frac{(1-\beta_0)}{\sigma_i} e^{\frac{I_i B_{i,2}}{\sigma_i^2}} \Psi\left(-\frac{B_{i,2}}{\sigma_i}\right) \tag{8}$$

$$p_{i,z_i=2} = \frac{\beta_0}{\mu} \frac{e^{\frac{I_i B_{i,1}}{\sigma_i^2}} \Psi\left(-\frac{B_{i,1}}{\sigma_i}\right) \Psi\left(\frac{\sigma_i}{\mu}+\frac{B_{i,1}}{\sigma_i}\right)}{\Psi\left(\frac{\sigma_i}{\mu}-\frac{I_i-B_{i,1}}{\sigma_i}\right)\left(\Psi\left(\frac{\sigma_i}{\mu}+\frac{B_{i,1}}{\sigma_i}\right)+\Psi\left(-\frac{B_{i,1}}{\sigma_i}\right)\right)} \tag{9}$$

$$p_{i,z_i=4} = \frac{\beta_0}{\mu} \frac{e^{\frac{I_i B_{i,2}}{\sigma_i^2}} \Psi\left(-\frac{B_{i,2}}{\sigma_i}\right) \Psi\left(\frac{\sigma_i}{\mu}+\frac{B_{i,2}}{\sigma_i}\right)}{\Psi\left(\frac{\sigma_i}{\mu}-\frac{I_i-B_{i,2}}{\sigma_i}\right)\left(\Psi\left(\frac{\sigma_i}{\mu}+\frac{B_{i,2}}{\sigma_i}\right)+\Psi\left(-\frac{B_{i,2}}{\sigma_i}\right)\right)} \tag{10}$$

Assuming that each pixel in the vicinity window exclusively belongs to only one background level, we got probability of background levels: Assuming that each pixel in the vicinity window exclusively belongs to only one background level, we got probability of background levels:

$$P_{j,1} = \sum_{i=1}^{N}\left((p_{i,1}+p_{i,2})\cdot p_i^{signal}\cdot p_j^{signal} + p_{i,1}\cdot\left(1-p_i^{signal}\cdot p_j^{signal}\right)\right)\exp\left(-\frac{(I_i-I_j)^2}{2\left(\sigma_i^2+\sigma_j^2\right)}\right) \tag{11}$$

$$P_{j,2} = \sum_{i=1}^{N}\left((p_{i,3}+p_{i,4})\cdot p_i^{signal}\cdot p_j^{signal} + p_{i,3}\cdot\left(1-p_i^{signal}\cdot p_j^{signal}\right)\right)\exp\left(-\frac{(I_i-I_j)^2}{2\left(\sigma_i^2+\sigma_j^2\right)}\right) \tag{12}$$

Finally consensus background intensity in the pixel of interest was calculated as weighted mean:

$$B = \frac{B_1 P_1 + B_2 P_2}{P_1 + P_2} \tag{13}$$

The estimation of variance of background intensity:

$$\sigma_b^2 = Var_{B_1}\left(\frac{P_1}{P_1+P_2}\right)^2 + Var_{B_2}\left(\frac{P_2}{P_1+P_2}\right)^2 + (B_1-B_2)^2\frac{P_1 P_2}{(P_1+P_2)^2}\frac{1}{N} \tag{14}$$

Final background was calculated by formula (14) (Figure 2b). Difference between original intensities and estimated backgrounds has residual "nuclei border artifact" within estimated uncertainty (Figure 2c).

Result of TBL algorithm is presented on Figure 3a,b.

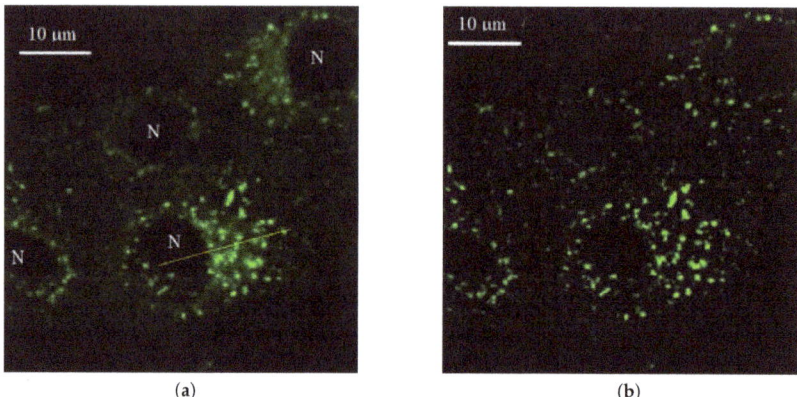

**Figure 3.** (a) Original image. (b) Image after subtraction background by TBL.

## 3. Conclusions

We developed a new algorithm (TBL) to estimate cytoplasm fluorescence (background) in conditions where high spatial frequency is present in both background and foreground signal. TBL avoids artefacts that are characteristic for state-of-the-arts background subtraction algorithms in presence inhomogeneous background with sharp transition between levels.

**Author Contributions:** Methodology Y.K.; software, H.M-N.; investigation, I.K.; supervision, M.Z.

**Funding:** This work was financially supported by the German Federal Ministry of Education and Research (BMBF) (LiSyM: grant #031L0038), European Research Council (ERC) (grant #695646) and the Max Planck Society (MPG).

**Acknowledgments:** We thank the Center for Information Services and High Performance Computing (ZIH) of the TU Dresden and the Light Microscopy Facility of the MPI-CBG.

**Conflicts of Interest:** The authors declare no conflict of interest. The funders had no role in the design of the study; in the collection, analyses, or interpretation of data; in the writing of the manuscript, or in the decision to publish the results.

## Appendix A. Estimation Prior Parameters $\mu$ and $\beta$

It is reasonable to assume that most of pixels have only one background level in their close vicinity. Fluorescence microscopy images of intracellular organelles have majority of pixels belonging to background. Therefore, median filter, which is insensitive to outliers (foreground), could be used as crude estimation of background. However, given that intensity is not normally distributed, median is shifted relative to mode. In our cropped Gaussian approximation, the median $M_i$ is solution of equation:

$$erf\left(\frac{1}{\sqrt{2}}\frac{M_i - B_i}{\sigma_i}\right) = \frac{1}{2}erfc\left(\frac{1}{\sqrt{2}}\frac{B_i}{\sigma_i}\right), \quad B_i \geq 0 \tag{A1}$$

Therefore, we first calculated median $\{M_i\}$ over vicinity window for image for image$\{J_i, \sigma_i\}$, where $J_i$ denote intensity after offset subtraction $J_i = I_i + \frac{\zeta}{\alpha}$, then calculated $\{B_i\}$ by numerical solution of equation A1 and, finally, constructed image $\{u_i = \frac{I_i - B_i}{\sigma_i}\}$, where in absence of foreground image has intensity distribution

$$p_b(u|b, \mu, \sigma) = \sqrt{\frac{2}{\pi}} \frac{e^{-\frac{1}{2}u^2}}{erfc\left(-\frac{b}{\sqrt{2}\sigma}\right)} \tag{A2}$$

In presence of foreground image has intensity distribution

$$p_f(u|\mu,\sigma,b) = \frac{\frac{\sigma}{\mu}e^{-\frac{\sigma}{\mu}u}erfc\left(\frac{1}{\sqrt{2}}\left(\frac{\sigma}{\mu}-u\right)\right)}{e^{\frac{b}{\mu}}erfc\left(\frac{1}{\sqrt{2}}\left(\frac{\sigma}{\mu}+\frac{b}{\sigma}\right)\right)+e^{-\frac{1}{2}\left(\frac{\sigma}{\mu}\right)^2}erfc\left(-\frac{b}{\sqrt{2}\sigma}\right)} \quad (A3)$$

Therefore ratio fraction of pixels with intensity in the interval $[t,\infty]$ to the number of pixels with intensity in the interval $[0,1]$ for distribution (A2) is:

$$b = \frac{\int_t^\infty p_b(u|b,\mu,\sigma)\,du}{\int_0^1 p_b(u|b,\mu,\sigma)\,du} = \frac{1-erf\left(\frac{t}{\sqrt{2}}\right)}{erf\left(\frac{1}{\sqrt{2}}\right)} \quad (A4)$$

For distribution (A3) the ration is:

$$f = \frac{\int_t^\infty p_f(u|b,\mu,\sigma)\,du}{\int_0^1 p_f(u|b,\mu,\sigma)\,du} = \frac{e^{-\frac{\sigma}{\mu}t}erfc\left(\frac{1}{\sqrt{2}}\left(\frac{\sigma}{\mu}-t\right)\right)+e^{-\frac{1}{2}\left(\frac{\sigma}{\mu}\right)^2}erfc\left(\frac{t}{\sqrt{2}}\right)}{erfc\left(\frac{1}{\sqrt{2}}\frac{\sigma}{\mu}\right)-e^{-\frac{\sigma}{\mu}}erfc\left(\frac{1}{\sqrt{2}}\left(\frac{\sigma}{\mu}-1\right)\right)+e^{-\frac{1}{2}\eta^2}erf\left(\frac{1}{\sqrt{2}}\right)} \quad (A5)$$

The integration $f$ with Jeffreys prior $\frac{1}{\mu}$ gives:

$$<f> = \lim_{A\to\infty}\frac{1}{A}\int_0^A f\frac{1}{\mu}d\mu \approx 1.47 - 0.009t \quad (A6)$$

Then we calculated ratio $R$ of number of pixels with intensities above $u > t$ to number of pixels with $u \le 1$ in the image:

$$R = \frac{b(1+<f>)+\beta(<f>-b)}{1+<f>-\beta(<f>-b)} \quad (A7)$$

The ratio (A7) was calculated for set of thresholds $t$ ($t = 1,2,3$). The resulting overdetermined system was solved in least square sense w.r.t $\beta$, which is probability of presence foreground signal in the pixel.

Expectations of mean value of pixels with intensities $u > t$ are:

$$\langle m_{b,u>t}\rangle = \frac{\int_t^\infty up_b(u|\mu,B,\sigma)\,du}{\int_1^\infty p_b(u|\mu,B,\sigma)\,du} = \sqrt{\frac{2}{\pi}}\frac{e^{-\frac{1}{2}t^2}}{erfc\left(\frac{t}{\sqrt{2}}\right)} \quad (A8)$$

$$\langle m_{f,u>t}\rangle = \frac{\int_t^\infty up_f(u|\mu,B,\sigma)\,du}{\int_1^\infty p_f(u|\mu,B,\sigma)\,du} = \frac{\left(\frac{\mu}{\sigma}+t\right)e^{\frac{1}{2}\left(\frac{\sigma}{\mu}-t\right)^2-\frac{1}{2}t^2}erfc\left(\frac{1}{\sqrt{2}}\left(\frac{\sigma}{\mu}-t\right)\right)+\sqrt{\frac{2}{\pi}}e^{-\frac{1}{2}t^2}+\frac{\mu}{\sigma}erfc\left(\frac{t}{\sqrt{2}}\right)}{e^{\frac{1}{2}\left(\frac{\sigma}{\mu}-t\right)^2-\frac{1}{2}t^2}erfc\left(\frac{1}{\sqrt{2}}\left(\frac{\sigma}{\mu}-t\right)\right)+erfc\left(\frac{t}{\sqrt{2}}\right)} \quad (A9)$$

Therefore expected mean value of pixels with intensities $u > t$ is: $\langle m_{u>t}\rangle = \beta\langle m_{f,u>t}\rangle + (1-\beta)\langle m_{b,u>t}\rangle$.

We calculated experimental mean value $M$ of pixels with intensities $u > t$ for a set of thresholds $t$ ($t = 1,2,3$):

$$M = \beta\frac{(\eta+t)e^{\frac{1}{2}\left(\frac{\sigma}{\mu}-t\right)^2-\frac{1}{2}t^2}erfc\left(\frac{t}{\sqrt{2}}(\eta-t)\right)+\sqrt{\frac{2}{\pi}}e^{-\frac{1}{2}t^2}+\eta erfc\left(\frac{t}{\sqrt{2}}\right)}{e^{\frac{1}{2}(\eta-t)^2-\frac{1}{2}t^2}erfc\left(\frac{1}{\sqrt{2}}(\eta-t)\right)+erfc\left(\frac{t}{\sqrt{2}}\right)}+(1-\beta)\sqrt{\frac{2}{\pi}}\frac{e^{-\frac{1}{2}t^2}}{erfc\left(\frac{t}{\sqrt{2}}\right)} \quad (A10)$$

and solved the overdetermined system in least square sense w.r.t. $\eta$. Then parameter $\mu$ was estimated as: $\mu = \eta \cdot \langle\sigma\rangle$

## References

1. Zimmer, M. Green Fluorescent Protein (GFP): Applications, Structure, and Related Photophysical Behavior. *Chem. Rev.* **2002**, *102*, 759–782.
2. Rink, J.C.; Ghigo, E.; Kalaidzidis, Y.L.; Zerial, M. Rab Conversion as a Mechanism of Progression from Early to Late Endosomes. *Cell* **2005**, *122*, 735–749.
3. Meijering, E.; Smal, I.; Danuser, G. Tracking in Molecular Bioimaging. *IEEE Signal Process. Mag.* **2006**, *23*, 46–53.
4. Sbalzarini, I.F.; Koumoutsakos, P. Feature point tracking and trajectory analysis for video imaging in cell biology. *J. Struct. Biol.* **2006**, *151*, 182–195.
5. Pfeffer, S.R. Rab GTPases: master regulators that establish the secretory and endocytic pathways. *Mol. Biol. Cell* **2017**, *28*, 712–715.
6. Pécot, T.; Bouthemy, P.; Boulanger, J.; Chessel, A.; Bardin, S.; Salamero, J.; Kervrann, C. Background Fluorescence Estimation and Vesicle Segmentation in Live Cell Imaging with Conditional Random Fields. *IEEE Trans. Image Process.* **2015**, *24*, 667–680.
7. Kalaidzidis, Y. Available online: http://motiontracking.mpi-cbg.de (Version 8.93.00, 27 June 2019).
8. Lee, H.-C.; Yang, G. Computational Removal of Background Fluorescence for Biological Fluorescence Microscopy. In Proceedings of the IEEE 11th International Symposium on Biomedical Imaging (ISBI), Beijing, China, 29 April–2 May 2014.
9. Yang, L.; Zhang,Y.; Guldner, I.H.; Zhang, S.; Chen, D.Z. Fast Background Removal in 3D Fluorescence Microscopy Images Using One-Class Learning. In *International Conference on Medical Image Computing and Computer-Assisted Intervention—ICCAI 2015, Part III*; Springer: Cham, Switzerland, 2015; pp. 292–299.
10. Sternberg, S.R. Biomedical Image Processing. *Computer (IEEE)* **1983**, *16*, 22–34.
11. Kalaidzidis, Y. Fluorescence Microscopy Noise Model: Estimation of Poisson Noise Parameters from Snap-Shot Image. In Proceedings of the International Conference on Bioinformatics and Computational Biology (BIOCOMP'17), Las Vegas, NV, USA, 17–20 July 2017; pp. 63–66.

© 2019 by the authors. Licensee MDPI, Basel, Switzerland. This article is an open access article distributed under the terms and conditions of the Creative Commons Attribution (CC BY) license (http://creativecommons.org/licenses/by/4.0/).

Article

# Bayesian Identification of Dynamical Systems [†]

Robert K. Niven [1,*], Ali Mohammad-Djafari [2], Laurent Cordier [3], Markus Abel [4,5] and Markus Quade [4]

[1] School of Engineering and Information Technology, The University of New South Wales, Canberra ACT 2600, Australia
[2] Laboratoire des signaux et systèmes (L2S), CentraleSupélec, 91192 Gif-sur-Yvette, France; Ali.mohammad-Djafari@l2s.centralesupelec.fr
[3] Institut Pprime, 86073 Poitiers Cedex 9, France; Laurent.Cordier@univ-poitiers.fr
[4] Ambrosys GmbH, 14469 Potsdam, Germany; markus.abel@ambrosys.de (M.A.); markus.quade@ambrosys.de (M.Q.)
[5] Institute for Physics and Astrophysics, University of Potsdam, 14469 Potsdam, Germany
\* Correspondence: r.niven@adfa.edu.au
[†] Presented at the 39th International Workshop on Bayesian Inference and Maximum Entropy Methods in Science and Engineering, Garching, Germany, 30 June–5 July 2019.

Published: 12 February 2020

**Abstract:** Many inference problems relate to a dynamical system, as represented by $dx/dt = f(x)$, where $x \in \mathbb{R}^n$ is the state vector and $f$ is the (in general nonlinear) system function or model. Since the time of Newton, researchers have pondered the problem of *system identification*: how should the user accurately and efficiently identify the model $f$ – including its functional family or parameter values – from discrete time-series data? For linear models, many methods are available including linear regression, the Kalman filter and autoregressive moving averages. For nonlinear models, an assortment of machine learning tools have been developed in recent years, usually based on neural network methods, or various classification or order reduction schemes. The first group, while very useful, provide "black box" solutions which are not readily adaptable to new situations, while the second group necessarily involve the sacrificing of resolution to achieve order reduction. To address this problem, we propose the use of an inverse Bayesian method for system identification from time-series data. For a system represented by a set of basis functions, this is shown to be mathematically identical to Tikhonov regularization, albeit with a clear theoretical justification for the residual and regularization terms, respectively as the negative logarithms of the likelihood and prior functions. This insight justifies the choice of regularization method, and can also be extended to access the full apparatus of the Bayesian inverse solution. Two Bayesian methods, based on the joint maximum *a posteriori* (JMAP) and variational Bayesian approximation (VBA), are demonstrated for the Lorenz equation system with added Gaussian noise, in comparison to the regularization method of least squares regression with thresholding (the SINDy algorithm). The Bayesian methods are also used to estimate the variances of the inferred parameters, thereby giving the estimated model error, providing an important advantage of the Bayesian approach over traditional regularization methods.

**Keywords:** Bayesian inverse problem; dynamical systems; system identification; regularization; sparsification

## 1. Introduction

Many problems of inference involve a dynamical system, as represented by:

$$\frac{d}{dt}x(t) = f(x(t)), \qquad (1)$$

where $x \in \mathbb{R}^n$ is the observable state vector, a function of time $t$ (and/or some other parameters), and $f \in \mathbb{R}^n$ is the (in general nonlinear) system function or model. Given a set of discrete time series data $[x(t_1), x(t_2), x(t_3), ...]$ from such a system, how should a user accurately and efficiently identify the model $f$? In dynamical systems theory, this is referred to as *system identification*, although for many problems of known mathematical structure, it can be simplified into a problem of *parameter identification*. The question then leads into deeper questions concerning the purpose of the prediction of $f$, and whether it is desired to reproduce a time series exactly, or more simply to extract its important mathematical and/or statistical properties.

For linear models, many methods are available for identification of the dynamical system (1), including linear regression, the Kalman filter and autoregressive moving averages. For nonlinear models, an assortment of machine learning tools have been developed in recent years, usually based on neural networks or evolutionary computational methods, or various classification or order reduction schemes. The first group, while very useful, provide "black box" solutions which are not readily adaptable to new situations, while the second group necessarily involve the sacrificing of resolution to achieve order reduction.

Very recently, a number of researchers in dynamical and fluid flow systems have applied sparse regression methods for system identification from time series data [e.g. 1–3]. The regression is used to determine a matrix of coefficients which – when multiplied by a matrix of functional operations – can be used to reproduce the time series. Such methods generally involve a regularization technique to conduct the sparse regression. However, both the regularization term and its coefficient are usually implemented in a heuristic or *ad hoc* manner, without much fundamental guidance on how they should be selected for any particular dynamical system.

In this study, we present a Bayesian framework for the system identification (or parameter identification) of a dynamical system using the Bayesian maximum *a posteriori* (MAP) estimate, which is shown to be equivalent to a variant of Tikhonov regularization. This Bayesian reinterpretation provides a rational justification for the choices of the residual and regularization terms, respectively as the negative logarithms of the likelihood and prior functions. The Bayesian approach can be readily extended to the full apparatus of the Bayesian inverse solution, for example to quantify the uncertainty in the model parameters, or even to explore the functional form of the posterior. In this study, we compare the prominent regularization method of least squares regression with thresholding (the SINDy algorithm) to two Bayesian methods, by application to the Lorenz system with added Gaussian noise. We demonstrate an advantage of the Bayesian methods, in their ability to calculate the variances of the inferred parameters, thereby giving the estimated model errors.

## 2. Theoretical Foundations

In recent years, a number of researchers have implemented sparse regression methods for the system identification of a variety of dynamical systems [e.g. 1–3]. The method proceeds from a recorded time series, which for $m$ time steps of an $n$-dimensional parameter $x$ is assembled into the $m \times n$ matrix:

$$X = \begin{bmatrix} x^\top(t_1) \\ \vdots \\ x^\top(t_m) \end{bmatrix} = \begin{bmatrix} x_1(t_1) & \cdots & x_n(t_1) \\ \vdots & & \vdots \\ x_1(t_m) & \cdots & x_n(t_m) \end{bmatrix}, \qquad (2)$$

and similarly for the time derivative:

$$\dot{X} = \begin{bmatrix} \dot{x}^\top(t_1) \\ \vdots \\ \dot{x}^\top(t_m) \end{bmatrix} = \begin{bmatrix} \dot{x}_1(t_1) & \cdots & \dot{x}_n(t_1) \\ \vdots & & \vdots \\ \dot{x}_1(t_m) & \cdots & \dot{x}_n(t_m) \end{bmatrix}. \qquad (3)$$

The user then chooses an alphabet of $c$ functions, which are applied to $X$ to populate a $m \times c$ matrix library, for example of the form:

$$\Theta(X) = \begin{bmatrix} 1 & X & X^2 & X^3 & \ldots & \sin(X) & \cos(X) & \ldots \end{bmatrix}, \tag{4}$$

in this case based on polynomial and trigonometric functions. The time series data for the dynamical system (1) are then analyzed by the matrix product:

$$\dot{X} = \Theta(X)\Xi, \tag{5}$$

in which $\Xi$ is a $c \times n$ matrix of coefficients $\xi_{ij} \in \mathbb{R}$. The matrix $\Xi$ is commonly computed by inversion of (5) using sparse regression. This generally involves a minimization equation of the form:

$$\hat{\Xi} = \arg\min_{\Xi} J(\Xi), \tag{6}$$

where $\hat{\ }$ indicates an inferred value, based on an objective function consisting of residual and regularization terms:

$$J(\Xi) = ||\dot{X} - \Theta(X)\Xi||_\beta^\alpha + \lambda ||\Xi||_\gamma^\alpha, \tag{7}$$

where $||\cdot||_p$ is the $p$ norm, $\lambda \in \mathbb{R}$ is the regularization coefficient and $\alpha, \beta, \gamma \in \mathbb{R}$ are constants. For dynamical system identification, (6)-(7) have been variously implemented with $\alpha \in \{1,2\}$, $\beta = 2$ and $\gamma \in \{0, [1,2]\}$ [e.g. 2–6]. Instead of (7), to enforce a sparse solution, some authors have implemented least squares regression with iterative thresholding, known as the sparse identification of nonlinear dynamics (SINDy) method [1]:

$$J(\Xi) = ||\dot{X} - \Theta(X)\Xi||_2^2 \quad \text{with} \quad |\xi_{ij}| \geq \lambda, \forall \xi_{ij} \in \Xi. \tag{8}$$

This has been shown to converge to (7) with $\alpha = \beta = 2$ and $\gamma = 0$ [7]. Other authors have implemented an objective function containing an information criterion, to preferentially select models with fewer parameters [2]. The above methods have been shown to have strong connections to the mathematical methods of singular value decomposition (SVD), dynamic mode decomposition (DMD) and Koopman analysis using various Koopman operators [e.g. 8–10].

In the Bayesian approach to this problem [e.g. 11–13], it is recognized that instead of (5), the time series decomposition should be written explicitly as:

$$\dot{X} = \Theta(X)\Xi + \epsilon, \tag{9}$$

where $\epsilon$ is a noise or error term, representing the uncertainty in the measurement data. The variables $\dot{X}, X, \Xi$ and $\epsilon$ are considered to be probabilistic, each represented by a probability density function (pdf) defined over their applicable domain. Instead of trying to invert (9), the Bayesian considers the posterior probability of $\Xi$ given the data, as given by Bayes' rule:

$$p(\Xi|\dot{X}) = \frac{p(\dot{X}|\Xi)p(\Xi)}{p(\dot{X})} \propto p(\dot{X}|\Xi)p(\Xi). \tag{10}$$

The simplest Bayesian method is to consider the maximum *a posteriori* (MAP) estimate of $\Xi$, given by maximization of (10):

$$\hat{\Xi} = \arg\max_{\Xi} p(\Xi|\dot{X}). \tag{11}$$

For greater fidelity, it is convenient to consider the logarithmic maximum instead of (11), hence from (10):

$$\hat{\Xi} = \arg\max_{\Xi} \left[\ln p(\Xi|\dot{X})\right] = \arg\max_{\Xi} \left[\ln p(\dot{X}|\Xi) + \ln p(\Xi)\right]. \tag{12}$$

If we now make the simple assumption of unbiased multivariate Gaussian noise with covariance matrix $\Gamma$, we have:

$$p(\epsilon|\Xi) = \mathcal{N}(0,\Gamma) = \frac{\exp\left(-\frac{1}{2}\epsilon^T \Gamma^{-1}\epsilon\right)}{\sqrt{(2\pi)^n \det \Gamma}}, \tag{13}$$

where det is the determinant. The numerator can be written as [13]

$$p(\epsilon|\Xi) \propto \exp\left(-\frac{1}{2}||\epsilon||^2_{\Gamma^{-1}}\right), \tag{14}$$

where $||\epsilon||^2_A = \epsilon^T A \epsilon$ is the norm defined by the $A$ bilinear product. From (9), this gives the likelihood

$$p(\dot{X}|\Xi) \propto \exp\left(-\frac{1}{2}||\dot{X} - \Theta(X)\Xi||^2_{\Gamma^{-1}}\right). \tag{15}$$

If we also assign a multivariate Gaussian prior with covariance matrix $\Sigma$

$$p(\Xi) = \mathcal{N}(0,\Sigma) \propto \exp\left(-\frac{1}{2}||\Xi||^2_{\Sigma^{-1}}\right), \tag{16}$$

then the MAP estimator (12) becomes [13]:

$$\hat{\Xi} = \arg\max_{\Xi}\left[\ln\exp\left(-\frac{1}{2}||\dot{X} - \Theta(X)\Xi||^2_{\Gamma^{-1}}\right) + \ln\exp\left(-\frac{1}{2}||\Xi||^2_{\Sigma^{-1}}\right)\right]$$

$$= \arg\max_{\Xi}\left[-\frac{1}{2}||\dot{X} - \Theta(X)\Xi||^2_{\Gamma^{-1}} - \frac{1}{2}||\Xi||^2_{\Sigma^{-1}}\right] \tag{17}$$

$$= \arg\min_{\Xi}\left[||\dot{X} - \Theta(X)\Xi||^2_{\Gamma^{-1}} + ||\Xi||^2_{\Sigma^{-1}}\right].$$

We see that the Bayesian MAP provides a minimization formula based on an objective function, which is remarkably similar to that used in the regularization method (6)-(7). Indeed, for isotropic variances of the noise $\Gamma = \sigma_\epsilon^2 I$ and prior $\Sigma = \sigma_\Xi^2 I$, where $I$ is the identity matrix, (17) reduces to the common regularization formula (6)-(7) with $\alpha = \beta = \gamma = 2$ and $\lambda = \sigma_\epsilon^2/\sigma_\Xi^2$ [11].

In Bayesian inference, any additional parameters can also be incorporated into the inferred posterior pdf. In the present study, the covariance matrices $\Gamma$ of the noise in (14) and $\Sigma$ of the prior in (16) are unknown. It is desirable to determine these directly from the Bayesian inversion process. Using the above simple model of isotropic variances, the posterior can be written as:

$$p(\Xi, \sigma_\epsilon^2, \sigma_\Xi^2 | \dot{X}) \propto p(\dot{X}|\Xi) p(\Xi|\sigma_\Xi^2) p(\sigma_\epsilon^2) p(\sigma_\Xi^2). \tag{18}$$

In the Bayesian joint maximum *a posteriori* (JMAP) algorithm, (18) is maximized with respect to $\Xi, \sigma_\epsilon^2$ and $\sigma_\Xi^2$, to give the estimated parameters $\hat{\Xi}, \hat{\sigma}_\epsilon^2$ and $\hat{\sigma}_\Xi^2$. In the variational Bayesian approximation (VBA), the posterior in (18) is approximated by $q(\Xi, \sigma_\epsilon^2, \sigma_\Xi^2) = q_1(\Xi) q_2(\sigma_\epsilon^2) q_3(\sigma_\Xi^2)$. The individual MAP estimates of each parameter are then calculated iteratively, using a Kullback-Leibler divergence $K = \int q \ln(q/p) \, d\Xi d\sigma_\epsilon^2 d\sigma_\Xi^2$ as the convergence criterion.

## 3. Application

To compare the traditional and Bayesian methods for dynamical system identification, a number of time series of the Lorenz system were generated and analyzed by several regularization methods, including SINDy, JMAP and VBA. The Lorenz system is described by the nonlinear equation [14]:

$$\frac{dx}{dt} = f(x) = [\sigma(y-x), x(\rho-z) - y, xy - \beta z]^T, \tag{19}$$

with parameter values $[\sigma, \rho, \beta]$ commonly assigned to $[10, \frac{8}{3}, 28]$ to generate chaotic behavior with a strange attractor. The analyses were conducted in Matlab 2018a on a MacBook Pro with 2.8 GHz Intel Core i7, with numerical integration by the ode45 function, using a time step of 0.01 and total time of 100. The calculated position data $X$ were then augmented by additive random noise, drawn from the standard normal distribution multiplied by a scaling parameter of 0.2. The regularization processes were then executed using a modified version of the published SINDy code and other utility functions [2], and modified forms of the JMAP and VBA functions implemented previously [11] with parameters $a_0 = 10^8$ and $b_0 = 10^{-8}$. For comparisons, the inferred parameters were then used to recalculate the time series and derivatives by a further function call. In the Bayesian algorithms, the estimated variances of the parameters and the prior were also calculated, assuming inverse gamma distributions for the variance priors; for JMAP this has an analytical solution, while for VBA the solution is found iteratively using a minimum Kullback-Leibler convergence criterion [11].

## 4. Results

The calculated noisy data for the Lorenz system are illustrated in Figure 1a,b, respectively for the parameter values and their derivatives. The calculated regularization results are then presented in Figures 2–4, respectively for the SINDy, JMAP and VBA methods. In each of these plots, the first subplot illustrates the difference in each inferred parameter (i.e., $\tilde{\zeta}_{ij} - \hat{\zeta}_{ij}$), while the second subplot gives the inferred time series of the parameters $X$, showing the noisy time series $x(t)$, the inferred series $\hat{x}(t)$ and their differences.

As evident in these plots, the three methods were approximately as effective in selection of the coefficients to recreate the Lorenz system. Of the other regularization methods published by [2], the iterative hard thresholding least squares and orthogonal matching pursuit also performed well, while the LASSO algorithm was unsuccessful for any system examined.

As noted, the two Bayesian methods also provided the variances of the predicted parameters, shown in Figures 3a and 4a as error bars corresponding to the standard deviations. These calculations indicate the inferred parameter errors to be larger than previously appreciated, for example $\pm 1.878 \times 10^{-10}$ in the coefficient of $x$ in all three series predicted by both JMAP and VBA. These values give a more realistic estimate of the inherent errors in the system identification method than suggested by the SINDy regularization.

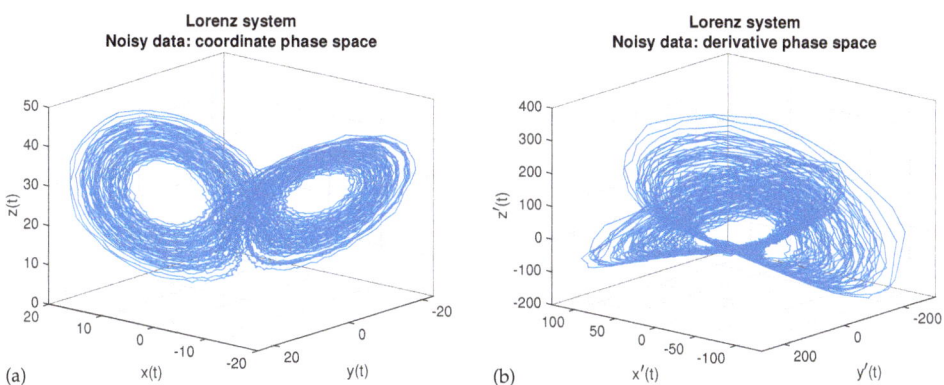

**Figure 1.** Calculated noisy data for the Lorenz system: (**a**) parameters $X$, and (**b**) derivatives $\dot{X}$.

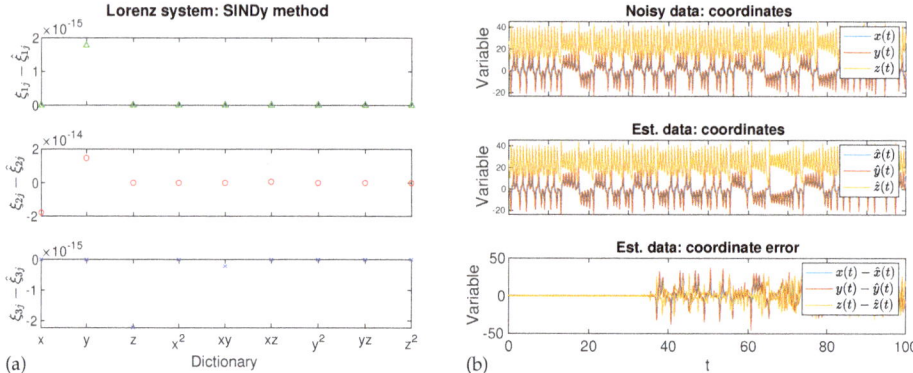

**Figure 2.** Output of SINDy regularization: (**a**) differences in predicted parameters $\zeta_{ij} - \hat{\zeta}_{ij}$, and (**b**) comparison of original and predicted time series $X$.

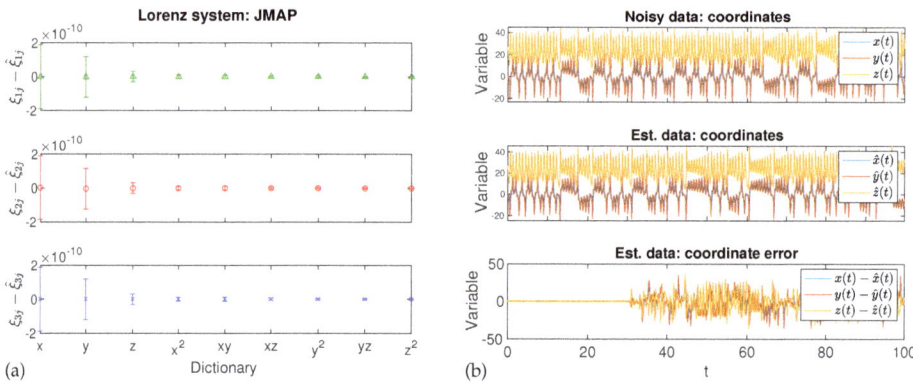

**Figure 3.** Output of JMAP regularization: (**a**) differences in predicted parameters $\zeta_{ij} - \hat{\zeta}_{ij}$ (the error bars indicate inferred standard deviations), and (**b**) comparison of original and predicted time series $X$

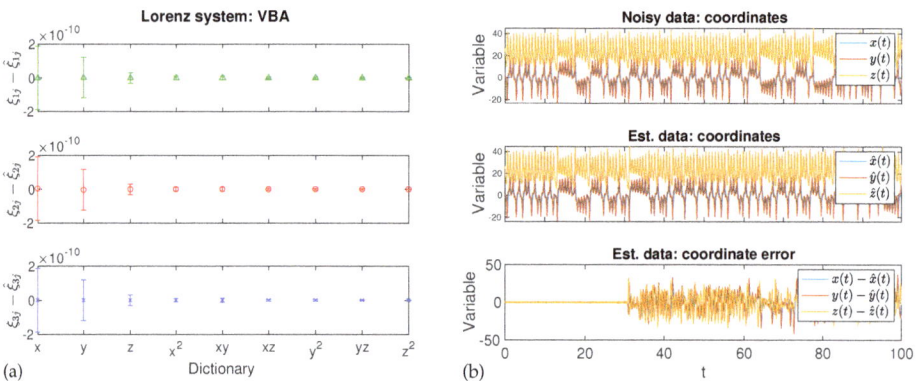

**Figure 4.** Output of VBA regularization: (**a**) differences in predicted parameters $\zeta_{ij} - \hat{\zeta}_{ij}$ (the error bars indicate inferred standard deviations), and (**b**) comparison of original and predicted time series $X$.

## 5. Conclusions

We examine the problem of system identification of a dynamical system, represented by a nonlinear equation system $dx/dt = f(x)$, from discrete time series data. For this, we present a

Bayesian inference framework based on the Bayesian maximum *a posteriori* (MAP) estimate, which for the assumption of Gaussian likelihood and prior functions, is shown to be equivalent to a variant of Tikhonov regularization. This Bayesian reinterpretation provides a clear theoretical justification for the choices of the residual and regularization terms, respectively as the negative logarithms of the likelihood and prior functions. The Bayesian approach is readily extended to the full apparatus of the Bayesian inverse solution, for example to quantify the uncertainty in the model parameters, or even to explore the functional form of the posterior pdf.

In this study, we compare the regularization method of least squares regression with thresholding (the SINDy algorithm) to two Bayesian methods JMAP and VBA, by application to the Lorenz system with added Gaussian noise. The Bayesian methods are shown to perform almost as effectively as SINDy for parameter estimation and reconstruction of the Lorenz time series. More importantly, the Bayesian methods also provide the variances – hence the standard deviations – of the inferred parameters, thereby giving a mathematical estimate of the system identification error. This is an important advantage of the Bayesian approach over traditional regularization methods.

**Funding:** This research was funded by the Australian Research Council Discovery Projects grant DP140104402, and also supported by French sources including Institute Pprime, CNRS, Poitiers, France, and CentraleSupélec, Gif-sur-Yvette, France.

**Conflicts of Interest:** The authors declare no conflict of interest. The funders had no role in the design of the study; in the collection, analyses, or interpretation of data; in the writing of the manuscript, or in the decision to publish the results.

## References

1. Brunton, S.L.; Proctor, J.L.; Kutz, J.N. Discovering governing equations from data by sparse identification of nonlinear dynamical systems. *Proc. Natl. Acad. Sci.* **2016**, *113*, 3932–3937.
2. Mangan, N.M.; Kutz, J.N.; Brunton, S.L.; Proctor, J.L. Model selection for dynamical systems via sparse regression and information criteria. *Roy. Soc. Proc. A* **2017**, *473*, 20170009.
3. Rudy, S.H.; Brunton, S.L.; Proctor, J.L.; Kutz, J.N. Data-driven discovery of partial differential equations. *Sci. Adv.* **2017**, *3*, e1602614.
4. Tikhonov, A.N. Solution of incorrectly formulated problems and the regularization method. *Dokl. Akad. Nauk SSSR* **1963**, *151*, 501–504. (In Russian)
5. Santosa, F.; Symes, W.W. Linear inversion of band-limited reflection seismograms. *SIAM J. Sci. Stat. Comp.* **1986**, *7*, 1307–1330.
6. Tibshirani, R. Regression shrinkage and selection via the Lasso. *J. R. Stat. Soc. B* **1996**, *58*, 267–288.
7. Zhang, L.; Schaeffer, H. On the convergence of the SINDy algorithm. *arXiv* **2018**, arXiv:1805.06445v1.
8. Brunton, S.L.; Brunton, B.W.; Proctor, J.L.; Kaiser, E.; Kutz, J.N. Koopman invariant subspaces and finite linear representations of nonlinear dynamical systems for control. *PLoS ONE* **2016**, *11*, e0150171.
9. Brunton, S.L.; Brunton, B.W.; Proctor, J.L.; Kaiser, E.; Kutz, J.N. Chaos as an intermittently forced linear system. *Nat. Comm.* **2017**, *8*, 19.
10. Taira, K.; Brunton, S.L.; Dawson, S.T.M.; Rowley, C.W.; Colonius, T.; McKeon, B.J.; Schmidt, O.T.; Gordeyev, S.; Theofilis, V.; Ukeiley, L.S. Modal analysis of fluid flows: An overview. *AIAA J.* **2017**, *55*, 4013–4041.
11. Mohammad-Djafari, A. Inverse problems in signal and image processing and Bayesian inference framework: From basic to advanced Bayesian computation. In Proceedings of the Scube Seminar, L2S, CentraleSupelec, At Gif-sur-Yvette, France, 27 March 2015.
12. Mohammad-Djafari, A. Approximate Bayesian computation for big data. In Proceedings of the Tutorial at MaxEnt 2016, Ghent, Belgium, 10–15 July 2016.
13. Teckentrup, A. Introduction to the Bayesian approach to inverse problems. In Proceedings of the MaxEnt 2018, Alan Turing Institute, UK, 6 July 2018.
14. Lorenz, E.N. Deterministic nonperiodic flow. *J. Atmos. Sci.* **1963**, *20*, 130–141.

 © 2020 by the authors. Licensee MDPI, Basel, Switzerland. This article is an open access article distributed under the terms and conditions of the Creative Commons Attribution (CC BY) license (http://creativecommons.org/licenses/by/4.0/).

*Proceedings*

# Estimating Flight Characteristics of Anomalous Unidentified Aerial Vehicles in the 2004 Nimitz Encounter †

**Kevin H. Knuth** [1,2,*], **Robert M. Powell** [2], **and Peter A. Reali** [2]

1. Department of Physics, University at Albany (SUNY), Albany, NY 12206, USA
2. Scientific Coalition for UAP Studies (SCU), Fort Myers, FL 33913, USA; robertmaxpowell@gmail.com (R.M.P.); preali@cableone.net (P.A.R.)
* Correspondence: kknuth@albany.edu
† Presented at the 39th International Workshop on Bayesian Inference and Maximum Entropy Methods in Science and Engineering, Garching, Germany, 30 June–5 July 2019.

Published: 16 December 2019

**Abstract:** A number of Unidentified Aerial Phenomena (UAP) encountered by military, commercial, and civilian aircraft have been reported to be structured craft that exhibit 'impossible' flight characteristics. We consider the 2004 UAP encounters with the *Nimitz* Carrier Group off the coast of California, and estimate lower bounds on the accelerations exhibited by the craft during the observed maneuvers. Estimated accelerations range from 75 g to more than 5000 g with no observed air disturbance, no sonic booms, and no evidence of excessive heat commensurate with even the minimal estimated energies. In accordance with observations, the estimated parameters describing the behavior of these craft are both anomalous and surprising. The extreme estimated flight characteristics reveal that these observations are either fabricated or seriously in error, or that these craft exhibit technology far more advanced than any known craft on Earth. In the case of the *Nimitz* encounters the number and quality of witnesses, the variety of roles they played in the encounters, and the equipment used to track and record the craft favor the latter hypothesis that these are technologically advanced craft.

**Keywords:** UAP; UAV; UFO

---

## 1. Introduction

Unidentified Aerial Phenomena (UAPs) partially identified as being unknown anomalous aircraft, referred to as Unidentified Anomalous Vehicles (UAVs) or Unidentified Flying Objects (UFOs), have been observed globally for some time [1]. Such phenomena were studied officially by the United States Air Force in a series of projects: Project Sign (1947), Project Grudge (1949) and Project Blue Book (1952–1969) [2]. Other nations, such as Australia, Brazil, Canada, Chile [3], Denmark, France, New Zealand, Russia (the former Soviet Union), Spain, Sweden, the United Kingdom, Uruguay, and the Vatican have also conducted studies, or are currently studying, UAPs [4]. In December of 2017 it was revealed that the United States government had been studying UAPs through at least one secret program called the Anomalous Aerospace Threat Identification Program (AATIP) [5], and that there have been times at which United States Naval pilots have had to deal with nearly daily encounters with UAVs [6,7]. These unidentified craft typically exhibit anomalous flight characteristics, such as traveling at extremely high speeds, changing direction or accelerating at extremely high rates, and hovering motionless for long periods of time. Furthermore, these craft appear to violate the laws of physics in that they do not have flight or control surfaces, any visible means of propulsion apparently violating Newton's Third Law, and can operate in multiple media, such as space (low Earth orbit), air, and water without apparent hindrance, sonic booms, or heat dumps [4].

The nature, origin, and purpose of these UAVs are unknown. It is also not known if they are piloted, controlled remotely, or autonomous. If some of these UAVs are of extraterrestrial origin, then it would be important to assess the potential threat they pose [4]. More interestingly, these UAVs have the potential to provide new insights into aerospace engineering and other technologies [8]. The potential of a serious threat as well as the promise of advancements in science and engineering, along with our evolving expectations about extraterrestrial life are important reasons for scientists to seriously study and understand these objects [9–13]. We carefully examine a series of encounters in 2004 by pilots and radar operators of the *Nimitz* carrier group, and estimate lower bounds on their accelerations. We demonstrate that the estimated accelerations are indeed extraordinary and surprising.

## 2. *Nimitz* Encounters (2004)

For a two week period in November of 2004, the U.S. Navy's Carrier Strike Group Eleven (CSG-11), which includes the USS *Nimitz* nuclear aircraft carrier and the Ticonderoga-class guided missile cruiser USS *Princeton*, encountered as many as 100 UAVs. We estimated the accelerations of UAVs relying on (1) radar information from USS *Princeton* former Senior Chief Operations Specialist Kevin Day; (2) eyewitness information from CDR David Fravor, commanding officer of Strike Fighter Squadron 41 and the other jet's weapons system operator, LCDR Jim Slaight; and (3) analyses of a segment of the Defense Intelligence Agency-released Advanced Targeting Forward Looking Infrared (ATFLIR) video. The following descriptions of the *Nimitz* encounters were summarized from the more detailed study published by the Scientific Coalition for UAP Studies (SCU) [14].

### 2.1. Senior Chief Operations Specialist Kevin Day (RADAR)

An important role of the USS *Princeton* is to act as air defense protection for the strike group. The *Princeton* was equipped with the SPY-1 radar system which provided situational awareness of the surrounding airspace. The main incident occurred on 14 November 2004, but several days earlier, radar operators on the USS *Princeton* were detecting UAVs appearing on radar at about 80,000+ feet altitude to the north of CSG-11 in the vicinity of Santa Catalina and San Clemente Islands. Senior Chief Kevin Day informed us that the Ballistic Missile Defense (BMD) radar systems had detected the UAVs in low Earth orbit before they dropped down to 80,000 feet [15]. The UAVs would arrive in groups of 10 to 20, subsequently drop down to 28,000 feet with a several hundred foot variation, and track south at a speed of about 100 knots [15]. Periodically, the UAVs would drop from 28,000 feet to sea level (approx. 50 feet), or under the surface, in 0.78 seconds. Without detailed radar data, it is not possible to know the acceleration of the UAVs as a function of time as they descended to the sea surface. However, one can estimate a lower bound on the acceleration by assuming that the UAVs accelerated at a constant rate halfway and then decelerated at the same rate for the remaining distance so that

$$\frac{1}{2}d = \frac{1}{2}a\left(\frac{t}{2}\right)^2. \tag{1}$$

The data consisted of the change in altitude $y \pm \sigma_y = 8530 \pm 90$ m ($-28,000$ ft $\pm 295$ ft) and the duration $t' \pm \sigma_t = 0.78 \pm 0.08$ s, where the goal was to estimate the acceleration, $a$. The dominant source of uncertainty in altitude was due to the observed variation in altitude among the observed UAVs, which was on the order of 200 to 300 ft.

In the first analysis, we assigned a joint Gaussian likelihood, $P(y, t|a, I)$ for the measured altitude change and the duration of the maneuver. Since the altitude change and the duration are independently measured, the joint likelihood is factored into the product of two likelihoods, and one can marginalize over the duration of the maneuver to obtain a likelihood for the altitude $y$

$$P(y\,|\,a,I) = \int_{-\infty}^{\infty} dt\, P(y,t\,|\,a,\sigma_y,t',\sigma_t,I) \tag{2}$$

$$= \int_{-\infty}^{\infty} dt\, P(y\,|\,a,t,\sigma_y,I)P(t\,|\,t',\sigma_t,I), \tag{3}$$

where the symbol $I$ represents the fact that these probabilities are conditional on all prior information. Assigning Gaussian likelihoods we have that

$$P(y\,|\,a,I) = \int_{-\infty}^{\infty} dt\, \frac{1}{\sqrt{2\pi}\sigma_y} \exp\left[-\frac{1}{2\sigma_y^2}\left(y+\frac{1}{4}at^2\right)^2\right] \frac{1}{\sqrt{2\pi}\sigma_t} \exp\left[-\frac{1}{2\sigma_t^2}(t-t')^2\right] \tag{4}$$

$$= \frac{1}{2\pi\sigma_y\sigma_t} \int_{-\infty}^{\infty} dt\, \exp\left[-\frac{1}{2\sigma_y^2}\left(y+\frac{1}{4}at^2\right)^2 - \frac{1}{2\sigma_t^2}(t-t')^2\right]. \tag{5}$$

The integrand is the exponential of a quartic polynomial in $t$, which was solved numerically. Assigning a uniform prior probability for the acceleration over a wide range of possible accelerations results in a posterior that is proportional to the likelihood (5) above resulting in a maximum likelihood analysis, which gave an estimate of $a = 5600^{+2270}_{-1190}$ g, as illustrated in Figure 1A.

As a second analysis, we employed sampling for which the change in altitude and the elapsed time were described by Gaussian distributions with $y \pm \sigma_y = 8530 \pm 90$ m and $t' \pm \sigma_t = 0.78 \pm 0.08$ s, respectively. The most probable acceleration was $5370^{+1430}_{-820}$ g while the mean acceleration was 5950 g (Figure 1B).

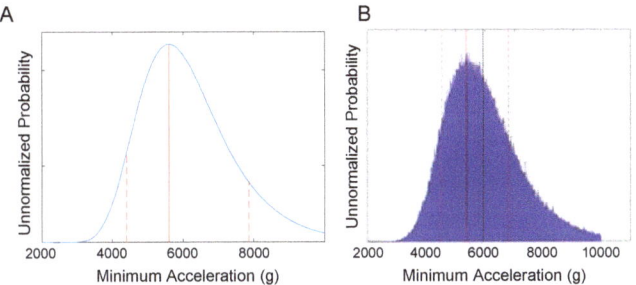

**Figure 1.** An analysis of Senior Chief Day's radar observations. (**A**) The posterior probability indicates the maximum likelihood estimate of the acceleration to be $5600^{+2270}_{-1190}$ g. (**B**) The accelerations obtained by sampling resulted in the most probable acceleration of $5370^{+1430}_{-820}$ g (red lines) while the mean acceleration is 5950 g (black dotted line).

With acceleration estimates in hand, we obtained a ballpark estimate of the power involved to accelerate the UAV. Of course, this required an estimate of the mass of the UAV, which we did not have. The UAV was estimated to be approximately the same size as an F/A-18 Super Hornet, which has a weight of about 32,000 lbs, corresponding to 14,550 kg. Since we want a minimal power estimate, we took the acceleration as 5370 g and assumed that the UAV had a mass of 1000 kg. The UAV would have then reached a maximum speed of about 46,000 mph during the descent, or 60 times the speed of sound, at which point the required power peaked at a shocking 1100 GW, which exceeds the total nuclear power production of the United States by more than a factor of ten. For comparison, the largest nuclear power plant in the United States, the Palo Verde Nuclear Generating Station in Arizona, provides about 3.3 GW of power for about four million people [16].

## 2.2. Commander David Fravor (PILOT)

On Nov. 14, 2004, CSG-11 was preparing for training exercises. Two F/A-18F Super Hornets were launched from the *Nimitz* for the air defense exercise to be conducted in an area 80–150 miles SSW of San Diego. Both planes, with call signs "FastEagle01" and "FastEagle02", had a pilot and a weapons system operator (WSO) onboard. VFA-41 Squadron Commanding Officer David Fravor was piloting FastEagle01 and LCDR Jim Slaight was the WSO of FastEagle02. CDR Fravor and his wingman were headed for the Combat Air Patrol (CAP) point, which is given by predefined latitude, longitude and altitude coordinates, where they would conduct the training exercises.

About a half-hour after take-off, Senior Chief Day operating the SPY-1 radar system on the *Princeton* detected UAVs entering the training area. The training exercise was delayed and FastEagle01 and FastEagle02 were directed to intercept a UAV at a distance of 60 miles and an altitude of 20,000 feet. As the F-18s approached *merge plot*, which is the point at which the radar could not differentiate the positions of the F-18s and the UAV, Fravor and Slaight noticed a disturbed patch of water, where it appeared as if there was a large object, possibly a downed aircraft, submerged 10 to 15 feet below the surface. As they observed the disturbance from 20,000 ft, all four pilots spotted a white UAV, shaped like a large cylindrical butane tank, or a Tic-Tac candy, moving erratically back and forth, almost like a bouncing ping-pong ball making instantaneous changes in direction without changing speed. The Tic-Tac UAV was estimated to be about the size of an F-18, about 40–50 feet in length and 10–15 feet wide, but had no apparent flight surfaces or means of propulsion, and its movement had no apparent effect on the ocean surface as one would expect from something like rotor wash from a helicopter.

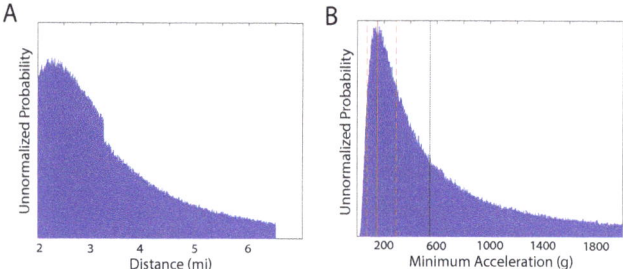

**Figure 2.** An analysis of CDR Fravor's encounter based on a Truncated Gaussian distribution ($1/30° \pm 1/60°$) of Fravor's visual acuity and a Truncated Gaussian distribution ($1 \pm 1\,s$) of elapsed time. A. Gaussian distribution of distances based on the visual acuity distribution. B. The distribution of accelerations has a maximum at $150\,^{+140}_{-80}$ g (red lines) and a mean of 550 g (black dotted line).

Fravor started a descent to investigate while his wingman kept high cover. As Fravor circled and descended, the UAV appeared to take notice of him and rose to meet him. The F-18 and the UAV circled one another. When Fravor reached the nine o'clock position, he performed a maneuver to close the distance by cutting across the circle to the three o'clock position. As he did so, the Tic-Tac UAV accelerated ([14], p.12) across Fravor's nose heading south. Fravor said that the UAV was gone within a second. As a comparison, Fravor noted that even a jet at Mach 3 takes 10 to 15 seconds to disappear from sight ([14], p.11). LCDR Slaight described the UAV as accelerating as if it was "shot out of a rifle" and that it was out of sight in a split second. ([14], p. 12).

The engagement lasted five minutes. With the Tic-Tac gone, the pilots turned their attention toward the large object in the water, but the disturbance has disappeared. The two FastEagles returned to the *Nimitz*, without sufficient fuel to attempt to pursue the Tic-Tac. On their way back, they received a call from the *Princeton* that the Tic-Tac UAV was waiting precisely at their CAP point. Senior Chief Day noted that this was surprising because those coordinates were predetermined and secret. Given that the CAP point was approximately $R = 60$ mi away, the probability of selecting the CAP point out of all the locations within the 60 mile radius, to within a one mile resolution (slightly more than the resolution of the radar system), is

$$P(x|I) = \frac{1}{\pi R^2} = \frac{1}{11310} = 0.0088\%, \qquad (6)$$

discounting the altitude. It appears that the Tic-Tac UAV intentionally went to that location, although it is not clear how this would be possible.

To obtain a lower bound on the acceleration, we assume that the UAV exhibited constant acceleration so that the distance traveled was given by

$$d = \frac{1}{2}at^2 \qquad (7)$$

during the elapsed time. The length of the Tic-Tac UAV was estimated to be about 40 ft with a cross sectional width of about $w = 10$ ft. Given that the acuity of human vision is about $\theta = 1/60°$ the UAV, at its narrowest, would be out of sight at a maximum distance of

$$d = \frac{w/2}{\tan(\theta/2)}, \qquad (8)$$

which is $d \approx 6.5$ mi. It is difficult to know what Fravor's acuity was given the viewing conditions. For this reason, we model the acuity conservatively as a truncated Gaussian distribution with a peak at $\theta = 1/30° \pm 1/60°$. The truncation at $\theta = 1/60°$ resulted in a discontinuity in the distribution of the distances (Figure 2A), which peaks around 2.25 mi.

The elapsed time is modeled as a Gaussian distribution with a mean of $1 \pm 1$ s and truncated for positive values of time. The resulting acceleration distribution was a skewed distribution of accelerations (Figure 2B) with a most probable acceleration of $150^{+140}_{-80}$ g, indicated in the figure by the red vertical lines and a mean acceleration of about 550 g indicated by the black vertical dotted line. Note that this is a lower bound, probably far below the observed acceleration if the UAV accelerated briefly as if "shot out of a rifle" and then traveled at a constant speed.

## 2.3. ATFLIR Video

Upon returning to the *Nimitz*, CDR Fravor requested that a crew equipped with the ATFLIR pod obtain videos of the Tic-Tac UAV. Two F/A-18Fs were launched under the guidance of an E-2 Hawkeye airborne radar plane. The two planes separated in search of the UAV, with one plane heading south toward the CAP point where the UAV was last seen on radar. That plane picked up a contact 33 miles to the south on the Range While Search (RWS) scan. This Tic-Tac UAV was filmed using the ATFLIR system, and the video was released to the public as the "*Nimitz* video" (Figure 3A).

We examined the last 32 frames of the *Nimitz* video in which the Tic-Tac UAV accelerated to the left and the targeting system lost lock. The video frame rate was 29.97 frames/s. As the UAV accelerates the image of the UAV becomes elongated and blurred. If the shutter speed was known, then this information could be used to better estimate the speed of the craft. This could be accomplished by treating the shutter speed as a model parameter, but such analysis is beyond the scope of this project. Instead, we concentrated on tracking the position of the right edge of the UAV and using those positions to estimate the kinematics. The left edge of the UAV was also estimated in the first frame to provide some information about the range, $z_o$, to the UAV given that that UAV was estimated to be about 40 feet in length. However, since the orientation was not known, this is modeled as a uniformly distributed unknown angular parameter $\phi \in \{0, 3\pi/8\}$, which allowed one to at least put an upper bound on the range $z_o$.

To estimate the position of the right edge of the craft in each frame (Figure 3A), the row of pixels for which the UAV has a maximum intensity was examined. The pixel intensities along that row at the right edge of the UAP were fit (maximum likelihood method with a Student-t likelihood) to half of a Gaussian curve. The center position of the Gaussian plus the standard deviation was used as the position of the right side of the UAV for that frame (Figure 3B).

Horizontal positions of the UAV are related to the pixel coordinates by noting that the entire field of view (approximately $FOV_{pix} = 606$ pixels) corresponds to an angular field of view of $0.7°$ in the narrow (NAR) mode, which is indicated in the upper left hand corner of the video. At the range, $z_o$, of the UAV this results in the proportion

$$X_{scale} = \frac{FOV_{pix}/2}{z_o \tan \frac{0.7°}{2}} \qquad (9)$$

where $X_{scale}$ has units $\frac{pixels}{m}$ when $z_o$ is in units of m. The ATFLIR has a zoom feature that can change the field of view. In the *Nimitz* video frames analyzed, the zoom is first set to unity in the NAR mode so that the angular field of view is $0.7°$. However, at frame 16, the zoom changes to two, so that the angular field of view in the NAR mode changes to $0.35°$. This appears as a discontinuity in the data ('+' symbols) illustrated in Figure 3C.

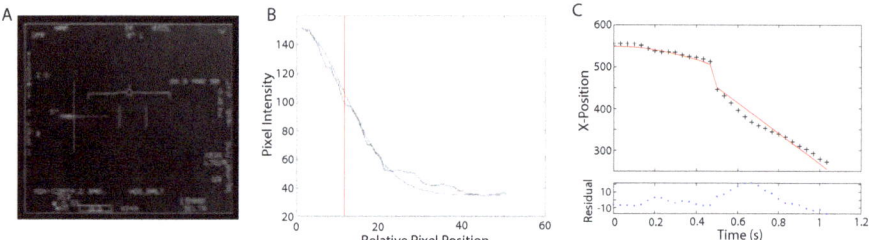

**Figure 3.** (**A**) Frame 19 of the last 32 frames of the *Nimitz* ATFLIR video. The narrow horizontal and vertical lines intersecting at the right edge of the UAP image indicate the position of the UAP. (**B**) The pixel intensities along a row of the frame are plotted along with the best Gaussian curve fit. The rightmost edge of the craft is defined as the center position of the Gaussian plus one standard deviation (indicated by the vertical red line). (**C**) This is an illustration the data (+), the most probable kinematic fit (solid curves) to the UAV positions in the *Nimitz* ATFLIR video, and the residuals (model minus data) for the model described by (11). Details can be found in Table 1.

We analyzed four different kinematic models using nested sampling, and statistically tested them by comparing the log Bayesian evidence. We used uniform prior probabilities for the kinematic parameters as well as a Student-t likelihood function, which is robust to outliers, such as those due to camera (airplane) motion. Model #1 considers constant acceleration to the left (-x direction). Model #2 considers constant acceleration both to the left (-x direction) and toward or away from the camera (z-direction). The forward model provides the position of the UAV as a function of time, where $t_i$ is the time of the $i^{th}$ video frame:

$$\text{Models \#1 and \#2} \quad \begin{cases} x(t_i) = \frac{1}{2}a_x t_i^2 + x_o \\ z(t_i) = \frac{1}{2}a_z t_i^2 + z_o \end{cases} \quad \text{const. accel.,} \qquad (10)$$

for which $a_x \in [-200, 0]$ g, $a_z \in [-100, 100]$ g, $x_o \in [-100, 100]$ px, $z_o \in [7.57, 75.75]$ mi, and Model #1 just considers the UAV's acceleration in the x-direction (to the left) so that $a_z \doteq 0$.

Models #3 and #4 describe the kinematics as constant acceleration followed by constant velocity motion after Frame 15:

$$\text{Models \#3 and \#4} \begin{cases} x(t_i) = \frac{1}{2}a_x t_i^2 + x_o & \text{for } t_i < t_{16} \\ x(t_i) = \frac{1}{2}a_x t_{15}^2 + a_x t_{15}(t_i - t_{15}) + x_o & \text{for } t_i \geq t_{16} \\ z(t_i) = \frac{1}{2}a_z t_i^2 + z_o & \text{for } t_i < t_{16} \\ z(t_i) = \frac{1}{2}a_z t_{15}^2 + a_z t_{15}(t_i - t_{15}) + z_o & \text{for } t_i \geq t_{16} \end{cases} \quad (11)$$

for which $a_x \in [-200, 0]$ g, $a_z \in [-100, 100]$ g, $x_o \in [-100, 100]$ px, $z_o \in [7.57, 75.75]$ mi, and Model #3 just considers the UAV's acceleration in the x-direction (to the left) so that $a_z \doteq 0$.

The models were analyzed using a nested sampling algorithm [17,18], which allowed for the estimation of the logarithm of the Bayesian evidence, logZ, as well as the logarithm of the likelihood, logL, and mean estimates of the model parameters. The analysis was performed for $N = 500$ samples and was run until the change in logZ from successive iterations was less than $10^{-5}$, ensuring a reliable estimate of the log evidence. Tests were performed to ensure that the trial-to-trial variations in parameter estimates were within the estimated uncertainties.

The results of the nested sampling analysis are listed in Table 1. The uncertainties in the logZ estimates (not listed) were on the order of one or less. Model 4, which describes the motion of the UAV as a constant acceleration to the left and away from the observer for the first 15 frames (approximately 0.53 s), is the most probable solution with acceleration components of $a_x = -35.64 \pm 0.08$ g and $a_z = 67.04 \pm 0.18$ g for a net acceleration of about $75.9 \pm 0.2$ g. The residuals indicate that a more precise model would consist of multiple episodes of acceleration during the maneuver. This was observed in SCU's analysis [14] where the accelerations were estimated to vary from around 40 to 80 g.

**Table 1.** Kinematic Models for the *Nimitz* Video Given the log evidence (logZ), Model 4 (**bold**) is most probable with a net acceleration of $75.9 \pm 0.2$ g.

| Model | logZ | LogL | $a_x$ (g) | $a_z$ (g) | $x_o$ (m) | $z_o$ (m) |
|---|---|---|---|---|---|---|
| Model 1 | $-253,640$ | $-253,614$ | $-71.1 \pm 0.7$ | – | $-15.40 \pm 0.04$ | $119,700 \pm 1200$ |
| Model 2 | $-236,950$ | $-236,287$ | $7.564 \pm 0.002$ | $99.994 \pm 0.005$ | $-13.36 \pm 0.04$ | $12,193 \pm 1$ |
| Model 3 | $-53,282$ | $-53,261$ | $-40.2 \pm 3.8$ | – | $-4.02 \pm 0.05$ | $49,700 \pm 4800$ |
| **Model 4** | $-52,084$ | $-52,031$ | $-35.64 \pm 0.08$ | $67.04 \pm 0.18$ | $-3.89 \pm 0.05$ | $43,870 \pm 110$ |

A more detailed analysis would involve modeling the motion of the UAV more precisely by modeling the pixel intensities on the video frames themselves. By considering the shutter speed, the blurring of the UAV image due to its motion would provide more information about its speed. In addition, the "change points" at which the accelerations changed could be treated as model parameters allowing for a more precise description of the UAV's behavior.

## 3. Discussion

In this paper, we have worked under the assumption that these UAPs were physical craft as described by the pilots. The fact that these UAPs exhibited astonishing flight characteristics leaves one searching for other possible explanations. One very clever explanation suggested by one of the reviewers was that these UAPs could have been generated by the intersection of two or more laser or maser beams ionizing the air, which could create a visual image, an infrared image, as well as a radar reflective region possibly explaining much of the observations.

While such an explanation could explain the visual, infrared and radar observations, it would not be able to explain either the suborbital radar returns from the ballistic missile defense (BMD) radar systems on the *Princeton* before the UAPs dropped to 80,000 ft, or the sonar returns when the TicTac UAPs went into the ocean [15], both of which are not as well substantiated or documented as the other observations.

More importantly, the distribution of the UAPs ranged from over 100 miles to the north over Catalina Island to about 70 miles to the west. This would require an array of widely distributed and coordinated lasers situated on multiple ships or aircraft. However, it is known that there were no other ships or airplanes in the area. In addition, the fact that the UAP reacted to CDR Fravor's maneuvers would require that radar be used to track the F-18s so that the laser-produced imagery could react to them. However, any such radar frequencies being used in the area would have been detected by the *Princeton*, the E-2 Hawkeye, and the F-18s themselves.

If any such system were being secretly tested against CSG-11, one would expect it to mimic real-life events, such as an enemy aircraft, drone, or missile launch. But the UAPs and their behavior were nothing like this. Furthermore, such powerful lasers might endanger the planes or personnel if anything went wrong in the testing, and the fact that the pilots were forced to take evasive maneuvers [19] reveals that they were being put in harms way. One wouldn't need to test a system in this manner, and if such a test did take place it would very likely have been illegal. Furthermore, such an explanation would have difficulty explaining the almost daily encounters experienced by pilots in the *Roosevelt* Carrier Group both off the coast of Virginia and during military operations in the Persian Gulf [6,7], or earlier encounters, such as that by Lt. Bethune in 1951, two years before the invention of the maser and nine years before the invention of the laser, which was analyzed in the extended version of this paper [20].

## 4. Conclusions

We have carefully considered a set of encounters between the *Nimitz* CSG-11 and UAPs of unknown nature and origin. Much of the information available consisted of eyewitness descriptions made by multiple trained witnesses observing in multiple modalities including visual contact from pilots, radar, and infrared video. While fabrication and exaggeration cannot be ruled out, the fact that multiple professional trained observers working in different modalities corroborate the reports greatly minimizes such risks.

The analysis aimed to estimate lower bounds on the acceleration. This was found by assuming that the UAVs accelerated a constant rate. We worked to obtain conservative estimates by assigning liberal uncertainties. It was found that the minimum acceleration estimates, ranging from about 70 g to well over 5000 g, far exceeded those expected for an aircraft (Table 2). For comparison, humans can endure up to 45 g for 0.044 s with no injurious or debilitating effects, but this limit decreases with increasing duration of exposure [21]. For durations more than 0.2 s the limit of tolerance decreases to 25 g and it decreases further still for longer durations [21].

**Table 2.** Summary of Estimated Accelerations ranging from about 75 g to over 5300 g. Detection Modalities refer to Multiple Pilots Visual Contact (Vs), Radar (R), Infrared Video (IR).

| Case | Detection Modalities | Kinematic Model | Figure | Min. Acceleration |
|---|---|---|---|---|
| Day | R | (1) | Figure 1B | $5370^{+1430}_{-820}$ g |
| Fravor | R,Vs | (7) | Figure 2C | $150^{+140}_{-80}$ g |
| ATFLIR | R,Vs,IR | (11) | Figure 3C | $75.9 \pm 0.2$ g |

These considerations suggest that these UAVs may not have been piloted, but instead may have been remote controlled or autonomous. However, it should be noted that even equipment can only handle so much acceleration. For example, the Lockheed Martin F-35 Lightning II has maintained structural integrity up to 13.5 g [22]. Missiles can handle much higher accelerations. The Crotale NG VT1 missile has an airframe capable of withstanding 50 g and can maintain maneuverability up to 35 g [23]. However, these accelerations are still only about half of lowest accelerations that we have estimated for these UAVs. The fact that these UAVs display no flight surfaces or apparent propulsion mechanisms, and do not produce sonic booms or excessive heat that would be released given the hundreds of GigaWatts of power that we expect should be involved, strongly suggests that these anomalous craft are taking advantage of technology, engineering, or physics that we are unfamiliar with. For example, the Tic-Tac UAV dropping from 28,000 ft to sea level in 0.78 s involved at least $4.3 \times 10^{11}$ J of energy (assuming a mass of 1000 kg), which is equivalent to about 100 tons of TNT, or the yield of 200 Tomahawk cruise missiles, released in $\frac{3}{4}$ of a second. One would have expected a catastrophic effect on the surrounding environment. This does not rule out the possibility that these UAVs have been developed by governments, organizations, or individuals on Earth, but it suggests that these UAVs and the technologies they employ may be of extraterrestrial origin. That being said, it should be strongly emphasized that proving that something is extraterrestrial would be extremely difficult, even if one had a craft in hand.

The purpose of this paper is to focus on the flight kinematics of these UAVs with the aim of building up a body of scientific evidence that will allow for a more precise understanding of their nature and origin.

As such, it is difficult to draw any useful conclusions at this point. We have characterized the accelerations of a number of UAVs and have demonstrated that if they are craft then they are indeed anomalous, displaying technical capabilities far exceeding those of our fastest aircraft and spacecraft. It is not clear that these objects are extraterrestrial in origin, but it is extremely difficult to imagine that anyone on Earth with such technology would not put it to use. Moreover, observations of similar UAPs go back to well before the era of flight [1]. Collectively, these observations strongly suggest that these UAVs should be carefully studied by scientists [9–13].

Unfortunately, the attitude that the study of UAVs (UFOs) is "unscientific" pervades the scientific community, including SETI (Search for Extraterrestrial Intelligence) [24], which is surprising, especially since efforts are underway to search for extraterrestrial artifacts in the solar system [25–29], in particular, on the Moon, Mars, asteroids [30], and at Earth-associated Lagrange points. Ironically, such attitudes inhibit scientific study, perpetuating a state of ignorance about these phenomena that has persisted for well over 70 years, and is now especially detrimental, since answers are presently needed [31–34].

**Author Contributions:** This work builds on analyses performed independently by K.H.K. and by R.M.P., P.A.R. and others [14]. For this work, K.H.K. determined the methodology, developed the software, performed the analysis, and wrote the original draft. R.M.P. and P.A.R. both reviewed and edited the work verifying correctness.

**Funding:** This research received no external funding.

**Acknowledgments:** The authors thank Kevin Day for discussing his experiences during the 2004 *Nimitz* encounters and patiently answering our numerous questions. KHK is especially grateful for the comments and suggestions made by John Skilling, as well as the careful and thoughtful recommendations made by Udo von Toussaint.

**Conflicts of Interest:** The authors declare no conflict of interest. Editorial decisions, including the decision to publish this work, were made by the MaxEnt 2019 Organizers.

## References

1. Vallee, J.; Aubeck, C. *Wonders in the Sky: Unexplained Aerial Objects from Antiquity to Modern Times*; Penguin: New York, NY, USA, 2010.
2. Unidentified Flying Objects and Air Force Project Blue Book. Available online: https://web.archive.org/web/20030624053806/http://www.af.mil/factsheets/factsheet.asp?fsID=188 (accessed on 9 June 2019).
3. CEFAA. Comité de Estudios de Fenómenos Aéreos Anómalos. Available online: http://www.cefaa.gob.cl/ (accessed on 27 July 2019).
4. Elizondo, L. The imminent change of an old paradigm: The U.S. government's involvement in UAPs, AATIP, and TTSA. In Proceedings of the Anomalous Aerospace Phenomena Conference (AAPC 2019) Presentation, Huntsville, AL, USA, 15–17 March 2019.
5. Cooper, H.; Blumenthal, R.; Kean, L. Glowing auras and "black money": The Pentagon's mysterious U.F.O. program. *The New York Times*, 2017.

6. Stieb, M. Navy pilots were seeing UFOs on an almost daily basis in 2014 and 2015: Report. *New York Magazine*, 2019. Available online: http://nymag.com/intelligencer/2019/05/navy-pilots-are-seeing-ufos-on-an-almost-daily-basis-report.html (accessed on 24 July 2019).
7. Rogoway, T. Recent UFO Encounters with Navy pilots occurred constantly across multiple squadrons. *The Drive*, 2019. Available online: https://www.thedrive.com/the-war-zone/28627/recent-ufo-encounters-with-navy-pilots-occurred-constantly-across-multiple-squadrons (accessed on 24 July 2019).
8. Monzon, I. Tech CEOs want to capture UFOs and reverse engineer them. *International Business Times*, 2019. Available online: https://www.ibtimes.com/tech-ceos-want-capture-ufos-reverse-engineer-them-2803920, (accessed on 24 July 2019).
9. Hynek, J.A. *The UFO Experience: A Scientific Inquiry*; Henry Regnery: Chicago, IL, USA, 1972.
10. Hill, P.R. *Unconventional Flying Objects: A Scientific Analysis*; Hampton Roads Publishing Co.: Charlottesville, VA, USA, 1995.
11. Sturrock, P.A. *The UFO Enigma: A New Review of the Physical Evidence*; Aspect: New York, NY, USA, 2000.
12. Knuth, K.H. Are we alone? The question is worthy of serious scientific study. *The Conversation*, 2018. Available online: https://theconversation.com/are-we-alone-the-question-is-worthy-of-serious-scientific-study-98843 (accessed on 24 July 2018).
13. Colombano, S.P. New Assumptions to Guide SETI Research. 2018. Available online: https://ntrs.nasa.gov/archive/nasa/casi.ntrs.nasa.gov/20180001925.pdf (accessed on 24 July 2018).
14. Powell, R.; Reali, P.; Thompson, T.; Beall, M.; Kimzey, D.; Cates, L.; Hoffman, R. A Forensic Analysis of Navy Carrier Strike Group Eleven's Encounter with an Anomalous aerial Vehicle. 2019. Available online: https://www.explorescu.org/post/nimitz_strike_group_2004 (accessed on 9 July 2018).
15. Day, K. (U.S. Navy (ret.)). Private Communication, 2019.
16. Palo Verde Nuclear Generating Station. Available online: https://en.wikipedia.org/wiki/Palo_Verde_Nuclear_Generating_Station (accessed on 8 August 2018).
17. Skilling, J. Nested sampling for general Bayesian computation. *Bayesian Anal.* **2006**, *1*, 833–859.
18. Sivia, D.S.; Skilling, J. *Data Analysis. A Bayesian Tutorial*, second ed.; Oxford University Press: Oxford, 2006.
19. 2004 Nimitz Pilot Report. 2017. Available online: https://thevault.tothestarsacademy.com/nimitz-report (accessed on 7 October 2018).
20. Knuth, K.H.; Powell, R.M.; Reali, P.A. Estimating Flight Characteristics of Anomalous Unidentified Aerial Vehicles. *Entropy* **2019**, *21*, 939.
21. Eiband, A.M. Human Tolerance to Rapidly Applied Accelerations: A Summary of the Literature. 1959. Available online: https://ntrs.nasa.gov/archive/nasa/casi.ntrs.nasa.gov/19980228043.pdf (accessed on 27 July 2019).
22. Kent, J. F-35 Lightning II News. 2010. Available online: http://www.f-16.net/f-35-news-article4113.html (accessed on 27 July 2019).
23. Army-Technology.com. Crotale NG Short Range Air Defence System. Available online: https://www.army-technology.com/projects/crotale/ (accessed on 27 July 2019).
24. Wright, J. Searches for technosignatures: The state of the profession. *arXiv* **2019**, arXiv:1907.07832.
25. Bracewell, R. Communications from superior galactic communities. *Nature* **1960**, *186*, 670–671.
26. Bracewell, R. Interstellar probes. In *Interstellar Communication: Scientific Perspectives*; Ponnamperuma, C., Cameron, A.G.W., Eds.; Houghton-Mifflin: Boston, MA, USA, 1974; pp. 141–167.
27. Freitas, R.A., Jr. The search for extraterrestrial artifacts (SETA). *J. Br. Interplanet. Soc.* **1983**, *36*, 501–506.
28. Tough, A.; Lemarchand, G. Searching for extraterrestrial technologies within our solar system. In *Symposium-International Astronomical Union*; Cambridge University Press: Cambridge, UK, 2004; Volume 213, pp. 487–490.
29. Haqq-Misra, J.; Kopparapu, R. On the likelihood of non-terrestrial artifacts in the Solar System. *Acta Astronaut.* **2012**, *72*, 15–20.
30. Kecskes, C. Observation of asteroids for searching extraterrestrial artifacts. In *Asteroids*; Badescu, V., Ed.; Springer: Berlin/Heidelberg, Germany, 2013; pp. 633–644.
31. Haines, R.F. *Aviation Safety in America: A Previously Neglected Factor*; NARCAP TR 01-2000; National Aviation Reporting Center on Anomalous Phenomena (NARCAP). 2000. Available online: http://www.noufors.com/Documents/narcap.pdf (accessed on 27 July 2019).

32. Bender, B.P. Senators Get Classified Briefing on UFO Sightings. Available online: https://www.politico.com/story/2019/06/19/warner-classified-briefing-ufos-1544273 (accessed on 27 July 2019).
33. Golgowski, N.H. Congress Briefed on Classified UFO Sightings as Threat to Aviator Safety, Navy Says. Available online: https://www.huffpost.com/entry/navy-briefs-congress-ufos_n_5d0baf79e4b06ad4d25cf1be (accessed on 27 July 2019).
34. Lutz, E.V.F. Congress Is Taking the UFO Threat Seriously. Available online: https://www.vanityfair.com/news/2019/06/congress-is-taking-the-ufo-threat-seriously (accessed on 27 July 2019).

© 2019 by the authors. Licensee MDPI, Basel, Switzerland. This article is an open access article distributed under the terms and conditions of the Creative Commons Attribution (CC BY) license (http://creativecommons.org/licenses/by/4.0/).

*Proceedings*

# Haphazard Intentional Sampling Techniques in Network Design of Monitoring Stations †

**Marcelo S. Lauretto [1], Rafael Stern [2], Celma Ribeiro [3] and Julio Stern [4,\*]**

1. School of Arts, Sciences and Humanities, University of Sao Paulo, 03828-000 Sao Paulo, Brazil; marcelolauretto@usp.br
2. Department of Statistics, Federal University of Sao Carlos, 13565-905 Sao Carlos, Brazil; rbstern@gmail.com
3. Polytechnic School, University of Sao Paulo, 05508-010 Sao Paulo, Brazil; celma@usp.br
4. Institute of Mathematics and Statistics, University of Sao Paulo, 05508-090 São Paulo, Brazil
\* Correspondence: jstern@ime.usp.br
† Presented at the 39th International Workshop on Bayesian Inference and Maximum Entropy Methods in Science and Engineering, Garching, Germany, 30 June–5 July 2019.

Published: 27 November 2019

**Abstract:** In empirical science, random sampling is the golden standard to ensure unbiased, impartial, or fair results, as it works as a technological barrier designed to prevent spurious communication or illegitimate interference between parties in the application of interest. However, the chance of at least one covariate showing a "significant difference" between two treatment groups increases exponentially with the number of covariates. In 2012, Morgan and Rubin proposed a coherent approach to solve this problem based on *rerandomization* in order to ensure that the final allocation obtained is balanced, but with an exponential computation cost in the number of covariates. Haphazard Intentional Sampling is a statistical technique that combines intentional sampling using goal optimization techniques with random perturbations. On one hand, it has all the benefits of standard randomization and, on the other hand, avoid exponentially large (and costly) sample sizes. In this work, we compare the haphazard and rerandomization methods in a case study regarding the re-engineering of the network of measurement stations for atmospheric pollutants. In comparison with rerandomization, the haphazard method provided groups with a better balance and permutation tests consistently more powerful.

**Keywords:** design of experiments; randomization; haphazard intentional sampling

---

## 1. Introduction

This paper addresses two related problems in the design of experiments: allocation and sampling.

The allocation problem can be illustrated with the classical example on clinical trials, see Fossaluza et al. [1]: Consider a research laboratory which wants to assess the effectiveness of a new drug for a particular disease. For this purpose, the laboratory may treat some patients with the new drug and others with a placebo. The problem of allocation consists of determining, for each patient in the trial, whether he/she will be treated with the new drug or the placebo. In order to obtain meaningful conclusions, researchers often wish the allocation to be balanced, in the sense that the distribution of some covariates (e.g., disease severity, gender, age, etc.) be the same among both treatment groups. This requirement is specially important to avoid spurious outcomes, such as different recovery rates due not to the effectiveness of each treatment, but to the imbalance in some of the covariates; for example, groups with a high proportion of patients with a mild form of the disease tend to have higher recovery rates than groups with a high proportion of patients with a severe form, even in the absence of treatment effect.

The sampling problem consists of drawing, from a (possibly large) set of sampling units or from a population, a subset for which some outcome variables shall be monitored. It is expected that the sample be a good *representative* of the whole original set, so that observations of outcomes in the sample can be used to make inferences about the whole set. As outcome variables may be influenced by some known covariates, a proxy to obtain such representative sampling is requiring that the distribution of these covariates be the same in the sample and in the remaining of complete set. In many practical applications, this problem may be considered analogous to the allocation problem, as it consists of partitioning the complete set of units into two groups—one composed by the sample and other the remaining (non monitored) units.

In both problems above, besides the requirement of obtaining well balanced groups, another fundamental requirement is that the allocation procedure be free of human ad-hoc interferences.

The standard solution for both problems is randomization, the golden standard to ensure unbiased, impartial, or fair results, see Pearl [2] and Stern [3]. Randomization works as a firewall, a technological barrier designed to prevent spurious communication of vested interests or illegitimate interference between parties in the application of interest, which may be a scientific experiment, a clinical trial, a legal case, an auditing process, or many other practical applications.

However, a common issue in randomized experiments is avoiding random allocations yielding groups that differ meaningfully with respect to relevant covariates. This is a critical issue, as the chance of at least one covariate showing a "significant difference" between two or more treatment groups increases exponentially with the number of covariates.

To overcome this issue, several authors suggest to repeat the randomization (i.e., to *rerandomize*) when it creates groups that are notably unbalanced on important covariates, see Sprott and Farewell [4], Rubin [5], Bruhn and McKenzie [6]. However, in the worst scenario, "ad hoc" rerandomization can be used to completely circumvent the haphazard, unpredictable or aimless nature of randomization, allowing a premeditated selection of a final outcome of choice, see Saa and Stern [7]. Another critique about rerandomization is that forms of analysis utilizing Gaussian distribution theory are no longer valid, as rerandomization changes the distribution of the test statistics, see Morgan and Rubin [8] and references therein.

As a response to these problems, Morgan and Rubin [8,9] proposed a coherent rerandomization approach in which the decision to rerandomize or not is based on a *pre-specified* criterion, e.g., a balance threshold. The inferential analysis of experimental data is based on a randomization test. The rerandomization procedure consists of the following steps:

1. Select units for the comparison of treatments, and collect covariate data on all units.
2. Define an explicit criterion for covariate balance.
3. Randomize the units to treatment groups.
4. Check covariate balance and return to Step 3 if the allocation is unacceptable according to the criterion specified in Step 2; continue until the balance is acceptable.
5. Conduct the experiment using the final randomization obtained in Step 4.
6. Perform inference (using a randomization test that follows exactly Steps 2–4).

Such approach aims to ensure balanced allocations, avoid subjective rejection criteria and provide sound inference procedures.

Despite the benefits of the above approach, it can be hard to use it in a way that yields a highly balanced allocation at a low computational cost. For example, in a problem of allocation into two groups, the probability that a simple random sampling generates an allocation that is significantly unbalanced (at level $\alpha$) for at least one out of $d$ covariates is proportional to $1 - (1-\alpha)^d$. As a result, the expected number of rerandomizations that are required in order for the sample to be balanced in every covariate grows exponentially with the number of covariates.

The Haphazard Intentional Sampling is a statistical technique developed with the specific purpose of yielding sampling techniques that, on one hand, have all the benefits of standard randomization and, on the other hand, avoid exponentially large (and costly) sample sizes. This approach, proposed by Lauretto et al. [10,11] and Fossaluza et al. [1], can be divided into a randomization and an optimization step. The randomization step consists of creating new artificial covariates that are distributed according to a standard multivariate normal. The optimization step consists of finding the allocation that minimizes a linear combination of the imbalances in the original covariates and in the artificial covariates.

In this article, we apply the Haphazard Intentional Sampling techniques to study how to rationally re-engineer networks of measurement stations for atmospheric pollution and/or gas emissions. We show how such re-engineering or re-design can substantially decrease the operation cost of monitoring networks while providing, at the same time, support for arriving at conclusions or taking decisions with the same statistical power as in conventional setups.

## 2. Haphazard Intentional Sampling

In this section, we present the formulation of Haphazard Sampling originally presented at Lauretto et al. [11]. Let $\mathbf{X}$ denote the covariates of interest. $\mathbf{X}$ is a matrix in $\mathbb{R}^{n \times d}$, where $n$ is the number of sampling units to be allocated and $d$ is the number of covariates of interest.

An allocation consists of assigning to each unit a group chosen from a set of possible groups, $\mathcal{G}$. We denote an allocation, $\mathbf{w}$, by a $1 \times n$ vector in $\mathcal{G}^n$.

For simplicity, we assume only two groups, that is, $\mathcal{G} = \{0, 1\}$. We also assume that the number of units assigned to each group is previously defined. That is, there exist integers $n_1$ and $n_0$ such that $n_1 + n_0 = n$, $\mathbf{1} \cdot \mathbf{w}^t = n_1$ and $\mathbf{1} \cdot (\mathbf{1} - \mathbf{w})^t = n_0$.

The goal of the allocation problem is to generate an allocation that, with high probability, is close to the infimum of the imbalance between groups with respect to individual covariate values, measured by a loss function, $L(\mathbf{w}, \mathbf{X})$.

An example of loss function is the Mahalanobis distance between the covariates of interest in each group [8], defined as follows. Let $\mathbf{A}$ be an arbitrary matrix in $\mathbb{R}^{n \times m}$. Furthermore, define $\mathbf{A}^* := \mathbf{A}\mathbf{L}$, where $\mathbf{L}$ is the Cholesky decomposition [12] of the inverse of covariance matrix of $\mathbf{A}$; that is, $\text{Cov}(\mathbf{A})^{-1} = \mathbf{L}^t \mathbf{L}$. For an allocation $\mathbf{w}$, let $\overline{\mathbf{A}^*}^1$ and $\overline{\mathbf{A}^*}^0$ denote the averages of each column of $\mathbf{A}^*$ over units allocated to, respectively, groups 1 and 0:

$$\overline{\mathbf{A}^*}^1 := \frac{\mathbf{w}}{n_1} \mathbf{A}^* \quad \text{and} \quad \overline{\mathbf{A}^*}^0 := \frac{(\mathbf{1} - \mathbf{w})}{n_0} \mathbf{A}^* . \tag{1}$$

The Mahalanobis distance between the average of the column values of $\mathbf{A}$ in each group specified by $\mathbf{w}$ is defined as:

$$M(\mathbf{w}, \mathbf{A}) := m^{-1} \| \overline{\mathbf{A}^*}^1 - \overline{\mathbf{A}^*}^0 \|_2 . \tag{2}$$

In this work, the haphazard allocation consists of finding the minimum of a noisy version of the Mahalabonis loss function. Let $\mathbf{Z}$ be an artificially generated matrix in $\mathbb{R}^{n \times k}$, with elements that are independent and identically distributed according to the standard normal distribution. For a given tuning parameter, $\lambda \in [0, 1]$, the haphazard allocation consists in solving the following optimization problem:

$$\begin{aligned}
\text{minimize} \quad & (1 - \lambda) M(\mathbf{w}, \mathbf{X}) + \lambda M(\mathbf{w}, \mathbf{Z}) \\
\text{subject to} \quad & \mathbf{1} \cdot \mathbf{w}^t = n_1 \\
& \mathbf{1} \cdot (\mathbf{1} - \mathbf{w})^t = n_0 \\
& \mathbf{w} \in \{0, 1\}^n
\end{aligned} \tag{3}$$

The parameter $\lambda$ controls the amount of perturbation that is added to the original Mahalanobis loss function, $M(\mathbf{w}, \mathbf{X})$. If $\lambda = 0$, then $w^*$ is the deterministic minimizer of $M(\mathbf{w}, \mathbf{X})$. If $\lambda = 1$, then $w^*$ is the minimizer of the unrelated random loss, $M(\mathbf{w}, \mathbf{Z})$. By choosing an intermediate value of $\lambda$ (as discussed in Section 4), one can obtain $w^*$ to be a random allocation such that, with a high probability, $M(w^*, \mathbf{X})$ is close to the infimum loss.

The formulation presented in Equation (3) is a Mixed-Integer Quadratic Programming Problem (MIQP) [13] and can be solved by the use of standard optimization software. As a MIQP may be computationally very expensive if $n$ and $d$ are large, a surrogate loss function that approximates $M(\mathbf{w}, \mathbf{A})$ is a linear combination of the norms $l_1$ and $l_\infty$ as follows [14]:

$$H(\mathbf{w}, \mathbf{A}) := m^{-1} \left( \|\overline{\mathbf{A}^*}^1 - \overline{\mathbf{A}^*}^0\|_1 + \sqrt{m} \ \|\overline{\mathbf{A}^*}^1 - \overline{\mathbf{A}^*}^0\|_\infty \right) \qquad (4)$$

The minimization of this *hybrid* norm yields a Mixed-Integer Linear Programming Problem (MILP), which is computationally much less expensive than a MIQP, see Murtagh [15], Wolsey and Nemhauser [13]:

$$\begin{aligned}
\text{minimize} \quad & (1-\lambda) \ H(\mathbf{w}, \mathbf{X}) + \lambda \ H(\mathbf{w}, \mathbf{Z}) \\
\text{subject to} \quad & \mathbf{1} \cdot \mathbf{w}^t = n_1 \\
& \mathbf{1} \cdot (\mathbf{1} - \mathbf{w})^t = n_0 \\
& \mathbf{w} \in \{0, 1\}^n
\end{aligned} \qquad (5)$$

## 3. Case Study

CETESB—The Environmental Company of Sao Paulo State, maintains a network of atmospheric monitoring stations, which provide hourly records of pollutant indicators and atmospheric parameters (Raw data are freely available at http://qualar.cetesb.sp.gov.br/qualar/home.do). The problem here addressed is to select 25 of 54 candidate stations to install additional pollutant sensors which, due to their costs, could not be installed in all monitoring stations.

Eight parameters were considered to compute (and control) the Mahalanobis distance between groups: Particulate matter 10 micrometers (PM10), Nitrogen monoxide (NO), Nitrogen dioxide (NO2), Nitrogen oxides (NOx), Ozon (O3), Air temperature (Temp), Relative humidity (RH) and wind speed (WS). An R routine was adapted from Amorim [16] to collect data from CETESB web site and build a dataset with one-year observations (August 2017–July 2018). Data was summarized by taking the medians of observations separately for rainy (october–march) and dry (april–september) seasons. Station coordinates (latitude and longitude) were also considered, to induce a suitable geographic representativeness in the selected subsample. Thus, our matrix data $\mathbf{X}$ has a total of 18 covariates—8 atmospheric summaries for each rain/dry season plus station coordinates.

In our empirical study, we explore the trade-off between randomization and optimization by using well calibrated values for the parameter $\lambda$, as defined in the next equation. The transformation between parameters $\lambda$ and $\lambda^*$ is devised to equilibrate the weights given to the terms of Equations (3) and (5) corresponding to the covariates of interest and artificial, which have distinct dimensions, $d$ and $k$.

$$\lambda = \lambda^* / [\lambda^*(1 - k/d) + k/d], \quad \text{where } \lambda^* \in \{0.05, 0.1, 0.2, 0.3, 0.4\}. \qquad (6)$$

For each value of $\lambda^*$, the haphazard allocation method was repeated 500 times (each time with a fresh random matrix of artificial covariates, $\mathbf{Z}$) with a fixed processing time $t = 120$ s.

For comparison, we drew 500 allocations using the rerandomization method proposed by Morgan and Rubin [8], which in its original version consists of repeatedly drawing random allocations until $M(\mathbf{w}, \mathbf{X})$ is below a given threshold $a$. Here we use a slightly modified *fixed-time* version of this method, that chooses the allocation which yields the lowest value for $M(\mathbf{w}, \mathbf{X})$ with a given processing time budget $t = 120$ s.

Finally, as a benchmark, we also drew 500 allocations using the standard (pure) randomization.

Computational tests were conducted on a desktop computer with a processor Intel I7-4930K (3.4 Ghz, 6 cores, 2 threads/core), Motherboard ASUS P9X79 LE, 24Gb RAM DDR3 and Linux Ubuntu Desktop v.18.04. The MILP problems were solved using Gurobi v.6.5.2 [17], a high performance solver that allows us to easily control all parameters of interest. Each allocation problem—among the batch of 500 allocations per allocation method, time budget and $\lambda$ value—was distributed to one of the 12 logical cores available. The computational routines were implemented in the R environment [18].

## 4. Results

### 4.1. Balance and Decoupling

Two performance criteria are analysed for each method:

1. The *balance* criterion, measured by the Mahalanobis distance between the covariates of interest, $M(\mathbf{w}^*, \mathbf{X})$. We computed the median and 95th percentile of $M(\mathbf{w}^*, \mathbf{X})$ over the 500 allocations yielded by each method.
2. The *decoupling* criterion, which concerns the absence of a systematic bias in allocating each pair of sampling units to the same group (positive association) or to different groups (negative association). For this purpose, we use the Yule's *coefficient of colligation* [19]: for each pair of units $(i,j) \in \{1, 2, \ldots, n\}^2, i < j$, and for each pair of groups $(r, s) \in \{0, 1\}^2$, let $z_{rs}(i,j)$ denote the number of times among the 500 allocations such that the units $i$ and $j$ are assigned, respectively, to groups $r$ and $s$. The Yule coefficient for the pair $(i, j)$ is computed as

$$Y(i,j) = \frac{\sqrt{z_{00}(i,j)z_{11}(i,j)} - \sqrt{z_{01}(i,j)z_{10}(i,j)}}{\sqrt{z_{00}(i,j)z_{11}(i,j)} + \sqrt{z_{01}(i,j)z_{10}(i,j)}}. \tag{7}$$

This coefficient ranges in the interval $[-1, 1]$ and measures how often the units $(i, j)$ are allocated to the same or to different groups. It equals zero when the numbers of agreements (allocations to the same group) and disagreements (allocations to different groups) are equal; and is maximum ($-1$ or $+1$) in the presence of total negative (complete disagreement) or positive (complete agreement) association.

The closer the $Y(i, j)$ to $+1$ or $-1$, the lower the decoupling provided by the allocation method with respect to $(i, j)$. So, for comparison purposes, we computed, for each method, the median and 95th percentile of $|Y(i, j)|$ among all pairs $(i, j)$.

Table 1 shows the median and 95th percentile for the Mahalanobis distances and absolute Yule coefficients for each method. As expected, the pure randomization method yields the highest Mahalanobis distances, as it does not take into account the balance between groups. For the haphazard method, the lower the $\lambda^*$ (and therefore, the lower the random component weight), the lower the Mahalanobis distance. It can be noticed that, for all values of $\lambda^*$ considered, the haphazard method yielded the lowest values for median and 95th percentile (outperforming the rerandomization method by a factor between 2 and 3). That means that the risk of getting a very bad allocation with the haphazard method is much smaller than using the rerandomization or pure randomization methods. Regarding the Yule coefficient, the pure randomization method is the benchmark for this parameter, as it naturally precludes any systematic association between individual allocations. For the haphazard allocation method, the Yule coefficient decreases as $\lambda^*$ increases.

The choice of the most suitable value of $\lambda^*$ among the candidate values in Table 1 is based on a graphical analysis, shown in Figure 1, in which we compare the variation rates of Mahalanobis distances and Yule coefficients with respect to $\lambda^*$. It can be noticed that, whereas the Mahalanobis distance increases almost linearly with $\lambda^* \in \{0.05, 0.1, 0.2, 0.3, 0.4\}$, the Yule coefficient decreases initially very fast for $\lambda^* \leq 0.2$ but afterward gets less sensitive with respect to $\lambda^*$. This suggests that, for our case study, $\lambda^* = 0.2$ is the most suitable choice, as values downward this point yield slightly lower Mahalanobis distances, but much higher Yule coefficients; conversely, values upward this point yield only slightly lower Yule coefficients, but considerably higher Mahalanobis distances.

In comparison with rerandomization, haphazard method set with $\lambda^* = 0.2$ yielded a 95th percentile for the Mahalanobis distances 140% better (0.20 vs. 0.48), with a 95th percentile for the Yule coefficient which is 73% higher (0.45 vs. 0.26).

**Table 1.** Mahalobis distances and absolute Yule coefficients yielded by the haphazard allocation, rerandomization and pure randomization methods (500 allocations for each method).

| Method | Mahalanobis Distance | | Yule Coefficient (Absolute Value) | |
|---|---|---|---|---|
|  | Median | 95th perc. | Median | 95th perc. |
| Haphazard ($\lambda^* = 0.05$) | 0.15 | 0.17 | 0.26 | 0.71 |
| Haphazard ($\lambda^* = 0.10$) | 0.16 | 0.18 | 0.16 | 0.51 |
| Haphazard ($\lambda^* = 0.20$) | 0.18 | 0.20 | 0.12 | 0.45 |
| Haphazard ($\lambda^* = 0.30$) | 0.18 | 0.21 | 0.12 | 0.44 |
| Haphazard ($\lambda^* = 0.40$) | 0.20 | 0.22 | 0.11 | 0.43 |
| Rerandomization | 0.44 | 0.48 | 0.07 | 0.26 |
| Pure random | 1.15 | 1.40 | 0.03 | 0.07 |

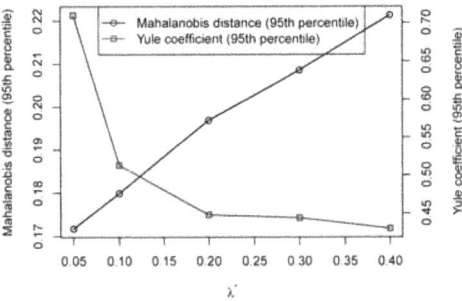

**Figure 1.** Mahalanobis distances and absolute Yule coefficients yielded by the haphazard allocation method with $\lambda^* \in \{0.05, 0.1, 0.2, 0.3, 0.4\}$.

Figure 2 illustrates the empirical distributions for the standardized difference in means for each covariate, each based on 500 simulated allocations per method. Each horizontal box plot represents, for each method and each covariate $j$, the empirical distribution of the statistics $(\overline{X}^1_{\cdot j} - \overline{X}^0_{\cdot j})/s_j$, where $\overline{X}^1_{\cdot j}$ and $\overline{X}^0_{\cdot j}$ denote the averages of the $j$-th column of $X$ over units allocated to, respectively, groups 1 and 0 (see Equation (1)); and $s_j$ is the reference scale given by the standard deviation of $\overline{X}^1_{\cdot j} - \overline{X}^0_{\cdot j}$ computed over 500 simple random allocations. It can be easily seen that differences are remarkably smaller in haphazard allocations than in rerandomization method that, in turn, are remarkably smaller than using pure randomization.

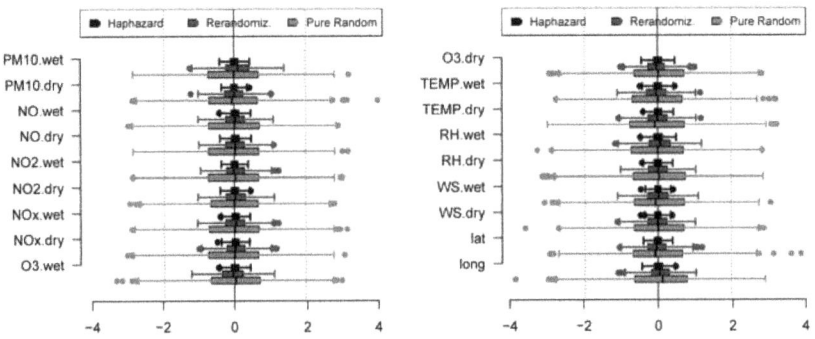

**Figure 2.** Difference between groups 0 and 1 with respect to average of standardized covariate values for each type of allocation (Adapted from Morgan and Rubin [9]).

## 4.2. Inference Power

The above measures can also be seen as a proxy for optimizing other statistical properties. For instance, one might be interested in testing the existence of a causal effect of the group assignment on a given response variable $Y$.

Consider that, for each $r \in \{0,1\}$, define $\boldsymbol{\mu}^r$ to be a $1 \times n$ vector where $\mu_i^r$ is the expected outcome for unit $i$ under treatment assign $r$, i.e., $\mu_i^r = E(Y_i | w_i = r)$. Define $\tau$ to be the true average treatment effect in the sample,

$$\tau = \frac{\mathbf{1} \cdot (\boldsymbol{\mu}^1)^t}{n} - \frac{\mathbf{1} \cdot (\boldsymbol{\mu}^0)^t}{n}. \tag{8}$$

Denoting by $\mathbf{w}$ the allocation and by $\mathbf{y}$ the vector of observations for $Y$ after units have received the corresponding treatments, $\tau$ can be estimated by:

$$\hat{\tau}_{\mathbf{w},\mathbf{y}} = \frac{\mathbf{w} \cdot \mathbf{y}^t}{n_1} - \frac{(\mathbf{1} - \mathbf{w}) \cdot \mathbf{y}^t}{n_0}. \tag{9}$$

Suppose the problem of interest is testing the hypothesis that treatment effect is null, that is, $H_0 : \tau = 0$.

A randomization test consists in simulating a reference distribution for $\hat{\tau}$ under the hypothesis $H_0$, and then estimating the probability of getting an estimate more extreme than the observed value of $\hat{\tau}_{\mathbf{w},\mathbf{y}}$. Considering a two-tailed test, a significance level $\alpha$ and the allocation method $\mathcal{M}(\mathbf{X})$ used to conduct the experiment, the randomization test follows the following steps:

1. Generate $B$ allocations $\mathbf{w}^{(1)}, \mathbf{w}^{(2)}, \ldots, \mathbf{w}^{(B)}$ using method $\mathcal{M}(\mathbf{X})$, constrained to $\mathbf{1} \cdot (\mathbf{w}^{(b)})^t = n_1$ and $\mathbf{1} \cdot (\mathbf{1} - \mathbf{w}^{(b)})^t = n_0$.
2. For each generated allocation $\mathbf{w}^{(b)}$, compute the corresponding $\hat{\tau}_{\mathbf{w}^{(b)},\mathbf{y}}$ according to Equation (9).
3. Estimate the $p$ value by

$$p \cong \frac{\sum_{b=1}^{B} I(|\hat{\tau}_{\mathbf{w}^{(b)},\mathbf{y}}| \geq |\hat{\tau}_{\mathbf{w},\mathbf{y}}|)}{B}, \tag{10}$$

   where $I(\cdot)$ is the indicator function.
4. $H_0$ is rejected if $p \leq \alpha$.

We performed a numerical experiment to assess the test power (i.e., the probability of rejecting $H_0$ when it is false) in the allocations obtained by each allocation method in this study. For this purpose, for each method $\mathcal{M}(\mathbf{X})$ and for each $\tau \in \{0, 0.1, 0.2, \ldots, 2\}$, we repeated 500 times the following procedure:

1. Generate an allocation $\mathbf{w}$ using the method $\mathcal{M}(\mathbf{X})$.
2. Simulate a response vector $\mathbf{y}$ in the following way:
   For $i \in \{1, \ldots, n\}$:
   - Draw a random number $\mu_i^0 \sim N(\theta, 1)$, where $\theta = \sum_j (X_{i,j} - \overline{X}_{.j}) / \operatorname{sd}(X_{.j})$ and $j$ indexes the columns of $\mathbf{X}$;
   - If $w_i = 0$, then set $y_i = \mu_i^0$; otherwise, set $y_i = \mu_i^0 + \tau$.
3. Apply the randomization test described above on $\mathbf{w}, \mathbf{y}$ to test $H_0 : \tau = 0$, with a significance level $\alpha = 0.05$ and $B = 500$ allocations.

For each value of $\tau$, the test power is estimated by the proportion of times the hypothesis $H_0$ has been rejected over the 500 repetitions of the procedure above. It is expected that this probability equals to $\alpha$ for $\tau = 0$, and approaches 1 as $\tau$ increases.

Figure 3 illustrates the difference of power in the allocations obtained by the haphazard ($\lambda^* = 0.2$), fixed-time rerandomization and pure randomization methods for the hypothesis $H_0 : \tau = 0$. The tests obtained using the haphazard allocations are consistently more powerful over $\tau$ than the ones obtained using the rerandomization allocations. Indeed, for $\tau \in [0.3, 1.2]$, the power yielded by haphazard allocations is more than twice the power yielded by rerandomized allocations, with a maximum factor of 3.9 for $\tau = 0.5$.

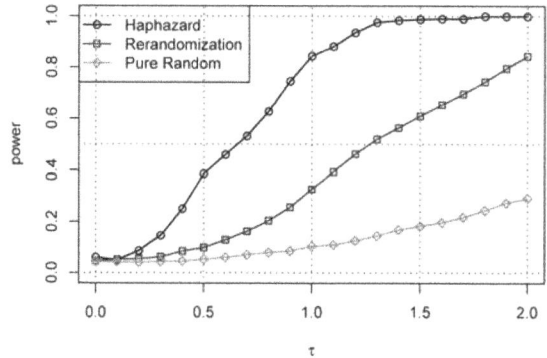

**Figure 3.** Power curves for each allocation method for testing the absence of treatment effect, $H_0 : \tau = 0$.

## 5. Conclusions

Results presented in this paper indicate that the haphazard intentional allocation method is a promising tool for design of experiments. In the numerical experiment conducted, the haphazard allocation method outperformed the alternative fixed-time rerandomization method by a factor 2.4 concerning the loss function of imbalance between the allocated groups. Besides, permutation tests using haphazard allocations are consistently more powerful than those obtained using the rerandomization allocations.

Future works shall explore the use of the Haphazard Intentional Allocation method and the Rerandomization method in applied problems in the fields of environmental monitoring, clinical trials, jurimetrics and audit procedures. We shall also explore the use of alternative surrogate Loss functions for balance performance, such as CVaR norms, Deltoidal norms and Block norms, see Pavlikov and Uryasev [20], Gotoh and Uryasev [21], Ward and Wendell [22].

**Author Contributions:** Conceptualization, J.S., M.L. and R.S.; Data acquisition preprocessing and analysis, C.R.; Programs implementation, M.L. and R.S.; Analysis of results, all authors.

**Funding:** This research was funded by FAPESP—the State of São Paulo Research Foundation (grants CEPID 2013/07375-0 and 2014/50279-4) and CNPq—the Brazilian National Counsel of Technological and Scientific Development (grant PQ 301206/2011-2).

**Acknowledgments:** The authors gratefully thank Kari L. Morgan (Penn State University) for the helpful comments and contributions to previous works, and for providing the R code of rerandomization method and useful graphs here adapted. The authors are also grateful for the support of the University of São Paulo (USP) and the Federal University of São Carlos (UFSCar).

**Conflicts of Interest:** The authors declare no conflict of interest.

## References

1. Fossaluza, V.; Lauretto, M.S.; Pereira, C.A.B.; Stern, J.M. Combining Optimization and Randomization Approaches for the Design of Clinical Trials. In *Interdisciplinary Bayesian Statistics*; Springer: New York, NY, USA, 2015; pp. 173–184.
2. Pearl, J. *Causality: Models, Reasoning, and Inference*; Cambridge University Press: Cambridge, UK, 2000.
3. Stern, J. Decoupling, Sparsity, Randomization and Objective Bayesian Inference. *Cybern. Hum. Knowing* **2008**, *15*, 49–68.
4. Sprott, D.A.; Farewell, V.T. Randomization in experimental science. *Stat. Pap.* **1993**, *34*, 89–94.

5. Rubin, D.B. Comment: The design and analysis of gold standard randomized experiments. *J. Am. Stat. Assoc.* **2008**, *103*, 1350–1353.
6. Bruhn, M.; McKenzie, D. In Pursuit of Balance: Randomization in Practice in Development Field Experiments. *Am. Econ. J. Appl. Econ.* **2009**, *1*, 200–232.
7. Saa, O.; Stern, J.M. Auditable Blockchain Randomization Tool. *arXiv* **2019**, arXiv:1904.09500.
8. Morgan, K.L.; Rubin, D.B. Rerandomization to improve covariate balance in experiments. *Ann. Stat.* **2012**, *40*, 1263–1282.
9. Morgan, K.L.; Rubin, D.B. Rerandomization to Balance Tiers of Covariates. *J. Am. Stat. Assoc.* **2015**, *110*, 1412–1421.
10. Lauretto, M.S.; Nakano, F.; Pereira, C.A.B.; Stern, J.M. Intentional Sampling by goal optimization with decoupling by stochastic perturbation. *Aip Conf. Proc.* **2012**, *1490*, 1490.
11. Lauretto, M.S.; Stern, R.B.; Morgan, K.L.; Clark, M.H.; Stern, J.M. Haphazard intentional allocation and rerandomization to improve covariate balance in experiments. *AIP Conf. Proc* **2017**, *1853*, 050003.
12. Golub, G.H.; Van Loan, C.F. *Matrix Computations*; JHU Press: Baltimore, MD, USA, 2012.
13. Wolsey, L.A.; Nemhauser, G.L. *Integer And Combinatorial Optimization*; John Wiley & Sons: Hoboken, NJ, USA, 2014.
14. Ward, J.; Wendell, R. Technical Note-A New Norm for Measuring Distance Which Yields Linear Location Problems. *Oper. Res.* **1980**, *28*, 836–844.
15. Murtagh, B.A. *Advanced Linear Programming: Computation And Practice*; McGraw-Hill International Book Co.: New York, NY, USA, 1981.
16. Amorim, W. *Web Scraping do Sistema de Qualidade do Ar da Cetesb*; R Foundation for Statistical Computing: Sao Paulo, Brazil, 2018.
17. Gurobi Optimization Inc. *Gurobi: Gurobi Optimizer 6.5 Interface*; R package version 6.5-0; Gurobi Optimization Inc.: Beaverton, OR, USA, 2015.
18. R Core Team. *R: A Language and Environment for Statistical Computing*; R Foundation for Statistical Computing: Vienna, Austria, 2018.
19. Yule, G.U. On the Methods of Measuring Association Between Two Attributes. *J. R. Stat. Soc.* **1912**, *75*, 579–652.
20. Pavlikov, K.; Uryasev, S. CVaR norm and applications in optimization. *Optim. Lett.* **2014**, *8*, 1999–2020.
21. Gotoh, J.Y.; Uryasev, S. Two pairs of polyhedral norms versus $l_p$-norms: proximity and applications in optimization. *Math. Program.* **2016**, *156*, 391–431.
22. Ward, J.; Wendell, R. Using Block Norms for Location Modeling. *Oper. Res.* **1985**, *33*, 1074–1090.

© 2019 by the authors. Licensee MDPI, Basel, Switzerland. This article is an open access article distributed under the terms and conditions of the Creative Commons Attribution (CC BY) license (http://creativecommons.org/licenses/by/4.0/).

*Proceedings*

# Determination of the Cervical Vertebra Maturation Degree from Lateral Radiography [†]

**Masrour Makaremi [1], Camille Lacaule [1] and Ali Mohammad-Djafari [2,*]**

[1] Department of Orthodontics, University of Bordeaux, 33000 Bordeaux, France; masrour@makaremi.fr (M.M.); camille_cml@hotmail.fr (C.L.)
[2] International Science Consulting and Training (ISCT), 10 rue de Montjay, 91440 Bures-sur-Yvette, France
* Correspondence: djafari@free.fr; Tel.: +33-6-22-95-42-33
[†] Presented at the 39th International Workshop on Bayesian Inference and Maximum Entropy Methods in Science and Engineering, Garching, Germany, 30 June–5 July 2019.

Published: 14 January 2020

**Abstract:** Many environmental and genetic conditions may modify jaws growth. In orthodontics, the right treatment timing is crucial. This timing is a function of the Cervical Vertebra Maturation (CVM) degree. Thus, determining the CVM is important. In orthodontics, the lateral X-ray radiography is used to determine it. Many classical methods need knowledge and time to look and identify some features to do it. Nowadays, Machine Learning (ML) and Artificial Intelligent (AI) tools are used for many medical and biological image processing, clustering and classification. This paper reports on the development of a Deep Learning (DL) method to determine directly from the images the degree of maturation of CVM classified in six degrees. Using 300 such images for training and 200 for evaluating and 100 for testing, we could obtain a 90% accuracy. The proposed model and method are validated by cross validation. The implemented software is ready for use by orthodontists.

**Keywords:** classification; orthodontics; Cervical Vertebra Maturation; Machin Learning; Artificial Intelligence; Deep Learning

---

## 1. Introduction

*1.1. Importance of the Work and Its Interest for Orthodontics Community*

Specialists in orthodontics are responsible for the treatment of dentofacial dysmorphisms, from different functional, genetical and morphological aetiologias. As a child or teenager is still growing, orthodontic treatment consists in a combination of orthodontics (about tooth position) and dentofacial orthopedics (about the guidance and stimulation of facial, maxilla and mandible growth in the three dimensions).

Many environmental and genetic conditions may induce upper or lower jaws lacks of growth. Classically, to handle a treatment properly, every etiological condition that can be modified or corrected, must be identified (diagnosis), normalized (treatment), and stabilized (retention). Specialists have to carefully examine and precisely analyze, all the medical, functional, clinical and radiographic data, in order to identify normal versus pathological conditions about tooth position, form or size, about lip, chin, cheeks, tongue and breathing functions, and about facial and jaws position and growing patterns. Adolescent orthodontic treatment also depends on proper management of jaws and facial growth, to allow a balanced jaws position, maximize the airway and improve the facial appearance [1]. Treatment planning in orthodontics depends on a systematic diagnosis and prognosis

Contemporary theories about craniofacial growth admit that the phenotype of the craniofacial complex is a result of a combination of genetic, epigenetic and environmental factors. The skeletal tissue of maxillomandibular complex is growing due to sutures and osteogenic cartilages proliferation depending on genetic, intrinsic and extrinsic environment. So facial growth can also be modified in amount and direction by extrinsic factors, including orthopedic and functional treatment. Thus, quantify facial and, in particular, mandibular growth remaining, influences diagnosis, prognosis, treatment goals and planning. Indeed, apart choosing the good appliance needed to change the rate and direction of jaw growth, the right treatment timing is crucial. If high growth rate is about to occur, orthopedic treatment may permit to correct jaws unbalanced, otherwise surgical correction of the jaws shift will be considered. The success of a dentofacial orthopedic treatment is linked to the determination of the best interventional frame (periods of accelerated or intense growth) to maximize the chances to reach skeletal goals, with adapted methods and devices, in an optimized duration.

The most common dentofacial dysmorphism, is the skeletal class II, corresponding to a short mandible. Study of normal mandibular growth and remodeling, has shown different ways of bone formation, that can be stimulated by functional and orthopedics treatments, in particular condylar growth responsible of 80% of the mandible growth. Numerous radiographic investigations have established that condylar/mandibular growth follows similar growth curve than statural growth [2]. This growth pattern is characterized by variations of growth rate in 4 stages: first a decrease of growth velocity from birth to 6 years old, then minor midgrowth spurt around 6 to 8 years, followed by a prepubertal plateau with decelerated growth rate, and finally the facial growth curve describe a peak of growth velocity corresponding at the pubertal growth spurt, which coincides, precedes or follows from 6 to 12 months the statural growth peak (controversial) [3]. This spurt occurs approximately two years earlier in girls than in boys [4].

To estimate mandibular growth potential left, the patient must be localized on is growth curve, and many biologic indicators have been proposed: increase in body height, menarche, breast and voice changes, dental development and eruption, middle phalanx maturation of the third finger, maturation of the hand and wrist, and cervical vertebral maturation [3,5–8].

*1.2. The Classical Radiographic Manual Methods*

1.2.1. Hand-Wrist Radiograph Method HWM

The comparison method describes in the Atlas of Greulich et Pyle in 1959 or the Fishman's method in 1982, permit to identify specific ossification stages occurring before, during, or after mandibular growth peak, on left hand and wrist radiographs [9,10]. The hand wrist radiographs have been used as a gold standard in the assessment of skeletal maturation for many decades, but presented several issues as: the additional X-ray exposure, the time spending and experience required (even if a digital software is now available [11]), and a sexual dimorphism and ethnic polymorphism in morphological modifications [12,13].

1.2.2. Vertebrae Maturation CVM

First who proposed to predict skeletal age and growth potential by cervical vertebrae maturation (CVM) method is LAMPARSKI in 1972. Cervical vertebrae are available on the lateral cephalometric radiographs, prescribed routinely by orthodontists for each patient diagnosis and treatment planning [14]. He has used measurements of mandibular length on several annual lateral cephalograms to describe individual mandibular growth curve, and correlated it with morphological description of vertebrae morphology at each stage. This method were modified several times first by Hassel and Farman (1995) [15], then twice by Baccetti et al. (2002 and 2005) for a more accurate assessment of cervical maturation, by 6 stages identified by morphological changes in the C2,C3,C4 vertebral bodies on a single lateral cephalogram, independently of patient gender [16].

This last version is the most used nowadays to detect the mandibular growth spurt, as it shows the best results in clinical applicability [17].

As every single bones of the human body, vertebrae growth and present maturational changes from birth to full maturity. Cervical vertebrae are the first seven pieces of the spinal column. Vertebral growth in the cartilaginous layer of the superior and inferior surfaces of each vertebrae, involves changes in size of vertebral bodies and shape of upper and lower borders of C2,C3,C4 vertebrae. These changes have been described into 6 stages, correlating with morphological modifications of the vertebral shapes and estimated time lapse from the mandibular growth peak. Both visual and cephalometric appraisals of morphological changes have been proposed.

Visual analysis [1]:

- Cervical stage 1 (CS1) = 2 years before mandibular growth peak:
  Lower borders of C2 to C4 vertebrae are flat. C3 and C4 superior borders are tapered from posterior to anterior.
- Cervical stage 2 (CS2) = 1 year before mandibular growth peak:
  Lower border of C2 presents a concavity. Bodies of C3 and C4 are the same.
- Cervical stage 3 (CS3) = during the year of the mandibular growth peak:
  Lower borders of C2 and C3 present concavities. Vertebrae are growing so C3 and C4 may be either trapezoid or rectangular shape, as superior borders are less and less tapered.
- Cervical stage 4 (CS4) = 1 or 2 years after mandibular growth peak:
  Lower borders of C2, C3 and C4 present concavities. Both C3 and C4 bodies are rectangular with horizontal superior borders longer than higher.
- Cervical stage 5 (CS5) = 1 year after the end of mandibular growth peak:
  Still concavities of lower borders of C2, C3 and C4. At least one of C3 or C4 bodies are squared and spaces between bodies are reduced.
- Cervical stage 6 (CS6) = 2 years after the end of mandibular growth peak:
  The concavities of lower borders of C2 to C4 have deepened. C3 and C4 bodies are both square or rectangular vertical in shape (bodies higher than wide).

1.2.3. Cephalometric appraisals:

Using the landmarks illustrated on Figure 1 (right), cephalometric analysis consists in the measurement of:

- The concavity depth of the lower vertebral border (estimated by the distance of the middle point (Cm) from the line connecting posterior to anterior points (Clp-Cla))
- The tapering of upper border of vertebral C3 and C4 bodies (estimated by the ratio between posterior and anterior bodies heights (Cup-Clp)/(Cua-Cla))
- The lengthening of vertebral bodies (estimated by the ratio between the bases length and anterior bodies borders height ( Clp-Cla)/Cua-Cla)

Many researchers found this method as valid and reliable as hand and wrist Xray [14]. The cervical vertebrae maturation stages have been demonstrated as a clinically useful maturation indicators for evaluation of pubertal growth height and mandibular velocities [18–20], by correlation between chronological age and cervical vertebrae maturation, between hand-wrist and cervical-vertebrae maturation [16,21–23].

Some studies underlined the need for association with other clinical assessments [24] in clinical practice, and a good reliability in differentiating pre and post mandibular growth spurt periods [25].

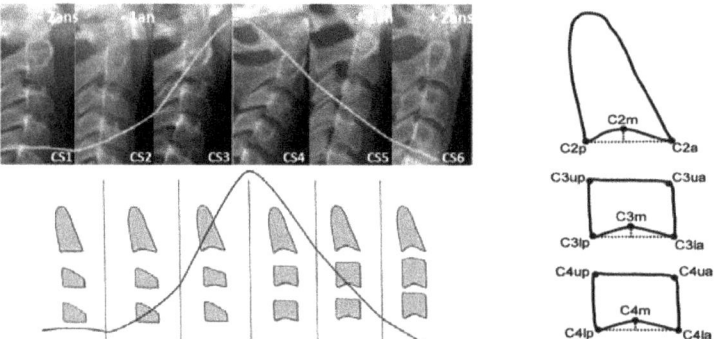

**Figure 1.** (**Left**) CVM radiological and morphological stages superposed with Björk growth curve [16], (**Right**) Cephalometric landmarks for CVM stages determination [1].

## 1.3. The Difficulties of the Labeling Task

Specific training is provided to assess CVM stages reliably, and repeatably at a satisfactory level [26,27]. Gabriel et al. minimized the risk of bias (radiographs without tracings, standardized training to private practice orthodontists...) and observed a moderate intra and inter-observer agreement (30 to 62% of cases). These results confirm the expertise required to proper determination of CVM stage, and may be explained by the use of a qualitative method of assessment, and the lack in detecting exceptional cases (individual variations in size and morphology, outside the norms defined by the method). Moreover, for orthodontists, the cervical vertebrae area on the lateral cephalograms is outside their expertise visual field. They have poor general knowledge and experience about vertebrae observation, as they focus on maxillomandibular bones and teeth at first glance. This would have been a difficulty in the labeling task of our radiographs. All lateral radiographs have been labeled by a radiologic technician, specialized in cephalometric tracing and over trained in CVM stages agreement (3 years full time), using a standardized morphologic and cephalometric protocol. Intra observer reproducibility must be estimated in further study.

## 1.4. The Need for Automatization and the Help Which It Brings

Estimation of CVM stage represents only one single element influencing the patient orthodontic treatment. The practitioner must master the entire clinical, functional, biomechanical and cephalometric data analysis in order to define proper diagnosis and treatment goals and planning. Even in being specialists, orthodontists require a very broad range of skills and a great deal of time for each patient complete diagnosis. Considering that reproducibility of classifying CVM stages is superior at 98% by trained examiners [1], automatization by expert eyes will provide time saving, efficiency, accuracy, repeatability in treatment planning and patient care.

Few studies have presented software programs for cephalometric determination of C2, C3 and C4 vertebrae proportions according reference points marked manually on the image, and automatically calculates the skeletal maturation stage. This computer-assisted analysis still depends on operator experience [28]. Padalino et al run a study comparing manual analysis of CVM stages and the analysis performed by a dedicated software. It has shown a concordance of 94% between the two methods but hand-tracing analysis was quicker of 28 seconds on average [29].

Deep learning conventional neural networks have already been used to diagnose metabolic disorders in pediatric endocrinology, in order to assess skeletal bone age on left hand-wrist radiographs. Deep learning approach proposes better accuracy than conventional methods in processing the image in less than 1 s. Our study aims to develop a fully automated deep learning assessment of CVM stages on lateral cephalograms in orthodontics.

## 2. Preprocessing of the Data

For this classification task, we had an image data base of 2000 X-ray radiographic images. Each image has a size of 2012 × 2012. These images are extracted from the patients files and are anonymized.

A selection of 600 images are studied and labelized by the experts in six classes (CVS1, ..., CVS6). These labelized data are divided in three sets of Training, Validation and Testing. We did different division of the data: First, we had started by 300, 200 and 100, respectively for Training, Validation and Testing. Then, we decided to divide them to 200, 200 and 200 and used Cross Validation technique by permutation of these sets.

Also, as these images are from the whole head, only a specific part of the image is usefull for this classification, we performed different preprocessing before feeding then to the DL input. In a preprocessing step, each original image is first cropped to the interesting part (Test1: size 488 × 488), then resized to (Test2: 244 × 244) or (Test3: 64 × 64) and after resizing to 244 × 244, they are Sobel filtered to enhance the contours of the image (Test 4). Figure 2 shows an example of these inputs.

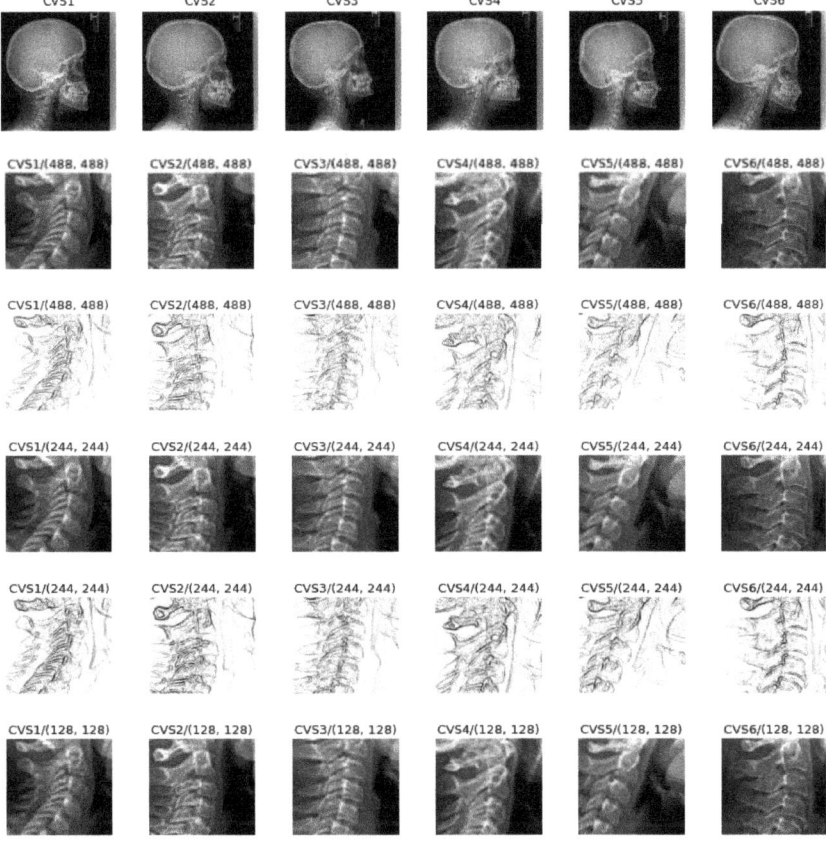

**Figure 2.** Originals and different preprocessing before training: (**a**) Originals (2012 × 2020), (**b**) test0: cropped images (488 × 488), (**c**) test1: cropped and sobel edge detector filter (488 × 488), (**d**) test2: cropped and resized (244 × 244), (**e**) test3: cropped, resized and sobel edge detector filter (244 × 244), (**f**) test4: cropped and resized (64 × 64).

## 3. Structure of the Deep Learning Network

In a preliminary study, we used different Deep Learning network structures for this classification task and finally we selected a Deep Learning structure (like resnet) which is adapted for our task.

We considered different classical networks:

- Resnet:
  Resnet was introduced in the paper "Deep Residual Learning for Image Recognition <https://arxiv.org/abs/1512.03385>". There are several variants with different output sizes, including Resnet18, Resnet34, Resnet50, Resnet101, and Resnet152, all of which are available from torchvision models. As our dataset is small, we used Resnet18 that we adapted in our case for 6 classes.
- Alexnet:
  Alexnet was introduced in the paper "ImageNet Classification with Deep Convolutional Neural Networks <https://papers.nips.cc/paper/4824-imagenet-classification-with-deep-convolutional-neural-networks.pdf>" and was the first very successful CNN on the ImageNet dataset.
- VGG:
  VGG was introduced in the paper "Very Deep Convolutional Networks for Large-Scale Image Recognition <https://arxiv.org/pdf/1409.1556.pdf>". Torchvision offers eight versions of VGG with various lengths and some that have batch normalizations layers.
- Squeezenet:
  The Squeznet architecture is described in the paper "SqueezeNet: AlexNet-level accuracy with 50× fewer parameters and <0.5 MB model size, <https://arxiv.org/abs/1602.07360>". It uses a different output structure than the other models mentioned here. Torchvision has two versions of Squeezenet. We used version 1.0.
- Densenet:
  Densenet was introduced in the paper "Densely Connected Convolutional Networks", <https://arxiv.org/abs/1608.06993>. Torchvision has four variants of Densenet. Here we used Densenet-121 and modified the output layer, which is a linear layer with 1024 input features, for our case.
- Inception v3:
  Inception v3 was first described in "Rethinking the Inception Architecture for Computer Vision", <https://arxiv.org/pdf/1512.00567v1.pdf>. This network is unique because it has two output layers when training. The second output is known as an auxiliary output and is contained in the AuxLogits part of the network. The primary output is a linear layer at the end of the network. Note, when testing we only consider the primary output.

As it can be seen from on Figure 3, the structure of Deep Learning model is composed of an input convolutional layer and three or four other convolutional nets (CNN) layers and a fully connected of (32 × 32) to 6 classes. Each of the three CNNs is followed by a normalization, pooling and dropout layers with different dropout coefficients.

The models are trained with different partitions of the images in Training, Validation and Testing sets. The following figure shows 300 images which have been prepared for the training, then validated on 200 images and saved to be used for testing step. A set of 120 images are used for testing step and the average score was 80 percent.

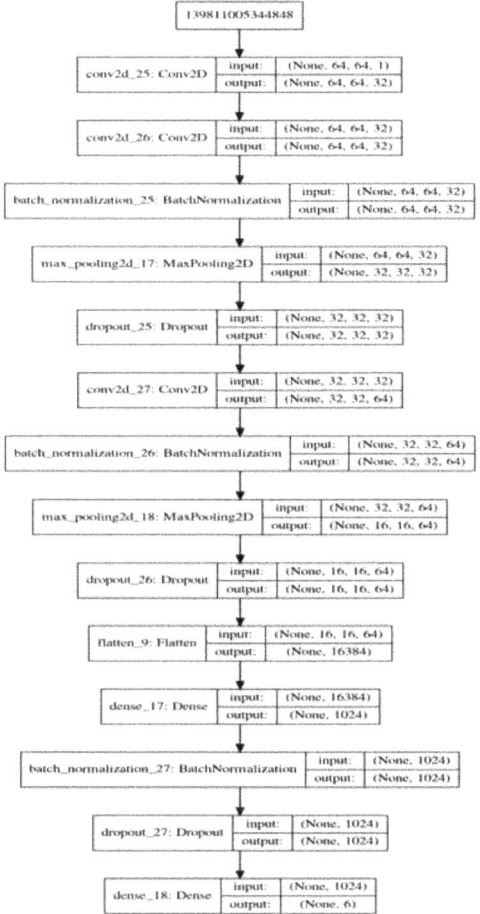

**Figure 3.** The structure of the proposed Deep Learning network.

## 4. Tools and Implementation

In this work, we used the following tools:

- SciKit-Learn: Data shuffling, Kmeans and Gaussian Mixture clustering, Principal Component Analysis and performance metrics.
- Keras wth tensorflow backend: VGG16, VGG19 and ResNet50 convolution network models with ImageNet weights

## 5. Prediction Results

With implemented DL structure, we used 300 images for the training step, 200 images for the validation step and finally 150 images for the testing step. We had to fix a great number of parameters such as dropout rates, optimization algorithms, regularization parameters, etc.

The following Figure 4 shows the evolution of the Loss function and the accuracy as a function of the epoch numbers for one of these different tests.

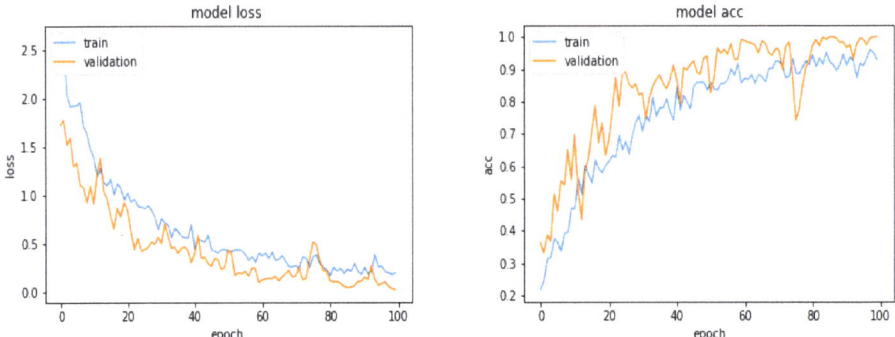

**Figure 4.** Loss function and the accuracy as a function of the epoch numbers.

Here, we show the prediction results obtained with different preprocessing of the data, both during the training and the testing

```
CVS1/ 0.83,   CVS2/ 0.85,   CVS3/ 0.72,   CVS4/ 0.72,   CVS5/ 0.79,   CVS6/ 0.82
```

## 6. Conclusions

In this work, we developed and presented a specifically designed classification method for classifying the lateral radiographs of a great number of patients with the objective of determining the cervical vertebra maturation degree of bones, which is an important parameter for the orthodontists. The proposed Deep Learning classification method is particularly adapted for this task. In a first step, we used 300 labeled images for training, 200 for validation and hyper parameter tuning and finally 100 for testing. Even if during the training and validation, we could obtain accuracies more than 95%, the accuracy for the testing images did not exceeded 85%. We think that with a greater number of training and validation images, this can be improved. Our plan is to use about 1000 images for training and 1000 for testing in near future.

## References

1. Baccetti, T.; Franchi, L.; McNamara, J.A. The Cervical Vertebral Maturation (CVM) Method for the Assessment of Optimal Treatment Timing in Dentofacial Orthopedics. *Semin. Orthod.* **2005**, *11*, 119–129.
2. Patcas, R.; Herzog, G.; Peltomäki, T.; Markic, G. New perspectives on the relationship between mandibular and statural growth. *Eur. J. Orthod.* **2016**, *38*, 13–21.
3. Moore, R.N.; Moyer, B.A.; DuBois, L.M. Skeletal maturation and craniofaciat growth. *Am. J. Orthod. Dentofac. Orthop.* **1990**, *98*, 33–40.
4. Hunter, C.J. The corelation of facial growth with body height and skeletal maturation at adolescence. *Angle Orthod.* **1966**, *36*, 44–54.
5. Raberin, M.; Cozor, I.; Gobert-Jacquart, S. Les vertèbres cervicales: Indicateurs du dynamisme de la croissance mandibulaire? *L'Orthodontie Française* **2012**, *83*, 45–58.
6. Franchi, L.; Baccetti, T.; McNamara, J.A. Mandibular growth as related to cervical vertebral maturation and body height. *Am. J. Orthod. Dentofac. Orthop.* **2000**, *118*, 335–340.
7. De Stefani, A.; Bruno, G.; Siviero, L.; Crivellin, G.; Mazzoleni, S.; Gracco, A. Évaluation radiologique de l'âge osseux avec la maturation de la phalange médiane du doigt majeur chez un patient orthodontique péripubertaire. *Int. Orthod.* **2018**, *16*, 499–513.
8. Krisztina, M.I.; Ogodescu, A.; Réka, G.; Zsuzsa, B. Evaluation of the Skeletal Maturation Using Lower First Premolar Mineralisation. *Acta Med. Marisiensis* **2013**, *59*, 289–292.
9. Pyle, S.I.; Waterhouse, A.M.; Greulich, W.W. Attributes of the radiographic standard of reference for the National Health Examination Survey. *Am. J. Phys. Anthropol.* **1971**, *35*, 331–337.

10. Loder, R.T.; Estle, D.T.; Morrison, K.; Eggleston, D.; Fish, D.N.; Greenfield, M.L.; Guire, K.E. Applicability of the Greulich and Pyle Skeletal Age Standards to Black and White Children of Today. *Am. J. Dis. Child.* **1993**, *147*, 1329–1333.
11. Bunch, P.M.; Altes, T.A.; McIlhenny, J.; Patrie, J.; Gaskin, C.M. Skeletal development of the hand and wrist: Digital bone age companion-a suitable alternative to the Greulich and Pyle atlas for bone age assessment? *Skelet. Radiol.* **2017**, *46*, 785–793.
12. Srinivasan, B.; Padmanabhan, S.; Chitharanjan, A.B. Constancy of cervical vertebral maturation indicator in adults: A cross-sectional study. *Int. Orthod.* **2018**, *16*, 486–498.
13. Shim, J.J.; Bogowicz, P.; Heo, G.; Lagravère, M.O. Interrelationship and limitations of conventional radiographic assessments of skeletal maturation. *Int. Orthod.* **2012**, *10*, 135–47.
14. O'Reilly, M.T.; Yanniello, G.J. Mandibular growth changes and maturation of cervical vertebrae—A longitudinal cephalometric study. *Angle Orthod.* **1988**, *58*, 179–184.
15. Hassel, B.; Farman, A.G. Skeletal maturation evaluation using cervical vertebrae. *Am. J. Orthod. Dentofac. Orthop.* **1995**, *107*, 58–66.
16. Elhaddaoui, R.; Benyahia, H.; Azaroual, F.; Zaoui, F. Intérêt de la méthode de maturation des vertèbres cervicales (CVM) en orthopédie dento-faciale: Mise au point. *Revue de Stomatologie, de Chirurgie Maxillo-faciale et de Chirurgie Orale* **2014**, *115*, 293–300.
17. Jaqueira, L.M.F.; Armond, M.C.; Pereira, L.J.; De Alcântara, C.E.P.; Marques, L.S. Determining skeletal maturation stage using cervical vertebrae: Evaluation of three diagnostic methods. *Braz. Oral Res.* **2010**, *24*, 433–437.
18. Uysal, T.; Ramoglu, S.I.; Basciftci, F.A.; Sari, Z. Chronologic age and skeletal maturation of the cervical vertebrae and hand-wrist: Is there a relationship? *Am. J. Orthod. Dentofac. Orthop.* **2006**, *130*, 622–628.
19. Hosni, S.; Burnside, G.; Watkinson, S.; Harrison, J.E. Comparison of statural height growth velocity at different cervical vertebral maturation stages. *Am. J. Orthod. Dentofac. Orthop.* **2018**, *154*, 545–553.
20. Perinetti, G.; Contardo, L.; Castaldo, A.; McNamara, J.A.; Franchi, L. Diagnostic reliability of the cervical vertebral maturation method and standing height in the identification of the mandibular growth spurt. *Angle Orthod.* **2016**, *86*, 599–609.
21. Mahajan, S. Evaluation of skeletal maturation by comparing the hand wrist radiograph and cervical vertebrae as seen in lateral cephalogram. *Indian J. Dent. Res.* **2011**, *22*, 309–316.
22. Sachan, K.; Tandon, P.; Sharma, V. A correlative study of dental age and skeletal maturation. *Indian J. Dent. Res.* **2011**, *22*, 882.
23. Danaei, S.M.; Karamifar, A.; Sardarian, A.; Shahidi, S.; Karamifar, H.; Alipour, A.; Boushehri, S.G. Measuring agreement between cervical vertebrae and hand-wrist maturation in determining skeletal age: Reassessing the theory in patients with short stature. *Am. J. Orthod. Dentofac. Orthop.* **2014**, *146*, 294–298.
24. Ball, G.; Woodside, D.; Tompson, B.; Hunter, W.S.; Posluns, J. Relationship between cervical vertebral maturation and mandibular growth. *Am. J. Orthod. Dentofac. Orthop.* **2011**, *139*, e455–e461.
25. Ballrick, J.W.; Fields, H.W.; Beck, F.M.; Sun, Z.; Germak, J. The cervical vertebrae staging method's reliability in detecting pre and post mandibular growth. *Orthod. Waves* **2013**, *72*, 105–111.
26. Perinetti, G.; Caprioglio, A.; Contardo, L. Visual assessment of the cervical vertebral maturation stages: A study of diagnostic accuracy and repeatability. *Angle Orthod.* **2014**, *84*, 951–956.
27. McNamara, J.A.; Franchi, L. The cervical vertebral maturation method: A user's guide. *Angle Orthod.* **2018**, *88*, 133–43.
28. Santiago, R.C.; Cunha, A.R.; Junior, G.C.; Fernandes, N.; Campos, M.J.S.; Costa, L.F.M.; Vitral, R.W.F.; Bolognese, A.M. New software for cervical vertebral geometry assessment and its relationship to skeletal maturation—A pilot study. *Dentomaxillofac. Radiol.* **2014**, *43*, 20130238.
29. Padalino, S.; Sfondrini, M.F.; Chenuil, L.; Scudeller, L.; Gandini, P. Fiabilité de l'analyse de la maturité squelettique selon la méthode CVM des vertèbres cervicales faite par un logiciel dédié. *Int. Orthod.* **2014**, *12*, 483–493.

© 2020 by the authors. Licensee MDPI, Basel, Switzerland. This article is an open access article distributed under the terms and conditions of the Creative Commons Attribution (CC BY) license (http://creativecommons.org/licenses/by/4.0/).

*Proceedings*

# Interaction between Model Based Signal and Image Processing, Machine Learning and Artificial Intelligence [†]

**Ali Mohammad-Djafari** [1,2]

[1] Laboratoire des Signaux et Systèmes, CNRS, CentraleSupélec-Univ Paris Saclay, 91192 Gif-sur-Yvette, France; djafari@free.fr
[2] International Science Consulting and Training (ISCT), 91440 Bures sur Yvette, France
[†] Presented at the 39th International Workshop on Bayesian Inference and Maximum Entropy Methods in Science and Engineering, Garching, Germany, 30 June–5 July 2019.

Published: 28 November 2019

**Abstract:** Signale and image processing has always been the main tools in many area and in particular in Medical and Biomedical applications. Nowadays, there are great number of toolboxes, general purpose and very specialized, in which classical techniques are implemented and can be used: all the transformation based methods (Fourier, Wavelets, ...) as well as model based and iterative regularization methods. Statistical methods have also shown their success in some area when parametric models are available. Bayesian inference based methods had great success, in particular, when the data are noisy, uncertain, incomplete (missing values) or with outliers and where there is a need to quantify uncertainties. In some applications, nowadays, we have more and more data. To use these "Big Data" to extract more knowledge, the Machine Learning and Artificial Intelligence tools have shown success and became mandatory. However, even if in many domains of Machine Learning such as classification and clustering these methods have shown success, their use in real scientific problems are limited. The main reasons are twofold: First, the users of these tools cannot explain the reasons when the are successful and when they are not. The second is that, in general, these tools can not quantify the remaining uncertainties. Model based and Bayesian inference approach have been very successful in linear inverse problems. However, adjusting the hyper parameters is complex and the cost of the computation is high. The Convolutional Neural Networks (CNN) and Deep Learning (DL) tools can be useful for pushing farther these limits. At the other side, the Model based methods can be helpful for the selection of the structure of CNN and DL which are crucial in ML success. In this work, I first provide an overview and then a survey of the aforementioned methods and explore the possible interactions between them.

**Keywords:** signal and image processing; transform based; model based; regularization; Bayesian inference; Gauss-Markov-Potts; variational Bayesian approach; Machine Learning; Artificial Intelligence

---

## 1. Introduction

Nowadays, there are great number of general purpose and very specialized toolboxes, in which, classical and advanced techniques of signal and image processing methods are implemented and can be used. Between them, we can mention all the transformation based methods (Fourier, Hilbert, Wavelets, Radon, Abel, ... and much more) as well as all the Model Based and iterative regularization methods. Statistical methods have also shown their success in some areas when parametric models are available.

Bayesian inference based methods had great success, in particular, when the data are noisy, uncertain, some missing and some outliers and where there is a need to account and to quantify uncertainties.

Nowadays, we have more and more data. To use these "Big Data" to extract more knowledge, the Machine Learning and Artificial Intelligence tools have shown success and became mandatory. However, even if in many domains of Machine Learning such as classification and clustering these methods have shown success, their use in real scientific problems are limited. The main reasons are twofold: First, the users of these tools can not explain the reasons when they are successful and when they are not. The second is that, in general, these tools can not quantify the remaining uncertainties.

Model based and Bayesian inference approach have been very successful in linear inverse problems. However, adjusting the hyper parameters is complex and the cost of the computation is high. The Convolutional Neural Networks (CNN) and Deep Learning (DL) tools can be useful for pushing farther these limits. At the other side, the Model based methods can be helpful for the selection of the structure of CNN and DL which are crucial in ML success. In this work, first I give an overview and a survey of the aforementioned methods and explore the possible interactions between them.

The rest of the paper is organized as follows: First a classification of signal and image processing methods is proposed. Then, very briefly, the Machine Learning tools are introduced. Then, through the problem of Imaging inside the body, we see the different steps from acquisition of the data, reconstruction, post-processing such as segmentation and finally the decision and conclusion of the user are presented. After mentioning some successful case studies in which the ML tools have been successful, we arrive at the main part of this paper: Looking for the possible interactions between Model based and Machine Learning tools. Finally, we mention the Open problems and challenges in both classical, model based and the ML tool.

## 2. Classification of Signal and Image Processing Methods

Signal and image processing methods can be classified in the following categories:

- Transform based methods
- Model based and inverse problem approach
- Regularisation methods
- Bayesian inference methods

In the first category, the main idea is to use different ways the signal and images can be represented in time, frequency, space, spacial frequency, time-frequency, wavelets, etc.

## 3. Transform Domain Methods

Figure 1 shows the main idea behind the transform based methods. Mainly, first a linear transform (Fourier, Wavelet, Radon, etc.) is applied to the signal or the image, then some thresholding or windowing is applied in this transform domain and finally an inverse transform is applied to obtain the result. Appropriate choices of the transform and the threshold or the widow size and shape are important for the success of such methods [1,2].

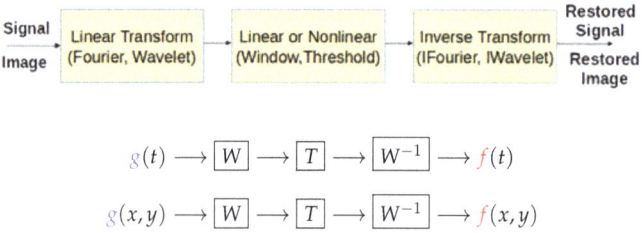

**Figure 1.** Transform methods.

## 4. Model Based and Inverse Problem Approach

The model based methods are related to the notions of forward model and inverse problems approach. Figure 2 shows the main idea:

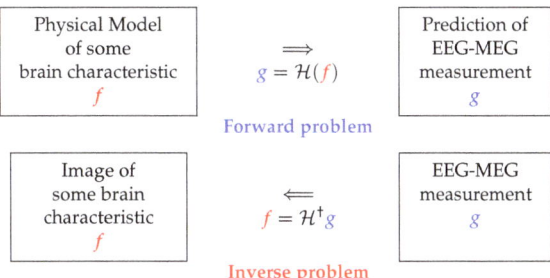

Figure 2. Model based methods.

Given the forward model $\mathcal{H}$ and the sources $f$, the prediction of the data $g$ can be done, either in a deterministic way: $g = \mathcal{H}(f)$ or via a probabilistic model: $p(g|f, \mathcal{H})$.

In the same way, given the forward model $\mathcal{H}$ and the data $g$, the estimation of the unknown sources $f$ can be done either via a deterministic method or probabilistic one. One of the deterministic method is the Generalized inversion: $f = \mathcal{H}^\dagger(g)$. A more general method is the regularization: $\hat{f} = \arg\min_f \{J(f)\}$ with $J(f) = \|g - \mathcal{H}(f)\|^2 + \lambda \mathcal{R}(f)$ [3].

As we will see later, the only probabilistic method which can be efficiently use for the inverse problems is the Bayesian approach.

## 5. Regularization Methods

Let consider the linear inverse problem:

$$g = Hf + \epsilon, \tag{1}$$

Then the basic idea in regularization is to define a regularization criterion:

$$J(f) = \frac{1}{2}\|g - Hf\|_2^2 + \lambda R(f) \tag{2}$$

and optimize it to obtain the solution [4]. The first main issue in such regularization method is the choice of the regularizer. The most common examples are:

$$R(f) = \left\{\|f\|_2^2, \|f\|_\beta^\beta, \|Df\|_2^2, \|Df\|_\beta^\beta, \sum_j \phi([Df]_j)\right\}, \ 1 \leq \beta \leq 2 \tag{3}$$

The second main issue in regularization is the the choice of appropriate optimization algorithm. Mainly, depending on the type of the criterion, we have:

- $R(f)$ quadratic: Gradient based, Conjugate Gradient algorithms are appropriate.
- $R(f)$ non quadratic, but convex and differentiable: Here too the Gradient based and Conjugate Gradient (CG) methods can be used, but there are also great number of convex criterion optimization algorithms.
- $R(f)$ convex but non-differentiable: Here, the notion of sub-gradient is used.

Specific cases are:

- L2 or quadratic: $J(f) = \frac{1}{2}\|g - Hf\|_2^2 + \lambda \|Df\|_2^2$.

  In this case we have an analytic solution: $\hat{f} = (H'H + \lambda D'D)^{-1}H'g$. However, in practice this analytic solution is not usable in high dimensional problems. In general, as the gradient $\nabla J(f) = -H'(g - Hf) + 2\lambda D'Df$ can be evaluated analytically, gradient based algorithms are used.

- L1 (TV): convex but not differentiable at zero: $J(f) = \frac{1}{2}\|g - Hf\|_2^2 + \lambda \|Df\|_1$.

  The algorithms in this case use the notions of Fenchel conjugate, Dual problem, sub gradient, proximal operator, ...

- Variable splitting and Augmented Lagrangian

$$(f, \hat{z}) = \arg\min_{f, z} \left\{ \frac{1}{2}\|g - Hf\|_2^2 + \lambda \|z\|_1 + q\|z\|_2^2 \right\} \quad \text{s.t.} \quad z = Df \tag{4}$$

A great number of optimization algorithms have been proposed: ADMM, ISTA, FISTA, etc. [5–7].

Main limitations of deterministic regularization methods are:

- Limited choice of the regularization term. Mainly, we have: a) Smoothness (Tikhonov), b) Sparsity, Piecewise continuous (Total Variation).
- Determination of the regularization parameter. Even if there are some classical methods such as L-Curve and Cross validation, there are still controversial discussions about this.
- Quantification of the uncertainties: This is the main limitation of the deterministic methods, in particular in medical and biological applications where this point is important.

The best possible solution to push further all these limits is the Bayesian approach which has: (a) Many possibilities to choose prior models, (b) possibility of the estimation of the hyper-parameters, and most important (c) accounting for the uncertainties.

## 6. Bayesian Inference Methods

The simple case of the Bayes rule is:

$$p(f|g, \mathcal{M}) = \frac{p(g|f, \mathcal{M}) \, p(f|\mathcal{M})}{p(g|\mathcal{M})} \quad \text{where} \quad p(g|\mathcal{M}) = \int p(g|f, \mathcal{M}) \, p(f|\mathcal{M}) \, df \tag{5}$$

When there are some hyper parameters which have also to be estimated, we have:

$$p(f, \theta|g, \mathcal{M}) = \frac{p(g|f, \theta, \mathcal{M}) \, p(f|\theta, \mathcal{M}) \, p(\theta|\mathcal{M})}{p(g|\mathcal{M})} \quad \text{where} \quad p(g|\mathcal{M}) = \int p(g|f, \theta, \mathcal{M}) \, p(f|\theta, \mathcal{M}) \, d\theta \, df \tag{6}$$

From that joint posterior distribution, we may also obtain the marginals:

$$p(f|g, \mathcal{M}) = \int p(f, \theta|g, \mathcal{M}) \, d f \quad \text{and} \quad p(\theta|g, \mathcal{M}) = \int p(f, \theta|g, \mathcal{M}) \, df \tag{7}$$

To be more specific, let consider the case of linear inverse problems $g = Hf + \epsilon_*$. Then, assuming Gaussian noise, we have:

$$p(g|f) = \mathcal{N}(g|Hf, v_\epsilon I) \propto \exp\left[\frac{-1}{2v_\epsilon}\|g - Hf\|_2^2\right] \tag{8}$$

Assuming a Gaussian prior:

$$p(f) \propto \exp\left[\frac{-1}{2v_f}\|f\|_2^2\right] \quad \text{or} \quad \exp\left[\frac{-1}{2v_f}\|Df\|_2^2\right], \tag{9}$$

Then, we see that the posterior is also Gaussian and the MAP and Posterior Mean (PM) estimates become the same and can be computed as the minimizer of : $J(f) = \|g - Hf\|_2^2 + \lambda R(f)$:

$$p(f|g) \propto \exp\left[\frac{-1}{2v_\epsilon}J(f)\right] \to \widehat{f}_{MAP} = \arg\max_f \{p(f|g)\} = \arg\min_f \{J(f)\} \tag{10}$$

In summary, we have:

$$\begin{cases} p(g|f) = \mathcal{N}(g|Hf, v_\epsilon I) \\ p(f) = \mathcal{N}(f|0, v_f I) \end{cases} \to \begin{cases} p(f|g) = \mathcal{N}(f|\widehat{f}, \widehat{\Sigma}) \\ \widehat{f} = [H'H + \lambda I]^{-1} H'g \\ \widehat{\Sigma} = v_\epsilon[H'H + \lambda I]^{-1}, \quad \lambda = \frac{v_\epsilon}{v_f} \end{cases} \tag{11}$$

For the case where the hyper parameters $v_\epsilon$ and $v_f$ are unknown (Unsupervised case), we can derive the following:

$$\begin{cases} p(g|f, v_\epsilon) = \mathcal{N}(g|Hf, v_\epsilon I) \\ p(f|v_f) = \mathcal{N}(f|0, v_f I) \\ p(v_\epsilon) = \mathcal{IG}(v_f|\alpha_{\epsilon_0}, \beta_{\epsilon_0}) \\ p(v_f) = \mathcal{IG}(v_f|\alpha_{f_0}, \beta_{f_0}) \end{cases} \to \begin{cases} p(f|g, v_\epsilon, v_f) = \mathcal{N}(f|\widehat{f}, \widehat{\Sigma}) \\ \widehat{f} = [H'H + \widehat{\lambda} I]^{-1} H'g \\ \widehat{\Sigma} = \widehat{v}_\epsilon[H'H + \widehat{\lambda} I]^{-1}, \quad \widehat{\lambda} = \frac{\widehat{v}_\epsilon}{\widehat{v}_f} \\ p(v_\epsilon|g, f) = \mathcal{IG}(v_\epsilon|\widetilde{\alpha}_\epsilon, \widetilde{\beta}_\epsilon) \\ p(v_f|g, f) = \mathcal{IG}(v_f|\widetilde{\alpha}_f, \widetilde{\beta}_f) \\ \widetilde{\alpha}_\epsilon, \widetilde{\beta}_\epsilon, \widetilde{\alpha}_f, \widetilde{\beta}_f \end{cases} \tag{12}$$

where the expressions for $\widetilde{\alpha}_\epsilon, \widetilde{\beta}_\epsilon, \widetilde{\alpha}_f, \widetilde{\beta}_f$ can be found in [8].

The joint posterior can be written as:

$$p(f, v_\epsilon, v_\xi|g) \propto \exp\left[-J(f, v_\epsilon, v_\xi)\right] \tag{13}$$

From this expression, we have different expansion possibilities:

- JMAP: Alternate optimization with respect to $f, v_\epsilon, v_f$:

$$J(f, v_\epsilon, v_f) = \frac{1}{2v_\epsilon}\|g - Hf\|_2^2 + \frac{1}{2v_f}\|f\|_2^2 + (\alpha_{\epsilon_0} + 1)\ln v_\epsilon + \frac{\beta_{\epsilon_0}}{v_\epsilon} + (\alpha_{f_0} + 1)\ln v_f + \frac{\beta_{f_0}}{v_f} \tag{14}$$

- Gibbs sampling MCMC:

$$f \sim p(f, v_\epsilon, v_f|g) \to v_\epsilon \sim p(v_\epsilon|g, f) \to v_f \sim p(v_f|g, f) \tag{15}$$

- Variational Bayesian Approximation: Approximate $p(f, v_\epsilon, v_f|g)$ by a separable one $q(f, v_\epsilon, v_f) = q_1(f)q_2(v_\epsilon)q_3(v_f)$ minimizing $KL(q|p)$ [8].

## 7. Imaging inside the Body: From Acquisition to Decision

To introduce the link between the different model based methods and the Machine Learning tools, let consider the case of medical imaging, from the acquisition to the decision steps:

- Data acquisition:

$$\text{Object } f \to \boxed{\text{Scanner}} \to \text{Data } g$$

- Reconstruction:

$$\text{Data } g \to \boxed{\text{Reconstruction}} \to \text{Image } \widehat{f}$$

- Post Processing (Segmentation):

$$\text{Image } \widehat{f} \to \boxed{\text{Segmentation}} \to \widehat{z}$$

- Understanding and Decision:

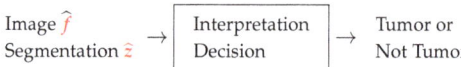

The questions now are: Can we join any of these steps? Can we go directly from the image to the decision? For the first one, the Bayesian approach can provide a solution:

Data $g \rightarrow$ [Reconstruction / Segmentation] $\rightarrow$ Reconstruction $\widehat{f}$ / $\rightarrow$ Segmentation $\widehat{z}$

The main tool here is to introduce a hidden variable which can represent the segmentation. A solution is to introduce a classification hidden variable $z$ with $z_j = \{1, 2, \cdots, K\}$. Then in Figure 3, we have in summary:

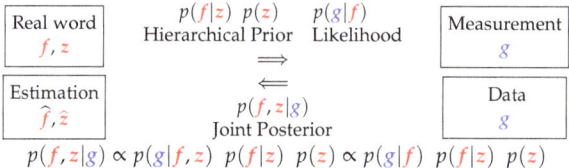

**Figure 3.** Bayesian approach for joint reconstruction and segmentation.

A few comments for these relations:

- $p(g|f,z)$ does not depend on $z$, so it can be written as $p(g|f)$.
- We may choose a Markovian Potts model for $p(z)$ to obtain more compact homogeneous regions [8,9].
- If we choose for $p(f|z)$ a Gaussian law, then $p(f,z|g)$ becomes a Gauss-Markov-Potts model [8].
- We can use the joint posterior $p(f,z|g)$ to infer on $(f,z)$: We may just do JMAP: $(\widehat{f},\widehat{z}) = \arg\max \{p(f,z|g)\}$ or trying to access to the expected posterior values by using the Variational Bayesian Approximation (VBA) techniques [8,10–13].

This scheme can be extended to consider the estimation of the hyper parameters too. Figure 4 shows this.

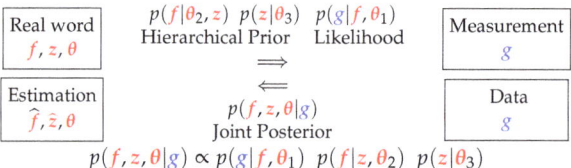

**Figure 4.** Advanced Bayesian approach for joint reconstruction and segmentation.

Again, here, we can use the joint posterior $p(f,z,\theta|g)$ to infer on all the unknowns [12].

## 8. Advantages of the Bayesian Framework

- Large flexibility of Prior models prior
  - Smoothness (Gaussian, Gauss-Markov)

- Direct Sparsity (Double Exp, Heavy-tailed distributions)
- Sparsity in the Transform domain (Double Exp, Heavy-tailed distributions on the WT coefficients)
- Piecewise continuous (DE or Student-t on the gradient)
- Objects composed of only a few materials (Gauss-Markov-Potts), ...
* Possibility of estimating hyper-parameters via JMAP or VBA
* Natural ways to take account for uncertainties and quantify the remaining uncertainties.

## 9. Imaging inside the Body for Tumor Detection

* Reconstruction and Segmentation

$$\text{Data } g \rightarrow \boxed{\text{Reconstruction}} \rightarrow \text{Image } \widehat{f} \rightarrow \boxed{\text{Segmentation}} \rightarrow \widehat{z}$$

* Understanding and Decision

$$\text{Image } \widehat{f},\ \text{Segmentation } \widehat{z} \rightarrow \boxed{\text{Interpretation Decision}} \rightarrow \text{Tumor or Not Tumor}$$

* Can we do all together in a more easily way?
* Machine Learning and Artificial Intelligence tools may propose solutions

$$\text{Data } g \rightarrow \boxed{\text{Machine Learning and Artificial Intelligence}} \rightarrow \text{Tumor or Not Tumor}$$

* **Learning from a great number of data**

From here, we may just do JMAP: $(\widehat{f}, \widehat{z}, \widehat{\theta}) = \arg\max \{p(f, z, \theta|g)\}$, or better, use the Variational Bayesian Approximation (VBA) to do inference.

## 10. Machine Learning Basic Idea

The main idea in Machine Learning is to learn from a great number of data: $(g_i, d_i), i = 1, \cdots N$:

$$\text{Data } (g_i, d_i)_{i=1}^{N} \rightarrow \boxed{\text{Machine Learning and Artificial Intelligence}} \rightarrow \text{Tumor or Not Tumor}$$

Between the basic tasks we can mention: (a) Classification (Tumor/Not Tumor), (b) Regression (Continuos parameter estimation) and (c) Clustering when the data have not yet labels.

Between the existing ML tools we may mention: Support Vector Machines (SVM), Decision-Tree learning (DT), Artificial Neural Networks (ANN), Bayesian Networks (BN), HMM and Random Forest (RF), Mixture Models (GMM, SMM, ...), KNN, Kmeans,...

The frontiers between Image processing, Computer vision, Machine Learning & Artificial intelligence are not very precise as it is shown in Figure 5.

| | |
|---|---|
| Image Processing | Acquisition, Representation, Compression, Transmission |
| Computer Vision | Enhancement, Restoration Segmentation |
| Machine Learning | Edge/Feature extraction Pattern matching Objects detection |
| Artificial Intelligence | Understanding, Recognition, 3D |

**Figure 5.** Frontiers between Image processing (IP), Computer vision (CV), Machine Learning (ML) and Artificial intelligence (AI).

Between the Machine learning tools, we may mention Neural Networks (NN), Artificial NN (ANN), Convolutional NN (CNN), Recurrent NN (RNN), Deep Learning (DL). This last one had shown great success in Speech Recognition, Computer Vision and specifically in Segmentation, Classification and Clustering and in Multi-modality and cross-domain information fusion [14–17]. However, there are still many limitations: Lack of interpretability, reliability and uncertainty and No reasoning and explaining capabilities [18]. To overcome, there still much to do with the Fundamentals.

## 11. Interaction between Model Based and Machine Learning Tools

To show the possibilities of the interaction between classical and machine learning, let consider a few examples. The first one is the case of linear inverse problems and quadratic regularization or the Bayesian with Gaussian priors. The solution has an analytic expression:

$$g = Hf + \epsilon \to \hat{f} = (HH^t + \lambda DD^t)^{-1} H^t g = BH^t g$$

which can be presented schematically as

$$g \to \boxed{H^t} \to \boxed{B} \to \hat{f} \text{ or directly } g \to \boxed{\text{CNN or DL}} \to \hat{f}$$

As we can see, this induces directly a linear NN structure. In particular, if $H$ represents a convolution operator, then $H^t$ and $H^t H$ are too and probably the operator $B$ can also be well approximated by a convolution and the whole inversion can be modelled by a CNN [19].

The second example is the denoising $g = f + \epsilon$ with L1 regularizer, or equivalently, the MAP estimator with a double exponential prior, where the solution can be obtained by a convolution followed by a thresholding [20,21].

$$g = f + \epsilon \to \hat{f} = \frac{1}{\lambda} H^t g \text{ followed by Thresholding}$$

$$g \to \boxed{H^t} \to \boxed{\text{Thresholding}} \to \hat{f} \text{ or directly } g \to \boxed{\text{CNN or DL}} \to \hat{f}$$

The third example is the Joint Reconstruction and Segmentation that was presented in previous sections. If we present the different steps of reconstruction, segmentation and parameter estimation, we can also compare it with some kind of NN (combination of CNN, RNN and GAN) [22,23].

$$g \to \boxed{\text{Reconstruction}} \to \hat{f} \to \boxed{\text{Segmentation}} \to \hat{z} \to \boxed{\text{Parameter estimation}} \to \hat{\theta}$$

## 12. Conclusions and Challenges

Signal and image processing, imaging systems and computer vision have made great progress in the last forty years. The first category of the methods was based on linear transformation followed by a thresholding or windowing and coming back. The second generation was model based: forward modeling and inverse problems approach. The main successful approach was based on regularization methods using a combined criterion. The third generation was model based but probabilistic and using the Bayes rule, the so called Bayesian approach. Nowadays, Machine Learning (ML), Neural Networks (NN), Convolutional NN (CNN), Deep Learning (DL) and Artificial Intelligence (AI) methods have obtained great success in classification, clustering, object detection, speech and face recognition, etc. But, they need great number of training data and lack still explanation and they may fail very easily. For signal and image processing or inverse problems, they need still progress. This progress is coming via their interaction with the model based methods. In fact, the successful of CNN and DL methods greatly depends on the appropriate choice of the network structure. This choice can be guided by the model based methods. For inverse problems, when the forward models are not available or too complex, NN and DL may be helpful. However, we may still need to choose the structure of the NN via approximate forward model and approximate Bayesian inversion.

## References

1. Carre, P.; Andres, E. Discrete analytical ridgelet transform. *Signal Process.* **2004**, *84*, 2165–2173.
2. Dettori, L.; Semler, L. A comparison of wavelet, ridgelet, and curvelet-based texture classification algorithms in computed tomography. *Comput. Biol. Med.* **2007**, *37*, 486–498.
3. Mohammad-Djafari, A. *Inverse Problems in Vision and 3D Tomography*; John Wiley & Sons: Hoboken, NJ, USA, 2010.
4. Idier, J. *Bayesian Approach to Inverse Problems*; John Wiley & Sons: New York, NY, USA, 2008.
5. Afonso, M.V.; Bioucas-Dias, J.M.; Figueiredo, M.A. Fast image recovery using variable splitting and constrained optimization. *IEEE Trans. Image Process.* **2010**, *19*, 2345–2356.
6. Bioucas-Dias, J.M.; Figueiredo, M.A. An iterative algorithm for linear inverse problems with compound regularizers. In Porceedings of the 2008 15th IEEE International Conference on Image Processing, San Diego, CA, USA, 12–15 October 2008; pp. 685–688.
7. Florea, M.I.; Vorobyov, S.A. A robust FISTA-like algorithm. In Porceedings of the 2017 IEEE International Conference on Acoustics, Speech and Signal Processing (ICASSP), New Orleans, LA, USA, 5–9 March 2017; pp. 4521–4525.
8. Chapdelaine, C.; Mohammad-Djafari, A.; Gac, N.; Parra, E. A 3D Bayesian Computed Tomography Reconstruction Algorithm with Gauss-Markov-Potts Prior Model and its Application to Real Data. *Fundam. Informaticae* **2017**, *155*, 373–405.
9. Féron, O.; Duchêne, B.; Mohammad-Djafari, A. Microwave imaging of inhomogeneous objects made of a finite number of dielectric and conductive materials from experimental data. *Inverse Probl.* **2005**, *21*, S95.
10. Chapdelaine, C.; Mohammad-Djafari, A.; Gac, N.; Parra, E. A Joint Segmentation and Reconstruction Algorithm for 3D Bayesian Computed Tomography Using Gauss-Markov-Potts Prior Model. In Porceedings of the 42nd IEEE International Conference on Acoustics, Speech and Signal Processing (ICASSP 2017), New Orleans, LA, USA, 5–9 March 2017.
11. Chapdelaine, C.; Gac, N.; Mohammad-Djafari, A.; Parra, E. New GPU implementation of Separable Footprint Projector and Backprojector: First results. In Porceedings of the 5th International Conference on Image Formation in X-ray Computed Tomography, Salt Lake City, UT, USA, 20–23 May 2018.
12. Ayasso, H.; Mohammad-Djafari, A. Joint NDT image restoration and segmentation using Gauss–Markov–Potts prior models and variational bayesian computation. *IEEE Trans. Image Process.* **2010**, *19*, 2265–2277.
13. Chapdelaine, C. Variational Bayesian Approach and Gauss-Markov-Potts prior model. *arXiv* **2018**, arXiv:1808.09552.
14. Szegedy, C.; Toshev, A.; Erhan, D. Deep Neural Networks for Object Detection. In Proceedings of the Advances in Neural Information Processing Systems 26 (NIPS 2013), Lake Tahoe, NV, USA, 5–10 December 2013; pp. 2553–2561.
15. Deng, L.; Yu, D. *Deep Learning: Methods and Applications*; NOW Publishers: Hanover, MA, USA, 2014.
16. Schmidhuber, J. Deep Learning in Neural Networks: An Overview. *Neural Netw.* **2015**, *61*, 85–117.

17. Schmidhuber, J. The New AI: General & Sound & Relevant for Physics. In *Artificial General Intelligence*; Goertzel, B., Pennachin, C., Eds.; Springer: Berlin/Heidelberg, Germany, 2006; pp. 175–198.
18. Chellappa, R.; Fukushima, K.; Katsaggelos, A.; Kung, S.Y.; LeCun, Y.; Nasrabadi, N.M.; Poggio, T.A. Applications of Artificial Neural Networks to Image Processing (guest editorial). *IEEE Trans. Image Process.* **1998**, *7*, 1093–1097.
19. Ciresan, D.C.; Meier, U.; Masci, J.; Gambardella, L.M.; Schmidhuber, J. Flexible, High Performance Convolutional Neural Networks for Image Classification. In Proceedings of the Twenty-Second International joint conference on Artificial Intelligence, Barcelona, Spain, 16–22 July 2011; pp. 1237–1242.
20. Krizhevsky, A.; Sutskever, I.; Hinton, G.E. ImageNet Classification with Deep Convolutional Neural Networks. In Proceedings of the Advances in Neural Information Processing Systems (NIPS 2012), Lake Tahoe, NV, USA, 3–6 December 2012; p. 4.
21. Zeiler, M.D.; Fergus, R. Visualizing and Understanding Convolutional Networks. *arXiv* **2013**, arXiv:1311.2901.
22. Masci, J.; Meier, U.; Ciresan, D.; Schmidhuber, J. Stacked Convolutional Auto-Encoders for Hierarchical Feature Extraction. In *Artificial Neural Networks and Machine Learning—ICANN 2011*; Honkela, T., Duch, W., Girolami, M.A., Kaski, S., Eds.; Lecture Notes in Computer Science; Springer: Berlin/Heidelberg, Germany, 2011; Volume 6791, pp. 52–59.
23. LeCun, Y.; Bengio, Y. Convolutional Networks for Images, Speech, and Time-Series. In *The Handbook of Brain Theory and Neural Networks*; Arbib, M.A., Ed.; MIT Press: Cambridge, MA, USA, 1995.

 © 2019 by the author. Licensee MDPI, Basel, Switzerland. This article is an open access article distributed under the terms and conditions of the Creative Commons Attribution (CC BY) license (http://creativecommons.org/licenses/by/4.0/).

*Proceedings*

# Radiometric Scale Transfer Using Bayesian Model Selection [†]

## Donald W. Nelson * and Udo von Toussaint

Longmont, Colorado, Max-Planck-Institut für Plasmaphysik, 85748 Garching, Germany; udo.von.toussaint@ipp.mpg.de
* Correspondence: donaldnelson@mac.com
† Presented at the 39th International Workshop on Bayesian Inference and Maximum Entropy Methods in Science and Engineering, Garching, Germany, 30 June–5 July 2019.

Published: 3 February 2020

**Abstract:** The key input quantity to climate modelling and weather forecasts is the solar beam irradiance, i.e., the primary amount of energy provided by the sun. Despite its importance the absolute accuracy of the measurements are limited—which not only affects the modelling but also ground truth tests of satellite observations. Here we focus on the problem of improving instrument calibration based on dedicated measurements. A Bayesian approach reveals that the standard approach results in inferior results. An alternative approach method based on monomial based selection of regression functions, combined with model selection is shown to yield superior estimations for a wide range of conditions. The approach is illustrated on selected data and possible further enhancements are outlined.

**Keywords:** broadband; irradiance; reference; solar radiation; climate modelling; pyrheliometer; Bayesian model comparison; evidence

## 1. Introduction

Broadband visible (0.295–3.5 microns) solar beam irradiance is measured using pyrheliometers. A concise history of solar beam irradiance measurements, beginning in the nineteenth century, is summarized in [1]. A pyrheliometer produces millivolt level output generated by a thermopile whose hot junctions are in contact with a black detector surface heated by incoming solar irradiance. The detector is situated behind a standardized view limiting aperture system with FOV (field of view) of five degrees. Thermopile voltage outputs must then be transformed into irradiance units of watts per square meter and this process requires a reference scale in irradiance units as well as a method to transform a raw voltage signal into an irradiance value from the reference scale. A history of radiometric reference scales in use during the twentieth century is discussed in [1–4]. The accuracy and precision of a radiometric reference scale transfer to operationally deployable pyrheliometers in use at the WMO regional level is the subject of this paper.

### 1.1. Current Practice

The World Meteorological Organization (WMO), since 1977, has sanctioned a reference scale for solar beam irradiance: the World Radiometric Reference (WRR) [5,6]. The WRR is maintained at the PMOD/WRC, (Physikalisch-Meteorologisches Observatorium Davos/World Radiation Center), located in Davos, Switzerland. At PMOD/WRC, a WSG (World Standard Group) of cavity radiometers calibrated by electrical substitution methods is used to create the WRR. At PMOD/WRC, the WSG has been in sustained use since the 1970s, and the WRR is defined as a weighted average of readings from the WSG [7].

Cavity type pyrheliometers are used as primary references for the WSG and in WMO regions. They are not used for operational monitoring sites due to their relatively high cost. Additionally, they are operated without a window to eliminate transmission and spectral effects on incoming solar irradiance so are vulnerable to ingestion of dust, rain, snow, insects etc. A standard field of view (FOV), 5 degrees, admits solar beam irradiance which enters a cavity shaped detector through a precision aperture whose area has been measured. The irradiance heats a detector surface in contact with a thermopile and an output signal on the order of millivolts is generated and recorded by an automated data acquisition system or by personnel manually recording readings from a voltmeter display. After a sequence of readings due to solar irradiance heating the detector, the cavity is blocked with a shutter mechanism and the detector is electrically heated to an output equivalent to the solar irradiance heating. The heater power is computed and the precision aperture allows the conversion into engineering units of watts per unit area. The cavity type radiometers are self calibrating, based on the ability of control circuitry to measure fundamental quantities: volts, ohms and amperes. The year over year sustained precision of ratios between cavity radiometers calibrated by electrical substitution is a reassuring achievement in accurate measurement of solar irradiance over the past forty years of the WRR.

The WSG at PMOD is routinely used to collect solar direct beam irradiance data throughout the year to ensure continuity of its group precision. At five year intervals, reference pyrheliometers and personnel from the seven worldwide WMO regions are invited to Davos Switzerland for an International Pyrheliometer Comparison (IPC). IPC-I, IPC-II and IPC-III were conducted in 1959, 1964 and 1970 respectively, and every five years since. The most recent, IPC-XII, was conducted in 2015, from 28 September to 16 October, and pyrheliometers from 15 Regional and 15 National Radiation Centers as well as 25 manufacturers and other institutions participated in the comparison and were represented by 111 individuals from 33 countries who operated 134 pyrheliometers of various configurations, design and manufacture [7].

During an IPC, clear sky periods occurring at Davos enable simultaneous measurements of solar beam irradiance by the WSG and each participating pyrheliometer, and a sufficient number of measurements are acquired for statistical analysis of the ratios of participating pyrheliometer irradiances to the WSG irradiances. A protocol for accepting or rejecting data points is adhered to [7]. After analysis, PMOD/WRC assigns a WRR factor to each pyrheliometer, which is its average ratio to the WRR during the IPC. Results for all pyrheliometers are summarized and published in a WMO publication [7]. In principle, this enables each IPC pyrheliometer to recreate the WRR as needed for reference scale transfer at the regional level. Self calibrating cavity type pyrheliometers are used as primary references for the WSG and in WMO regions. The righthand panel of Figure 1 illustrates precision achievable for a group of three participating cavity radiometers from the North American Region (WMO Region IV) in IPC-XII.

During IPC-XII, the self calibrating cavity type pyrheliometers exhibited standard deviations of their ratios to the WRR on the order of hundreds of parts per million, or less that 0.1 percent, as the three North American cavities show in Figure 1.

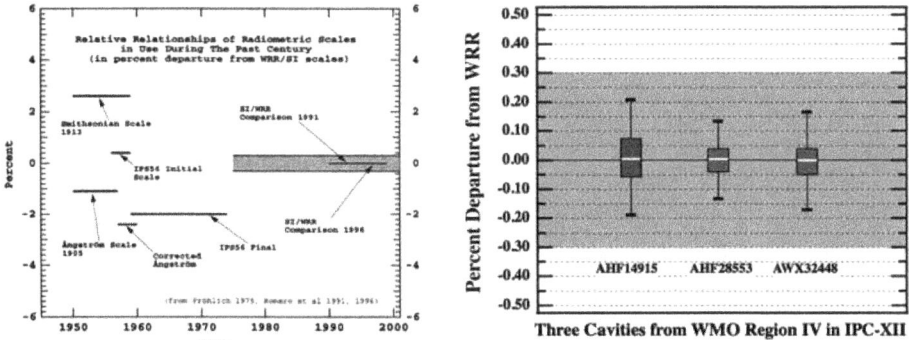

**Figure 1.** Left panel [8] shows efforts to establish a radiometric scale during the 20th century. The shaded band marks the adoption of the WRR. Relative relationships of the scales with respect to the current WRR are shown. Right panel illustrates the results for three cavities from WMO Region IV (North American Region) which participated in IPC-XII in 2015. Box plot data: box midlines are medians, box lower and upper boundaries are at the 25th and 75th percentiles and the box whisker caps are at the 2nd and 98th percentiles. Shaded bands in both panels denote the accepted uncertainty in the WRR. The scale in righthand panel is expanded by a factor of 12 with respect to panel on left.

*1.2. Implementing the WRR Within WMO Regions*

Throughout the seven WMO world regions, routine solar beam measurements are typically made using pyrheliometers that must be able to withstand exposure to extreme environments in which they may be used. The environments range from the South Pole in Antarctica to mid latitude sites, maritime South Pacific locations, equatorial continental sites, high mountain sites (e.g., Mauna Loa Observatory in Hawaii) and the Arctic. This requires a windowed hermetically sealed housing with the same field of view as a cavity but without the ability of self calibration. Detectors are thermopile based as in cavity type pyrheliometers but are not capable of self calibration and these pyrheliometers will be referred to as working class pyrheliometers and are mounted on solar tracking devices which maintain continuous alignment to the sun. An additional requirement is they must be economically feasible for purchase by a global community of users in need of reliable and accurate measurements of solar irradiance as well as being able to withstand exposure to weather and climate extremes.

A standard field of view, depending on vintage and manufacturer of the instrument, admits solar beam irradiance which heats a detector surface in contact with thermopile and an output signal on the order of millivolts is generated and recorded by a data acquisition system. Conversion of the original millivolt level signals from a working class pyrheliometer into engineering units of watts per square meter requires that it be calibrated against a radiometric reference scale traceable to the WRR.

However, currently achievable precision demonstrated at recent IPCs through ratioing self calibrating cavities to a WSG/WRR is not realized for working class pyrheliometers. A group of three working class pyrheliometers from WMO Region VI were participants in IPC-XII. Figure 2 below illustrates precision achievable for a group of three participating cavity radiometers from the North American Region (WMO Region IV) in IPC-XII, and three working class pyrheliometers from a North American Region manufacturer currently used in European Region VI. The degradation of precision is considerable. The purpose of this paper is to present results of an alternative method to calibrate working class pyrheliometers. The goal is to achieve a reduction in the uncertainty of an assigned irradiance from the WRR scale to an individual working class pyrheliometer output voltage.

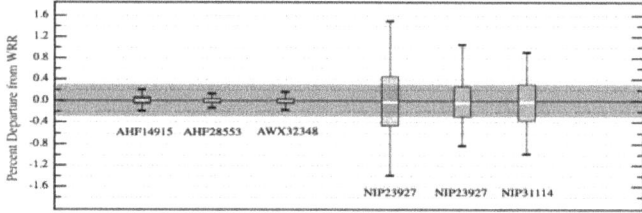

**Figure 2.** Box and whisker plots from IPC-XII Results for three North American Region cavity type pyrheliometers and a group of three working class pyrheliometers typically used at monitoring sites in WMO regions. Whisker caps are at 2nd and 98th percentiles. Box boundaries at 25th and 75th percentiles and median. Shaded band is stated uncertainty of the WRR. Precision losses in WRR ratios can range from factors of ten up to forty [7].

## 2. The Physical Model

### 2.1. Regional Calibration Procedure

Solar radiometry metrology has advanced considerably since 1970. Discussions of metrology of solar radiometry can be found in [9–17]. The advances in metrology of solar radiometry have been driven by the requirements of more accurate determination of the TSI, (total solar irradiance) measured by orbiting satellites equipped with self calibrating cavity radiometers. Ground based measurements have benefitted from these advances, and in particular, the cavity radiometers calibrated by electrical substitution.

IPC participants with self calibrating cavity radiometers are in position to reproduce the precision achievable at an IPC. For example, in Region IV (North America, Central America and the Caribbean), an annual ad hoc pyrheliometer comparison is conducted at NREL (National Renewable Energy Laboratory) in Golden Colorado. These NPC (NREL Pyrheliometer Comparison) events are conducted every fall during non-IPC years. A surrogate WSG, based on a group of participating cavities from the most recent IPC, is used to create a reference irradiance scale. NPC participants unable to attend a recent IPC, can compare their cavities to this surrogate WSG and realize a precision of ratios comparable to those achievable at an IPC. Figure 3 illustrates the precision of the same cavities displayed in Figure 2, but for their results from an NPC conducted in 2018 at NREL [18], three years after their most recent participation in IPC-XII. Cavities that participate in an NPC but not the most recent IPC are able to create their own surrogate WRR and establish traceability to the WSG/WRR in Davos.

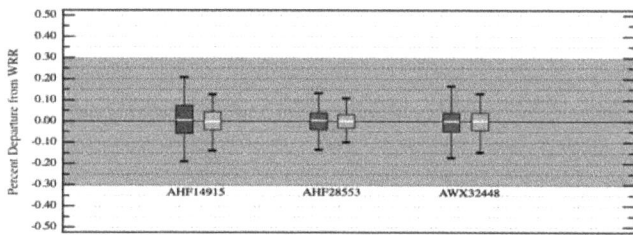

**Figure 3.** Results for three North American Region cavity type pyrheliometers. Darker shaded box plot data: IPC-XII 2015. Lighter shaded box plot data: NPC-2018. The shaded band across the graph is the historical assigned uncertainty of the WRR. Box and whisker information is the same as in Figures 1 and 2.

A working class pyrheliometer at the regional level is usually calibrated by operating it side by side with a reference cavity traceable to the most recent IPC or NPC and calibrations can be performed on an as needed basis. Typically, a group of working class pyrheliometers are calibrated together using a WRR or NPC traceable cavity. A protocol similar to an IPC is used for collecting measurements of direct beam solar irradiance from the reference cavity and output voltages. Data are collected at chosen time intervals and the voltage readings from a pyrheliometer under test are ratioed to the concurrent reference cavity irradiance values. Unlike an IPC or NPC, the pyrheliometers under test are not self-calibrating. The ratios formed by dividing readings from the pyrheliometers under test by the irradiances measured with the cavity have units of microvolts per watt per square meter and are referred to as responsivities. These ratios are collected over time periods that can vary from hours to days and weeks, depending on frequency of clear sky conditions during the calibration period and the judgment of personnel performing the calibration. Orthodox statistical techniques are used to process the set of ratios from individual pyrheliometers and one ratio value is generated and assigned as its responsivity. The assigned ratio is the chosen model for transforming voltage readings from a working class pyrheliometer into irradiance values traceable to the surrogate WRR generated by the reference cavity. In various forms, depending on the history of pyrheliometer design, this has been the model for assigning a reference scale irradiance to a given output from a working class pyrheliometer and has been used for the past century. In contrast, a typical IPC lasts for three weeks and usually clear sky periods occur such that enough readings are recorded to confidently produce WRR correction factors for all participating instruments. Protocols for IPC comparisons impose strict constraints on when data is officially recorded for determination of WRR factors but the IPCs are only scheduled every five years. At the regional level, working class pyrheliometers are utilized in long term monitoring networks, renewable energy applications, efficiency monitoring of solar power generation sites, commercial calibration services and research institutions. These pyrheliometers may also be installed at remote field sites and operate unattended and continuously for months and years. Periodically they require recalibration and are replaced by a more recently calibrated unit. This is the reality of establishing and maintaining long term monitoring networks for measurement of solar radiation.

## 2.2. Standard Calibration Procedure

As outlined before the standard calibration procedure consists of assigning the responsivity (i.e., the conversion factor from the measured voltage $v$ to the irradiance $P$). As shown in Figure 4 the assumption of a linear relation $P = \alpha v$ holds to a large extent.

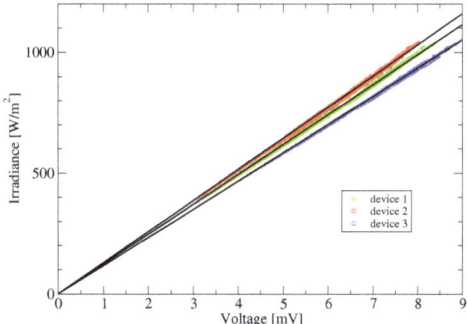

**Figure 4.** Results for three working class pyrheliometers used to collect clear sky solar irradiance data from 16 May 2013 to 12 August 2013. Sampling rate was 1 hertz and data were processed into one minute averages. A total of 14914 data points are included in the data set. The solid lines are given by linear regression of $P = \alpha_i \times v$ to the measured data for each of the three devices.

However, closer inspection of the residuals (c.f. Figure 5) reveals some remaining structure, thus indicating that the difference between data and model is not purely stochastic.

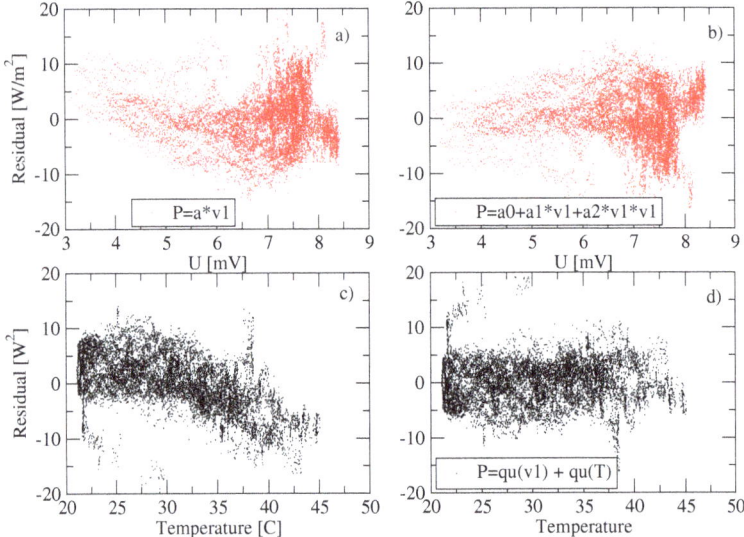

**Figure 5.** Panel (**a**) displays the difference between the data and the linear model for one of the data sets displayed in Figure 4. Panel (**b**) provides the residuum if instead of a model linear in the measured voltage a quadratic relationship is being assumed. The overall magnitude of the residuum in panel (**b**) is visibly smaller compared with panel (**a**). However, if the residuum is plotted as function of the device temperature (panel (**c**)) again a non-stochastic behavior is evident. Instead, panel (**d**) displays a bivariate regression function i.e., a sum of a quadratic function of the voltage $v_1$ and a quadratic function of the temperature $T$ appears to capture the data reasonably well. The resulting residuum shows no clear non-stochastic behavior.

In the panels of Figure 5 different residuals are displayed. Panel (a) displays the difference between the data and the linear model for one of the data sets displayed in Figure 4. Panel (b) provides the residuum if instead of a model linear in the measured voltage a quadratic relationship is being assumed. The overall magnitude of the residuum in panel b is visibly smaller compared with panel a. However, if the residuum is plotted as function of the device temperature (panel (c)) again a non-stochastic behavior is evident. Instead, a bivariate regression function i.e., a sum of a quadratic function of the voltage and a quadratic function of the temperature appears to capture the data reasonably well: the resulting residuum shows no clear systematic. But of course this may not be the optimal regression function. In any case the results so far indicate that the evaluation of the pyrheliometers may benefit from a multivariate regression approach, considering also other factors besides the measured voltage. Based on the design of the devices at least two additional parameters could be of importance besides the voltage $v$: the temperature $T$, potentially affecting the electronics or the device geometry (by thermal expansion), thermopile temperature dependence and the cosine of the solar zenith angle $c$.

## 2.3. Multivariate Linear Regression

The preceding discussion results in three likely parameters for the model: $f = f(T,c,v)$. However, the functional form of $f$ is unknown. Based on the experience of the change of the residuals using low order polynomials it appears reasonable to express the model function as a sum of multivariate monomials

$$f(T_i, c_i, v_i) = \sum_{k=1}^{E} x_{ik} a_k = \sum_{k=1}^{E} T_i^{l_k} c_i^{m_k} v_i^{q_k} \cdot a_k, \tag{1}$$

where $E$ is the expansion order of the model. As basis sets we consider the set of all monomials in these variables up to a total degree (the sum of all three exponents $l_k + m_k + q_k$) of 3, corresponding to 20 different basis functions, for example monomials like $T^2 c^0 v^0$, $T^1 c^1 v^1$ or $T^1 c^0 v^2$. Depending on the expansion order out of these 20 basis functions $E$ are chosen and used for a multivariate linear regression to the data. This yields for a given expansion order ($\binom{20}{E}$) possible models. Since a priori we neither know the most adequate expansion order $E$ nor the best set of monomials for a given expansion order we compute the model evidence in the MAP approximation for all possible combinations up to $E = 10$, resulting in the comparison of more than $10^6$ different models. Our approach employs standard Bayesian model comparison as outlined e.g., in [19]. For the large number of models it is only possible because most of the necessary numerics to compute the evidence can be done analytically, as is shown next.

If we assume that $\langle \varepsilon_i \rangle = 0$ and $\langle \varepsilon_i^2 \rangle = \sigma_i^2$ and negligible uncertainty in the voltage measurement the data $y$ are given by

$$y_i = f(T_i, c_i, v_i) + \varepsilon_i ,\qquad(2)$$

and the likelihood for $N$ independent measurements becomes

$$p(y|x, a, \sigma, E, \mathcal{I}) = \frac{(2\pi)^{-\frac{N}{2}}}{\prod_i \sigma_i} \exp\left\{ -\frac{1}{2} \sum_{i=1}^{N} \frac{(y_i - \sum_{k=1}^{E} x_{ik} a_k)^2}{\sigma_i^2} \right\}$$

$$= \frac{(2\pi)^{-\frac{N}{2}}}{\prod_i \sigma_i} \exp\left\{ -\frac{1}{2} \phi \right\} .\qquad(3)$$

The notation simplifies if we introduce the vectors $y$ and $a$ and the matrices $X = \{x_{ik}\}$ and $S^{-2} = \mathrm{diag}(\sigma_i^{-2})$ in the argument $\phi$ of the exponential. Then

$$\phi = (y - Xa)^T S^{-2} (y - Xa) .$$

It is convenient to introduce new variables $y' = S^{-1} y$ and $X' = S^{-1} X$ in the exponent, resulting in

$$\phi = (y' - X'a)^T (y' - X'a) = y'^2 - 2a^T X'^T y' + a^T X'^T X' a ,\qquad(4)$$

The maximum of the likelihood is achieved for

$$0 = \nabla \phi = -2 X'^T y' + 2 X'^T X' a ,$$
$$\Rightarrow \qquad (X'^T X') a^{\mathrm{ML}} = X'^T y' .$$

In the following, we assume that the inverse of the matrix $X'^T X'$ exists. A necessary condition is $N \geq E$, i.e., we have at least as many measurements as there are parameters. If the inverse matrix exists then

$$a^{\mathrm{ML}} = (X'^T X')^{-1} X'^T y' .$$

Since $\phi$ is quadratic in $a$ we can rewrite $\phi$ in Equation (4) as a complete square in $a$ plus a residue

$$\phi = R + (a - a_0)^T Q (a - a_0) .\qquad(5)$$

This form is achieved via a Taylor expansion about $a = a_{\mathrm{ML}}$. The second derivatives (Hessian) provide

$$Q = \frac{1}{2} \nabla \nabla^T \phi(a) \bigg|_{a^{\mathrm{ML}}} = X'^T X' .\qquad(6)$$

The matrix on the right-hand-side also shows up in the maximum likelihood solution

$$a^{\mathrm{ML}} = Q^{-1} X'^T y' .\qquad(7)$$

The residue $R$ in Equation (5) is the constant $R = \phi(a^{\text{ML}})$ of the Taylor expansion, that can be transformed into

$$R = y'^T \left(1 - X'Q^{-1}X'^T\right) y' . \tag{8}$$

This general result can be cast in an advantageous form employing singular value decomposition [20] of the matrix $X'$:

$$X' = UDV^T . \tag{9}$$

The sizes of the matrices $U$, $D$, and $V$ are $(N \times E)$, $(E \times E)$, and $(E \times E)$ respectively. The transposed matrix $X'^T$ is simply

$$X'^T = VDU^T \tag{10}$$

and the product $X'^T X'$ becomes, using the unitarity of $U$

$$Q = X'^T X' = VDU^T UDV^T = VD^2 V^T \tag{11}$$

The last equation is also known as the spectral decomposition of the real symmetric matrix $X'^T X'$. We assume that the singular values are strictly positive, in order that the inverse of $Q$ exists. The virtue of the spectral decomposition is that it yields immediately the inverse $Q^{-1}$ as

$$Q^{-1} = VD^{-2}V^T . \tag{12}$$

It is easily verified that the matrix product of Equations (11),12 yields the identity matrix $1$, which is a consequence of the left unitarity of $U$. The maximum likelihood estimate $a_{\text{ML}}$ is given by

$$a^{\text{ML}} = Q^{-1} X'^T y' = VD^{-2} V^T VDU^T y' = VD^{-1} U^T y' = \sum_i \left(\frac{u_i^T y'}{\lambda_i}\right) v_i , \tag{13}$$

where $u_i$ ($v_i$) are the column vectors of $U$ ($V$). The maximum likelihood estimate $a^{\text{ML}}$ is thereby expanded in the basis $\{v_i\}$ with expansion coefficients $(u_i^T y' / \lambda_i)$.

We now turn to the Bayesian estimation of $a$ [19]. The full information on the parameters $a$ is contained in the posterior distribution

$$p(a|y, X, \sigma, E, \mathcal{I}) = \frac{1}{Z} p(a|E, \mathcal{I}) \, p(y|X, a, \sigma, E, \mathcal{I}) , \tag{14}$$

from which we can determine for example the maximum posteriori (MAP) solution via

$$p(a|E, \mathcal{I}) \nabla_a p(y|X, a, \sigma, E, \mathcal{I}) + p(y|X, a, \sigma, E, \mathcal{I}) \nabla_a p(a|E, \mathcal{I}) = 0 .$$

For a sufficient number of well determined data the likelihood is strongly peaked around its maximum $a^{\text{ML}}$, while the prior will be comparatively flat, in particular if it is chosen uninformative. A reliable approximate solution will then be obtained from

$$p(a|E, \mathcal{I}) \nabla p(y|X, a, \sigma, E, \mathcal{I}) = 0 , \tag{15}$$

which is the maximum likelihood estimate. Similar arguments hold for posterior expectation values of any function $f(a)$

$$\langle f(a) \rangle = \frac{1}{Z} \int d^E a \, f(a) \, p(a|E, \mathcal{I}) \, p(y|X, a, \sigma, E, \mathcal{I}) ,$$
$$Z = \int d^E a \, p(a|E, \mathcal{I}) \, p(y|X, a, \sigma, E, \mathcal{I}) . \tag{16}$$

For general $p(a|E,\mathcal{I})$ the integrals can only be performed numerically. Using the fact that the likelihood is generally precisely localized compared to the diffuse prior. This suggests, that we replace the prior $p(a|E,\mathcal{I})$ by $p(a^{\text{ML}}|E,\mathcal{I})$ and take it out of the integrals

$$\langle f(a) \rangle \approx \frac{p(a^{\text{ML}}|E,\mathcal{I})}{Z} \int d^E a\, f(a)\, p(y|X,a,\sigma,E,\mathcal{I}),\tag{17}$$

$$Z \approx p(a^{\text{ML}}|E,\mathcal{I}) \int d^E a\, p(y|X,a,\sigma,E,\mathcal{I}).\tag{18}$$

Now, since $p(y|X,a,\sigma,E,\mathcal{I})$ is a multivariate Gaussian in $a$ the expectation value $\langle f(a) \rangle = a_i$ and the covariance $\langle f(a) \rangle = \text{cov}(a_i,a_j)$ can easily be determined

$$\langle a \rangle = a_{\text{ML}} \tag{19}$$

$$\text{cov}(a_i,a_j) = Q_{ji}^{-1} \tag{20}$$

We have derived a reasonable approximation for the normalization $Z$, also called the »prior predictive value« or the »evidence«. $Z$ represents the probability for the data, given the assumed model. The question at hand in the present problem is of course whether the data require really a high expansion order $E$ or are they also satisfactorily explained by some lower order $E' < E$. For these problems the full expression for $Z$ is required including the prior factor. The remaining Gaussian integral in Equation (18) can be performed easily resulting in

$$Z \approx p(a^{\text{ML}}|E,\mathcal{I}) \frac{(2\pi)^{\frac{E-N}{2}}}{\prod_i \sigma_i} |Q|^{-1/2} \exp\left\{-\frac{1}{2}y'^T (1 - X'Q^{-1}X'^T) y'\right\}.$$

The singular value decomposition of the argument of the exponential yields

$$1 - X'Q^{-1}X'^T = 1 - UDV^T VD^{-2}V^T VDU^T = 1 - UU^T = 1 - \sum_{i=1}^{E} u_i u_i^T$$

and the evidence finally reads

$$p(y|X,\sigma,E,\mathcal{I}) \approx p(a^{\text{ML}}|E,\mathcal{I}) \frac{(2\pi)^{\frac{E-N}{2}}}{\prod_k^E \lambda_k \prod_i^N \sigma_i} \exp\left\{-\frac{1}{2}y'^T \left(1 - \sum_{i=1}^{E} u_i u_i^T\right) y'\right\}.\tag{21}$$

The exponent in Equation (21) represents that part in the data that is due to noise or a different model.

It remains to assign a prior distribution to the linear parameters $a$ of the model. For simplicity we assign a normalized uniform prior for all components $p(a_i) = 1/400$ in the range $[-200, 200]$ and $p(a_i) = 0$ outside. Thus each additional parameter reduces the log-evidence by $\log(400) \simeq 6$. More refined prior distributions based on Maximum Entropy concepts are possible [19] but have not been considered in this work. The standard deviation of the power measurements was assumed to be $\sigma = 1\frac{W}{m^2}$ throughout—this is presumably too low but we did not want to mask systematic trends by assuming a too large noise level.

## 3. Results

The key results of this massive model comparison scanning over the expansion order and searching the best monomial combination for each expansion are displayed in Figure 6. The model evidence indicates that the data mandate an expansion order of around 8. Here also the improvement of the misfit with increasing number of parameters levels off. Inspection of the selected terms reveals that consistently - besides the powers of the measured voltage mixed terms of the form $T^k v^q$ and $c^l v^q$

were present in the best regression model. It is noteworthy that the key monomials $(v, v^3, cv, Tcv, T^2c)$ were the same for all three investigated devices. The increased order of the regression function reduced the deviation between the data and the model by more than 20% without any indications of overfitting. For expansion orders above 10 the condition number of the design matrix exceeded $10^5$, indicating that the available basis set becomes increasingly colinear and the fit less reliable. Therefore results obtained with $E > 10$ were not considered further.

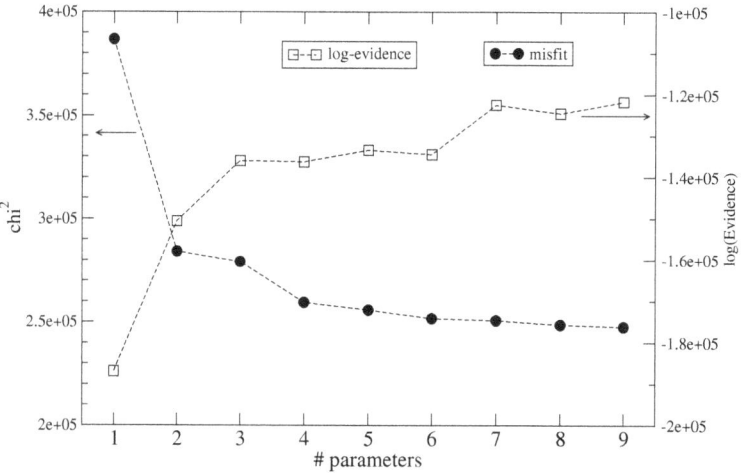

**Figure 6.** Misfit and evidence as function of the number of parameters. For each number of parameters the corresponding best fit is provided. On the left hand ordinate the misfit ($\chi^2$-value) between data and model is given. On the ordinate on the right hand side the logarithm of the model evidence is given. Higher values indicate a more probable model. As can be clearly seen from both misfit and model evidence the commonly used 1 parameter regression is inadequate. The evidence obtains nearly identical peak value maxima for 7 and 9 parameters where also the improvement for the $\chi^2$-values levels off.

Based on these results the calibration of pyrheliometer should take into account the temperature and the angle of irradiance as well instead of relying on a single calibration factor. This can significantly improve the accuracy. This is illustrated in Figure 7. A data set was collected using a group of three working class pyrheliometers in use for decades at a WMO Region IV RRC (Regional Radiation Center) located at the NOAA/Earth Systems Research Laboratory in Boulder, Colorado. One minute averages of data sampled at one hertz over the time period from 16 May–12 August, 2013 were used to create a data set for analysis. Clear sky periods were chosen and no lower limits were imposed on irradiance values (An IPC only uses irradiance values above 700 watts/sq.m). All solar zenith angles were used for the analysis subject to a constraint that the skies were categorized as clear. A total of 14914 one minute averages were used in the analysis. The darker shaded box and whisker plots summarize residuals after subtracting the pyrheliometer irradiances computed using its assigned responsivity value from the surrogate WRR reference. The lighter shaded box plots summarize the residuals after fitting the three parameter linear model using 8 monomials. There is measurable reduction in the uncertainty.

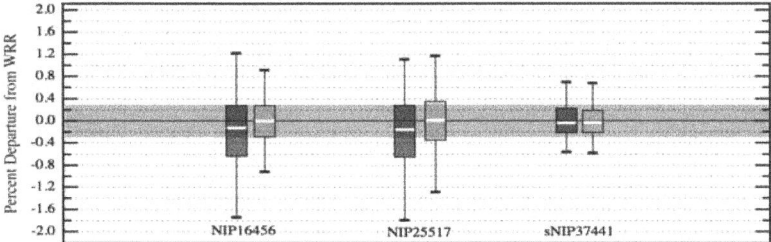

**Figure 7.** Results for three working class pyrheliometers used to collect clear sky solar irradiance data from 16 May 2013 to 12 August 2013. Sampling rate was 1 hertz and data were processed into one minute averages. A total of 14914 data points are included in the data set. Solar zenith angle range: 16–84 degrees Irradiance range: 375–1050 watts per square meter. Darker shaded boxes are residual summaries of ratio responsivity scaling. The lighter shaded boxes are residual summaries of a linear model scaling of pyrheliometer outputs. Box and whisker information is the same as in Figures 1–3.

## 4. Conclusion and Outlook

The precise estimation of solar irradiation is a key factor for climate modelling. The measurements are challenging due to a wide variety of measurement conditions, the need for stability on an absolute scale and the high precision requirements. The present results - based on an exhaustive Bayesian model comparison on more than $10^6$ different models- indicate that with relatively minor changes to the calibration protocol the achievable measurement accuracy can be significantly increased. It is highly beneficial to consider also the temperature and the angle of incidence for the calibration of the devices, improving the accuracy by more than 20% on the devices. Some further improvements appear possible: The data were analyzed as provided without any preprocessing. The residuals, however, display the presence of a small fraction of outliers which are not in agreement with the Gaussian likelihood used for the analysis. This may distort the estimate of the best fit parameters. Thus the application of a robust estimation method may improve the accuracy of the measurements even further. It could also be of interest to check if additional factors (besides temperature and angle) affect the device performance. And finally it may also be worthwhile - besides a self-consistent estimation of $\sigma$ - to validate the estimated evidence values for the selected best fit models using thermodynamic integration or nested sampling as function of expansion order to check for a potential bias in the MAP based estimation of the evidence.

**Author Contributions:** D.W.N. performed the experiments and provided the data; U.v.T. analyzed the data and contributed analysis tools. The paper was written jointly.

**Conflicts of Interest:** The authors declare no conflict of interest.

## References

1. Fröhlich, C. History of Solar Radiometry and the World Radiometric Reference. *Metrologia* **1991**, *28*, 111–115.
2. Latimer, J. On the Angström and Smithsonian absolute pyrheliometric scales and the International Pyrheliometric Scale 1956. *Tellus* **1973**, *15*, 586–592.
3. Hickey, J. Laboratory Methods of Experimental Radiometry Including Data Analysis. *Adv. Geophys.* **1970**, *14*, 227–267.
4. Smithsonian Institution Radiation Biology Laboratory. In Proceedings of the Symposium on Solar Radiation Measurements and Instrumentation, Washington, DC, USA, 13–15 November 1973.
5. Brusa, R.; Fröhlich, C. *Realisation of the Absolute Scale of Total Irradiance*; Technical Report; Swiss Meteorological Institute: Zürich, Switzerland, 1975.

6. Fröhlich, C. *World Radiometric Reference, Final Report*; Technical Report WMO 490; World Meteorological Organization: Geneva, Switzerland, 1977.
7. Finsterle, W. *The International Pyrheliometer Comparison, Final Report*; Technical Report WMO IOM 124; World Meteorological Organization: Davos, Switzerland, 2016.
8. Nelson, D. *The NOAA Climate Monitoring and Diagnostic Laboratory Solar Radiation Facility*; Technical Report Technical Memorandum OAR CMDL-15; Climate Monitoring and Diagnostics Laboratory: Boulder, CO, USA, 2000.
9. Kendall, J.; Burdahl, C. Two Blackbody Radiometers of High Accuracy. *J. Appl. Opt.* **1970**, *9*, 1082–1091.
10. Brusa, R.; Fröhlich, C. Absolute radiometers (PMO6) and their experimental characterisation. *J. Appl. Opt.* **1986**, *25*, 4173–4180.
11. Romero, J.; Fox, N.P.; Fröhlich, C. First compariosn of the solar and SI Radiometric Scale. *Metrologia* **1991**, *28*, 125–128.
12. Romero, J.; Fox, N.P.; Fröhlich, C. Improved Comparison of the World Radiometric Reference and the SI Radiometric Scale. *Metrologia* **1995**, *32*, 523–524.
13. Romero, J.; Fox, N.P.; Fröhlich, C. *Maintenance of the World Radiometric Reference*; Technical Report International Pyrheliometer Comparison IPCVIII: Working Report 188; Swiss Meteorological Organization: Zurich, Switzerland, 1996.
14. Finsterle, W.; Blattner, P.; Moebus, S.; Rüedi, I.; Wehrli, C.; White, M.; Schmutz, W. Third comparison of the World Radiometric Reference and the SI radiometric scale. *Metrologia* **2008**, *45*, 377–381. doi:10.1088/0026-1394/45/4/001.
15. Fehlmann, A. Metrology of Solar Irradiance. Ph.D. Thesis, University of Zurich, Zurich, Switzerland, 2011.
16. Kopp, G.; Fehlmann, A.; Finsterle, W.; Harber, D.; Heuerman, K.; Willson, R. Total solar irradiance data record accuracy and consistency improvements. *Metrologia* **2012**, *49*, S29–S33. doi:10.1088/0026-1394/49/2/s29.
17. Suter, M. Advances in Solar Radiometry. Ph.D. Thesis, University of Zurich, Zurich, Switzerland, 2014.
18. Reda, I.M.; Dooraghi, M.R.; Andreas, A.M.; Grobner, J.; Thomann, C. *NREL Comparison Between Absolute Cavity Pyrgeometers and Pyrgeometers Traceable to World Infrared Standard Group and the InfraRed Integrating Sphere*; Technical Report NREL/TP-1900-72633; National Renewable Energy Laboratory: Golden, CO, USA, 2018. doi:10.2172/1480239.
19. von der Linden, W.; Dose, V.; von Toussaint, U. *Bayesian Probability Theory: Applications in the Physical Sciences*; Cambridge University Press: Cambridge, UK, 2014. doi:10.1017/CBO9781139565608.
20. Press, W.H.; Teukolsky, S.A.; Vetterling, W.T.; Flannery, B.P. *Numerical Recipes 3rd Edition: The Art of Scientific Computing*, 3rd ed.; Cambridge University Press: Cambridge, UK, 2007.

© 2020 by the authors. Licensee MDPI, Basel, Switzerland. This article is an open access article distributed under the terms and conditions of the Creative Commons Attribution (CC BY) license (http://creativecommons.org/licenses/by/4.0/).

*Proceedings*

# 2D Deconvolution Using Adaptive Kernel [†]

**Dirk Nille * and Udo von Toussaint**

Max-Planck-Institut für Plasmaphysik, Boltzmannstrasse 2, 85748 Garching, Germany; Udo.v.Toussaint@ipp.mpg.de
* Correspondence: Dirk.Nille@ipp.mpg.de
† Presented at the 39th International Workshop on Bayesian Inference and Maximum Entropy Methods in Science and Engineering, Garching, Germany, 30 June–5 July 2019.

Published: 21 November 2019

**Abstract:** An analysis tool using Adaptive Kernel to solve an ill-posed inverse problem for a 2D model space is introduced. It is applicable for linear and non-linear forward models, for example in tomography and image reconstruction. While an optimisation based on a Gaussian Approximation is possible, it becomes intractable for more than some hundred kernel functions. This is because the determinant of the Hessian of the system has be evaluated. The SVD typically used for 1D problems fails with increasing problem size. Alternatively Stochastic Trace Estimation can be used, giving a reasonable approximation. An alternative to searching for the MAP solution is to integrate using Marcov Chain Monte Carlo without the need to determine the determinant of the Hessian. This also allows to treat problems where a linear approximation is not justified.

**Keywords:** inverse problem; regularisation; Adaptive Kernel

## 1. Introduction

An Adaptive Kernel model formulated in 2D is introduced, with the application of analysing data from an infrared camera system in order to determine surface heat loads. A key ingredient to solve this problem efficiently is the use of automatic differentiation (AD). The fast availability of gradients increases the efficiency and reliability of the optimisation significantly. However, the combination of matrix operations and AD results in a poor scaling of the demanded memory for increasing system size. In extending the model from 1D to 2D, this is the mayor obstacle. Stochastic Trace Estimation as alternative way to deal with large matrices is investigated and together with the SVD compared against results obtained by Marcov Chain Monte Carlo.

The forward model is based on a non-linear heat diffusion solver and the measurement system in form of Planck's Law. The classic version of the numerical tool THEODOR—a solver for the heat diffusion equation—used at ASDEX Upgrade and other machines solves the heat diffusion equation in two dimensions—1D surface and into the depth, further more referred to as 1D. Numerical tools for the general 2D case—2D surface plus depth—exist, e.g. [1], but are only used for deterministic calculations. A former contribution [2] introduced a Bayesian approach with THEODOR as forward model, called Bayesian THEODOR (BayTH). Its capabilities were extended from 1D to 2D data. This includes the forward model based on THEODOR and the Aadaptive Kernel (AK) model. The latter is used to describe our quantity of interest, the heat flux impinging on the surface.

The shape and temporal evolution of the heat flux pattern of a magnetically confined plasma onto the first wall is of great interest for fusion research. Heat flux densities of several MW/m$^2$ poses a threat to the exposed material [3,4]. The heat flux distribution is a footprint of the transport in the plasma edge [5,6]. Understanding the transport in the plasma edge is important to predict the behaviour of larger devices, aiming for a future fusion power plant. No direct measurement of the heat flux in the plasma is available. A method with sufficient spatial and temporal resolution to analyse many effects is to measure the thermal response of the target material, where the plasma deposits thermal energy. The impinging heat raises the temperature of the material, which itself transports the heat via conduction into the bulk. From the measured temporal evolution of the surface temperature, the heat flux into the material is deduced.

## 2. Forward Model

### 2.1. Heat Diffusion

The forward model for the heat transport in the target material is based on the THEODOR code, as described in [7]. Figure 1 shows an example: a rectangular cross section through the material, the colour-coding represents the temperature distribution in the tile.

The heat transport in the divertor target is described by heat diffusion, with a non-linear diffusion coefficient $\kappa$ with respect to the temperature.

$$\frac{\partial T}{\partial t} \rho c_p = \nabla \cdot (\kappa(T) \nabla T) \ . \tag{1}$$

Here $\rho$ and $c_p$ are the mass density and specific heat capacity of the material. The temperature $T$ is furthermore substituted by the heat potential $\kappa$

$$u(\kappa) = \int_0^T \kappa(T') dT' \tag{2}$$

leading to the semi-linear differential equation

$$\frac{\partial u}{\partial t} = \frac{1}{\rho c_p} \chi(u) \Delta u \ . \tag{3}$$

With the diffusivity $\chi$ beeing related to the conductivity $\kappa$ via

$$\chi = \frac{\kappa}{\rho c_p} \ . \tag{4}$$

This system is solved using the finite difference implicit Euler scheme with operator splitting. The spatial derivative is split into three parts: two along the surface—$\Delta_x$ and $\Delta_z$—and a part into the depth of the tile $\Delta_y$. This leads to three tridiagonal systems, which are solved successively using the Thomas Algorithm [8].

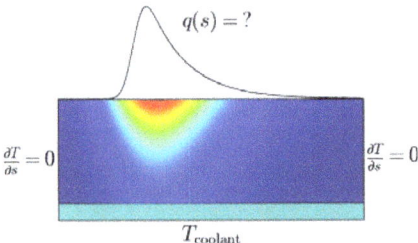

**Figure 1.** Sketch of the cross section of the target material with the temperature encoded in the colour. From measured surface temperatures the spatially resolved heat flux density $q(s)$ impinging onto the surface has to be deduced. The lateral boundary conditions allow no heat transport while the back side is in contact with a coolant.

## 2.2. Measurement System—Infrared Thermography

The second part of the forward calculation connects the behaviour of the target with the measurement system. Infrared (IR) thermography is based on Planck's Law, describing radiation emitted from surfaces with finite temperature. emitted from the surface of materials with finite temperature. Knowing material parameters like the emissivity allows to deduce the surface temperature from measuring the emitted photon flux and vice versa [2]. The modelled surface temperature translates into the photon rate emitted by the surface. For given integration times of the sensor, the photon rate observed through the aperture is translated into the counted photons. For systems where electronic noise can be neglected, the uncertainty of the signal is dominated by the photon statistics. The signal to compare to are *counts*, integers obtained by an analog-to-digital-converter, which are typically a fraction of the observed photons.

Figure 2 shows an image of the IR camera in ASDEX Upgrade (AUG) for the upper divertor during a discharge. The corners of the tile are clearly warmer than most of the exposed tile surface. This highlights the need of a 2D evaluation.

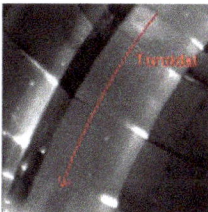

**Figure 2.** IR camera image in AUG during discharge 34549 on the upper divertor. The signal at the edges differs from the bulk and even neighbouring tiles show different responses. The toroidal symmetry approximation breaks.

## 3. Heatflux Model: Adaptive Kernel

To describe the surface heat flux distribution a multi-resolution model is used. A detailed description for the 1D variant can be found in [2,9]. Another approach is introduced in [10], including an iterative method to reduce the number of kernels—respective pixon in the reference.

The noise level for our application is expected to vary in time and space, as the amount of emitted radiation depends on the temperature of the surface area, which can be strongly peaked. Instead of using a global regularisation term, the Aadaptive Kernel approach allows a self consistent determination of the *best* resolution.

For an inverse problem with measured data $\vec{d}$ and linear model **A**—for this application a discrete formulation of equation (3)—we want to deduce our quantity of interest $\vec{u}$ from

$$\vec{d} = \mathbf{A} \cdot \vec{u} + \vec{\epsilon} . \tag{5}$$

Here $\vec{\epsilon}$ is the *noise* or *uncertainty* of the measurement—e.g., zero mean normal distributed. For the forward modelling we describe our quantity of interest $\vec{u}$ with Aadaptive Kernel. They are expressed in terms of another linear system

$$\vec{u} = \mathbf{B} \cdot \vec{h} \tag{6}$$

with a hidden image $\vec{h}$ and a smoothing matrix **B** depending on another set of parameters, the kernel widths $\vec{b}$. In total we find the model prediction for the likelihood

$$\vec{m} = \mathbf{AB} \cdot \vec{h} \tag{7}$$

where $\vec{h}$ can have an arbitrary resolution and the regularisation **B** is part of the inference.

As kernel function the normal distribution is used, characterised by the distance to its centre $x$ and its width $\sigma$. As upper limit for the kernel widths the system size or a physical reasonable limit may be used. For the lower limit the distance between the kernels is a good measure. For a normal distribution it is recommended to evaluate the error function for the contributions instead of a point wise evaluation of the exponential function. This ensures normalisation and a more smooth transition for small kernel widths, say $\sigma \leq 3$:

$$B_{i,j} = \frac{1}{2}\left[\operatorname{erf}\left(\frac{x_i - x_j + 0.5}{\sqrt{2}\sigma_j}\right) - \operatorname{erf}\left(\frac{x_i - x_j - 0.5}{\sqrt{2}\sigma_j}\right)\right] \quad (8)$$

For the 2D case the kernel function has to be generalised from an 1D to a 2D distribution. The simplest approach for any 1D distributions is to use the outer product of two independent distributions along the two axes. Figure 3 shows an example for two perpendicular, independent normal distributions located at different positions along their axis. This is justified for signals with a *fast axis*, which is aligned to one of the kernel axis. A more general treatment can be desirable and would include an angle between the elliptical kernel and the image or the kernel expressed via a covariance matrix with three independent entries.

For this work the outer product of two Gaussian kernels is used, which allows faster computation and normalisation than a rotated kernel system. Therefore, per kernel there are three parameters: amplitude $h_{i,j}$, kernel widths $\sigma_{x,i,j}$ and $\sigma_{y,i,j}$. In addition to these $3 \times N_k$ parameters for $N_k$ kernel, there is a weight for the entropic prior $\alpha$ for the hidden image [11].

The wider—hence smoother—the kernel, the smoother the resulting function. By using not a fixed width, but treating the width of every kernel as hyper parameter, the *best* resolution is found via model selection. This works implicitly, as Bayes Theorem acknowledges the increasing anti-correlation of neighbouring kernels for increasing kernel width.

This approach has been shown to work well for positive additive distributions (PADs) [9] like spectroscopy and depth profiles.

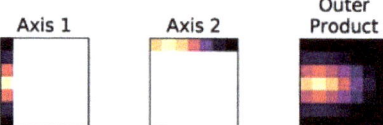

**Figure 3.** Example for the 2D kernel function as outer product of two Gaussians.

### 3.1. Model Selection and Effective Number of Degrees of Freedom (eDOF)

The degrees of freedom are an important quantity for model comparison, as additional DOFs typically improve the likelihood, while not necessarily gaining more information about the system. For the Aadaptive Kernel, the complexity is described by the transfer matrix, mapping the hidden image into the model space. In the simplest case **B** is the unit matrix, corresponding to a 1 to 1 map from hidden image to model function $f$ without smoothing. Note that this matrix is not in general square, e.g., for more or less kernels than cells in the model space for over- or under-sampling.

A more detailed explanation of the effective degrees of freedom can be found in [9]. Applying the evidence approximation to the probability distribution of the kernel widths $b$ given the data $P(b|d)$, assuming the distributions is peaked around a set of optimal widths $\hat{b}$

$$P(b|d) \approx \delta(b - \hat{b}) , \quad (9)$$

leaves us with the marginal

$$P(b|d) \propto \int d^N h P(d|b,h) P(b) P(h) . \quad (10)$$

Expanding its logarithm to second order around the *best* kernel weights $\vec{h}^\star$ leads to

$$P(b|d) \approx P(b)\ P(d|b,\vec{h}^\star)P(\vec{h}^\star)\det{}^{-1/2}(\mathbf{H}) \tag{11}$$

with $\mathbf{H}$ the hessian of $\log\left(P(d|h,b)\cdot P(h)\right)$. For a linear operator $\mathbf{A}$ the result is

$$\vec{d} = \mathbf{A}\cdot\mathbf{B}\vec{h}^\star \tag{12}$$

$\mathbf{H}$ is expressed as

$$\mathbf{H} = \mathbf{B}^T\mathbf{A}^T\mathbf{diag}(1/\vec{\sigma}^2)\mathbf{AB} + \mathbf{diag}\left(\frac{\alpha}{\vec{h}^\star}\right). \tag{13}$$

Here $\mathbf{diag}(1/\vec{\sigma}^2)$ is a diagonal matrix containing the inverse of the—uncorrelated—data uncertainties $\vec{\sigma}$. The term $\mathbf{diag}\left(\frac{\alpha}{\vec{h}^\star}\right)$ is the contribution from the entropic prior on $\vec{h}$. Expressing the determinant with the product of the eigenvalues of this matrix, the *model weight* enters the log probability via the sum of the logarithm of its eigenvalues:

$$\log\left(\det\left(\mathbf{B}^T\mathbf{A}^T diag(1/\vec{\sigma}^2)\mathbf{AB} + \mathbf{diag}\left(\frac{\alpha}{\vec{h}^\star}\right)\right)\right) = \sum_i^N \log\lambda_i \tag{14}$$

A straight forward approximation for the logdet of $\mathbf{H}$ is to evaluate the pseudo-determinants, formed from the singular values. For large systems and the use of an automatic differentiation library the SVD leads to before mentioned problems.

When using MCMC—or similar integration techniques—no explicit model selection is necessary. When a flat model is able to describe the data, neighbouring amplitudes become stronger anti-correlated for larger kernel widths, increasing the prior volume in the high-likelihood region. Hence, large kernel widths are favoured when compatible with the data. For ∼100 kernels an optimisation routine searching for the MAP solution is faster, in which case the *weight* for the anti-correlated amplitudes for a set of kernel widths has to be taken into account.

## 4. Exploring the Parameter Space

For optimisation the routine *e04wdc* from NAG [12] is used, facilitating gradient information for the cost function. The C++ code uses the *adept* [13] library to efficiently determine the gradient vector of the posterior with respect to all input parameters.

### 4.1. Automatic Differentiation

Optimisation algorithms generally benefit from gradient information of the cost function. Straight forward is the use of finite difference *FD*, where usually one or two additional function calls are made with perturbed input parameters to estimate the gradient.

An alternative is given by automatic differentiation (AD), also known as algorithmic differentiation. An overview of libraries for various languages can be found on http://www.autodiff.org/. Here, the gradient is internally calculated via the exact differential based on algebraic equations in the algorithm.

When the functions involved map from a $N$ dimensional space to a scalar—like the posterior—a single backward propagation—called inverse-mode differentiation—is enough to obtain the full Jacobian vector. This takes about 3–10 times the computation time compared to a simple function call. This becomes beneficial for functions depending on several—say more than 5—parameters with the additional benefit of more precise differentials. Also, no finite step width like for FD has to be chosen to find an optimum between numerical cutoff and approximation of the slope.

A drawback is the increased memory consumption, as most libraries create what is called a *tape* to store the path from input variables to cost function. This allows to calculate the contribution of each expression to the overall gradient as well as to use conditionals. As the problem size increases, the matrices and number of operations increase, affecting the size of the tape.

Using stochastic algorithms instead of the SVD to obtain the logdet—described in more details in Section 5—scales less strong in computation time as well as in memory consumption with the problem size, which allows to use the AD implementation for larger problems.

*4.2. Computation Time*

For the shown example with 1200 parameters the number of function calls from a *standard* parameter distribution—constant values for hidden image and kernel widths, initial likelihood about $3.4 \times 10^4$ for 400 data points—is on the order of some thousand. On the shown example, the bottle neck is the evaluation of the log determinant, independent of the method used—with some seconds for the full evaluation. The solving time is therefore on the order of 5 min to 60 min per frame. For similar consecutive frames, where the last parameter set is a good starting point for the optimisation, this can drop to about 1 min.

An alternative is to use Marcov Chain Monte Carlo MCMC to explore the parameter space. This circumvents the calculation of the logdet all together. The run-time of the forward model for a reasonable system size is on the order of 1 s—without the need to compute the kernel matrix explicitly and taking the system response into account. However, for $20 \times 20$ kernels we already have to deal with 1200 parameter. Using 1000 sweeps—each representing a sequential scan through the parameters—and 10 bins leads to about $10^3 \times 10^3 \times 10 = 10^7$ function evaluations. Assuming a run-time of 1 s per call, this sums up to 116 days of computation. For the shown example, the forward model evaluation took about 2.5 ms, which corresponds to about 7 h.

On this scale, minimisation seems to be the only feasible way, although the logdet determination becomes cumbersome. Alternatively Hamiltonian Monte Carlo can make use of the gradient information, which speeds up the process significantly. However, the comparison presented is based on results from classic MCMC.

## 5. Comparison of SVD, STE and MCMC in Regard to Model Selection

In this section, the model selection described in Section 3.1 is discussed for large problems. Large in this sense means $\geq 1000$ parameter, which is expected for our 2D Data.

Figure 4a shows the reference heat flux density for the further comparison with $20 \times 20$ pixels, on which a dense set of kernels—one for every pixel—will be applied. Starting with a tile at equilibrium at 80 °C, the resulting synthetic data after 50 ms is shown in Figure 4b.

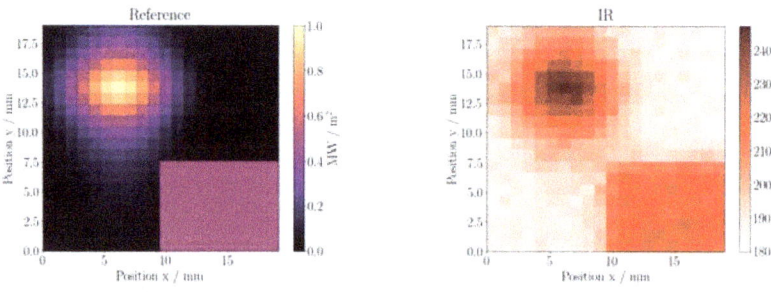

(a) Reference Heat Flux Pattern  (b) Synthetic Measurement

**Figure 4.** (a) Reference heat flux pattern. Peak: along x-axis a Cauchy distribution with width of 5 pixel, along y-axis a gaussian with width of 6 pixel. Peak set to 1 MW m$^{-2}$. Bottom right corner overwritten with a plateau of 0.5 MW m$^{-2}$. (b) Resulting IR data—in counts—after 50 ms exposure to the heat flux pattern shown in (a).

## 5.1. MCMC

Classic Metropolis Hastings Monte Carlo serves as reference for the distributions of the amplitudes and kernel widths. For 1200 parameters the integration is still feasible, see Section 4.2. Figure 5a shows the reconstructed heat flux pattern—top left—the spatial amplitude distribution—top right—and in the bottom the distributions for the kernel width along the horizontal axis—$\sigma_x$ and the vertical axis $\sigma_y$. At the edges of the step function the Aadaptive Kernel width represent the sharp transition.

## 5.2. SVD

Instead of determining the eigenvalues of Equation (14), the singular values composition is used, which is available as robust algorithm. Run time scales typically with $O(n^3)$, which is feasible for *small* systems of around 100 kernels—resulting in a $100 \times 100$ matrix—with calculation times on the millisecond scale. For the shown test system with $20 \times 20$ kernels, the SVD for the $400 \times 400$ matrix takes about 100 ms. The distribution of the kernel width is sharper at the edges of the step in the bottom right than for the MC result. As the kernel width acts as regularisation and the resulting heat flux pattern is virtually identical, the logdet evaluation is justified. Especially given the speedup from about 10 hours to about 10 min per frame.

Total memory including adjoints: 5.0 GB and time per function call including gradient evaluation is 1600 ms to 2000 ms. Without the gradient information, the values are 70 MB and 100 ms.

By increasing the image resolution by a factor of 2—leading to $40 \times 40$ pixels and kernels—the computation time without gradients increases to 13 s and the overall memory demand is 730 MB. With gradients, the memory demand exceeds 120 GB, which is the upper limit on the used computer system.

## 5.3. Stochastic Trace Estimation—STE

For larger matrices, the SVD becomes too expensive in terms of time and memory consumption. An alternative is to sample the matrix with test-vectors, in order to estimate the result of a function—like the logarithm—applied the eigenvalues of the matrix. This is known as Stochastic Trace Estimation STE. This is based on moments gained by matrix-vector multiplications, see e.g., [14,15]. The resulting distributions are shown in Figure 5c, which come close to the results of the SVD. The downside is, that the number of test-vectors and the order of the expansion has to be set a-priori for the optimisation procedure. For the shown example the expansion order is 10 and 50 test-vectors have been used.

Using adjoints the total memory demand is 4.7 GB and time per function call is 2000 ms. Without the gradient information, the values are 70 MB and 100 ms. The computation time and memory demand scale linear with both, the number of test-vectors and expansion order.

Even for larger matrices, the memory demand and computation time can be controlled by the order of expansion and test vectors used. Used on a case with $40 \times 40$ pixels and kernels—overall 4801 parameters—the time raises to 10 s and the memory to 730 MB without gradients. Including gradients, the computation time increases to 20 s for 10 vectors and expansion order 10, the memory demand is about 200 MB per test vector and expansion order. Intermittent evaluation of the gradients is possible to free the memory, as the results are independent. For 20 vectors and 10 orders, the memory demand is just above 100 GB.

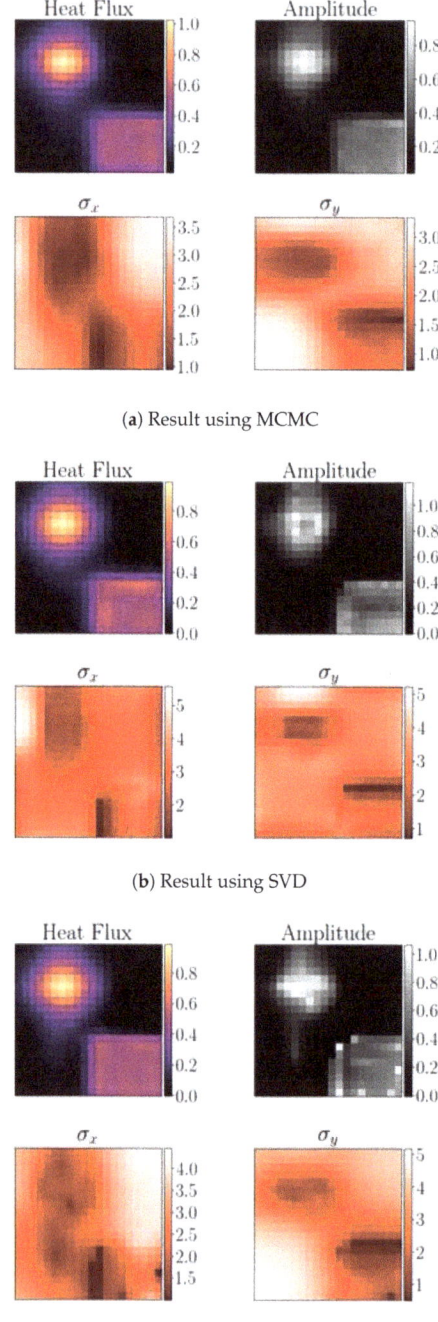

**Figure 5.** Comparison between the three approaches. Structure of subplots: top left—Heat Flux Pattern. Top right—amplitudes of each kernel. Bottom left—kernel width along x-axis. Bottom right—kernel width along y-axis.

### 5.3.1. Remark 1

For large matrices, the matrix matrix product $\mathbf{B}^T\mathbf{B}$ turns out to be memory consuming in the adjoint formulation. More memory-efficient implementations become necessary for increasing matrix sizes beyond $1000 \times 1000$—here about 20 GB are reserved for the matrix multiplication.

In addition to the SVD and STE, a conjugate gradient method was tested as well. However, due to the iterative nature the determination of the gradients mostly fails. In addition, although the precision of the result can be controlled, jumps in the logdet for slight changes to the kernel matrix prohibit the use in an optimisation routine. This limits the alternatives to the SVD for large matrices—to the authors knowledge—to the STE approach.

### 5.3.2. Remark 2

For the IR system it seems natural to place a kernel for every pixel. This however is probably not needed for most cases, especially when the profile has some known decay lengths. An example is the heat flux which may vary fast—on a few pixels basis—along the poloidal orientation, but slow along the toroidal orientation. For application to large data sets the number of kernels needed should be investigated beforehand to reduce the problem size.

## 6. Conclusions

A 2D formulation of the Adaptive Kernel model is introduced, including the means of using it in combination with an optimisation routine respecting the model selection via logdet. As the use of the SVD is limited in regard of the system size, the Stochastic Trace Estimation is suggested as alternative and shows comparable results.

Treating 2D distributions with the Aadaptive Kernel model is viable with modern techniques and computation power. While long time series of time dependent data stay challenging to analyse, single images or short sequences can be treated with this probabilistic framework. Key-ingredient in this presentation is gradient information, to navigate efficiently in high-dimensional spaces.

Integration with Monte Carlo Methods is generally possible, with the benefit not to have to calculate the determinant. However, for many parameters and a non-trivial forward model—with respect to the computation time—this analysis can be infeasibly slow. For smaller test cases it is the method of choice, to check the results of the logdet evaluation, which includes some approximations.

**Author Contributions:** "Conceptualization, D.N. and U.vT.; Methodology, D.N. and U.vT.; Software, D.N.; Validation, D.N.; Formal Analysis, D.N., U.vT.; Investigation, D.N.; Resources, D.N. and U.vT.; Data Curation, D.N.; Writing – Original Draft Preparation, D.N.; Writing – Review & Editing, D.N.; Visualization, D.N.; Supervision, D.N. and U.vT.; Project Administration, D.N.; Funding Acquisition, U.vT."

**Funding:** This work has been carried out within the framework of the EUROfusion Consortium and has received funding from the Euratom research and training programme 2014–2018 under grant agreement No. 633053. The views and opinions expressed herein do not necessarily reflect those of the European Commission.

**Conflicts of Interest:** The authors declare no conflict of interest.

## Abbreviations

The following abbreviations are used in this manuscript:

| | |
|---|---|
| AD | Automatic Differentiation |
| AK | Adaptive Kernel |
| AUG | ASDEX Upgrade |
| MCMC | Marcov Chain Monte Carlo |
| STE | Stochastic Trace Estimation |
| SVD | Singular Value Decomposition |

## References

1. Kang, C.S.; Lee, H.H.; Oh, S.; Lee, S.G.; Wi, H.M.; Kim, Y.S.; Kim, H.S. Study on the heat flux reconstruction with the infrared thermography for the divertor target plates in the KSTAR tokamak. *Rev. Sci. Instrum.* **2016**, *87*, 083508. doi:10.1063/1.4961030.
2. Nille, D.; von Toussaint, U.; Sieglin, B.; Faitsch, M. Probabilistic Inference of Surface Heat Flux Densities from Infrared Thermography. In *Bayesian Inference and Maximum Entropy Methods in Science and Engineering*; Polpo, A., Stern, J., Louzada, F., Izbicki, R., Takada, H., Eds.; Springer: Cham, Switzerland, 2018; pp. 55–64. doi:10.1007/978-3-319-91143-4_6.
3. Bazylev, B.; Janeschitz, G.; Landman, I.; Pestchanyi, S.; Loarte, A.; Federici, G.; Merola, M.; Linke, J.; Zhitlukhin, A.; Podkovyrov, V.; et al. ITER transient consequences for material damage: Modelling versus experiments. *Phys. Scr.* **2007**, *2007*, 229. doi:10.1088/0031-8949/2007/T128/044.
4. Li, M.; You, J.H. Interpretation of the deep cracking phenomenon of tungsten monoblock targets observed in high-heat-flux fatigue tests at 20 MW/m$^2$. *Fusion Eng. Des.* **2015**, *101*, 1–8. doi:10.1016/j.fusengdes.2015.09.008.
5. Goldston, R. Heuristic drift-based model of the power scrape-off width in low-gas-puff H-mode tokamaks. *Nucl. Fusion* **2012**, *52*, 013009. doi:10.1088/0029-5515/52/1/013009.
6. Stangeby, P.; Chankin, A. Simple models for the radial and poloidal E × B drifts in the scrape-off layer of a divertor tokamak: Effects on in/out asymmetries. *Nucl. Fusion* **1996**, *36*, 839. doi:10.1088/0029-5515/36/7/I02.
7. Herrmann, A.; Junker, W.; Günther, K.; Bosch, S.; Kaufmann, M.; Neuhauser, J.; Pautasso, G.; Richter, T. Energy flux to the ASDEX-Upgrade diverter plates determined by thermography and calorimetry. *PPCF* **1995**, *37*, 17. doi:10.1088/0741-3335/37/1/002.
8. Press, W.; Teukolsky, S.A.; Vetterling, W.T.; Flannery, B.P. *Numerical Recipes*, 3rd ed.; Cambridge University Press: Cambridge, UK, 2007.
9. Fischer, R.; Mayer, M.; von der Linden, W.; Dose, V. Enhancement of the energy resolution in ion-beam experiments with the maximum-entropy method. *Phys. Rev.* **1997**, *55*, 6667–6673. doi:10.1103/PhysRevE.55.6667.
10. Puetter, R.C. Pixon-based multiresolution image reconstruction and the quantification of picture information content. *Int. J. Imaging Syst. Technol.* **1999**, *6*, 314–331, doi:10.1002/ima.1850060405.
11. Skilling, J. Classic Maximum Entropy. In *Maximum Entropy and Bayesian Methods: Cambridge, England, 1988*; Springer: Dordrecht, The Netherlands, 1989; pp. 45–52. doi:10.1007/978-94-015-7860-8_3.
12. The NAG C Library, The Numerical Algorithms Group (NAG), Oxford, United Kingdom. Available online: www.nag.com (accessed on 15 July 2016).
13. Hogan, R. Fast Reverse-Mode Automatic Differentiation Using Expression Templates in C++. *ACM Trans. Math. Softw.* **2014**, *40*, 26:1–26:24. doi:10.1145/2560359.
14. Skilling, J. The Eigenvalues of Mega-dimensional Matrices. In *Maximum Entropy and Bayesian Methods: Cambridge, England, 1988*; Springer: Dordrecht, The Netherlands, 1989; pp. 455–466. doi:10.1007/978-94-015-7860-8_48.
15. Fitzsimons, J.K.; Granziol, D.; Cutajar, K.; Osborne, M.A.; Filippone, M.; Roberts, S.J. Entropic Trace Estimates for Log Determinants. *CoRR* **2017**, abs/1704.07223.

© 2019 by the authors. Licensee MDPI, Basel, Switzerland. This article is an open access article distributed under the terms and conditions of the Creative Commons Attribution (CC BY) license (http://creativecommons.org/licenses/by/4.0/).

 *proceedings*

*Proceedings*

# Bayesian Determination of Parameters for Plasma-Wall Interactions [†]

**Roland Preuss \*, Rodrigo Arredondo and Udo von Toussaint**

Max-Planck-Institut für Plasmaphysik, 85748 Garching, Germany; Rodrigo.Arredondo@ipp.mpg.de (R.A.); udt@ipp.mpg.de (U.v.T.)
\* Correspondence: preuss@ipp.mpg.de; Tel.: +49-89-3299-1202
† Presented at the 39th International Workshop on Bayesian Inference and Maximum Entropy Methods in Science and Engineering, Garching, Germany, 30 June–5 July 2019.

Published: 18 December 2019

**Abstract:** Within a Bayesian framework we propose a non-intrusive reduced-order spectral approach (polynomial chaos expansion) to assess the uncertainty of ion-solid interaction simulations. The method not only reduces the number of function evaluations but provides simultaneously a quantitative measure for which combinations of inputs have the most important impact on the result. It is applied to the ion-solid simulation program SDTrimSP with several uncertain and Gaussian distributed input parameters, i.e., angle $\alpha$, projectile energy $E_0$ and surface binding energy $E_{sb}$. In combination with recently acquired experimental data the otherwise hardly accessible model parameter $E_{sb}$ can now be estimated.

**Keywords:** uncertainty quantification; non-intrusive; spectral expansion; plasma-wall interactions; Bayesian analysis

## 1. Introduction

Plasma-wall interactions are of crucial importance in the design of future fusion reactors, since they determine the replacement cycle for the plasma exposed components of the wall. In order to estimate the life-time of those wall components atomistic simulations are essential. Almost all computer codes for the simulation of ion-solid interactions [1] rely on a large number of input parameters, e.g., surface binding energies, composition, energy distribution etc. However, many of these parameters are uncertain and a proper comparison with experimental data or other models requires the quantification of the uncertainty of the result. Unfortunately, the computational demand of single simulation runs often severely restricts the quantification of output uncertainties by full-grid or simple sampling (e.g., Monte Carlo sampling) based approaches due to the curse of dimensionality for more than a very limited number of uncertain input parameters. Therefore, to reduce the computational effort we propose a non-intrusive reduced-order model approach (polynomial chaos expansion), which not only reduces the number of function evaluations but provides simultaneously a quantitative measure of which combinations of inputs have the most important impact on the result, i.e., it yields a sensitivity analysis and the associated Sobol coefficients.

## 2. Bayesian Uncertainty Quantification

Based on the Bayesian framework we employ a spectral expansion to quantify the propagation of uncertainty through the model. First introduced by Wiener [2] in the context of Hermite basis functions it was termed 'polynomial chaos expansion' at his time. Nowadays the notion of 'chaos' has shifted and the use of the term 'spectral expansion' is more appropriate. Once successfully achieved, the spectral representation is capable of quantifying the uncertainty for any point in model space or to serve as a surrogate model.

Since we calculate the sought-for spectral coefficients from a discrete set of collocation points in the space of the random variable, our approach is non-intrusive, but approximate. The emerging integrals in the calculation of the coefficients are evaluated by Gaussian quadrature which identifies the collocation points with those of the quadrature. Moreover, we assume mutually independent normally distributed random variables. The adjunctive set of orthonormal basis functions in such a case are Hermite polynomials.

To quantify the uncertainty of we seek the appropriate function $g(\xi)$, such that $R$ will have the required distribution of the model response, $R = g(\xi)$. As for all random variables with finite variance it is possible to find an infinite expansion

$$g(\xi) = \sum_{k=0}^{\infty} a_k \psi_k(\xi) \approx \sum_{k=0}^{P} a_k \psi_k(\xi) \quad , \tag{1}$$

which we limit to polynomial order $P$ since the contributions of higher orders become numerically insignificant. The coefficients are given by

$$a_k = \frac{\langle g(\xi), \psi_k(\xi) \rangle}{\langle \psi_k(\xi), \psi_k(\xi) \rangle} \quad , \quad \text{with} \quad \langle g(\xi), \psi(\xi) \rangle = \int g(\xi) \psi(\xi) p(\xi) d\xi \quad . \tag{2}$$

We assume Gaussian character for the random variable, so the density $p(\xi)$ is distributed according to the normal (probability) distribution

$$p(\xi) = \frac{1}{\sqrt{2\pi}} \exp\left\{-\frac{\xi^2}{2}\right\} \quad . \tag{3}$$

The adjunctive set of orthonormal basis functions is given by the so-called *probabilist* Hermite functions, which read up to fourth order

$$\psi_0(\xi) = 1, \quad \psi_1(\xi) = \xi, \quad \psi_2(\xi) = \xi^2 - 1, \quad \psi_3(\xi) = \xi^3 - 3\xi, \quad \psi_4(\xi) = \xi^4 - 6\xi^2 + 3. \tag{4}$$

It turns out that for the model simulations under consideration this polynomial order is sufficient since contributions from higher orders become numerically insignificant for the result. With these definitions the normalization constants in Equation (2) are readily

$$\langle \psi_k, \psi_k \rangle = \int \psi_k(\xi) \psi_k(\xi) p(\xi) d\xi = k! \quad . \tag{5}$$

Due to the Gaussian nature of the probability function omnipresent in the integrals above, it is beneficial to use Gauss-Hermite quadrature for the evaluation

$$\langle g(\xi), \psi(\xi) \rangle \stackrel{\text{G.H.}}{=} \sum_{l=0}^{L} g(\xi_l) \psi(\xi_l) w_l \quad , \tag{6}$$

where the weights $w_l$ and the abscissas $\xi_l$ are for instance provided by Numerical Recipes [3]. Eventually, by exploiting the properties of the orthogonal Hermite polynomials the expectation value of the model outcome and its variance can be assigned to the spectral coefficients in Equation (2)

$$\langle R \rangle = a_0 \quad , \quad \text{var}(R) = \langle R^2 \rangle - \langle R \rangle^2 = \sum_{k=1}^{P} a_k^2 k! \quad . \tag{7}$$

In order to provide a measure for the influence of the uncertainty of input variables on the above variance we employ Sobol coefficients [4]. They are defined by

$$S_i = \frac{D_i}{\text{var}(R)}, \quad S_{ij} = \frac{D_{ij}}{\text{var}(R)}, \quad \ldots, \quad (8)$$

where the evaluation of the integrals

$$D_i = \int g_i^2(\tilde{\zeta}_i) d\tilde{\zeta}_i,$$
$$D_{ij} = \int\int g_{ij}^2(\tilde{\zeta}_i, \tilde{\zeta}_j) d\tilde{\zeta}_i d\tilde{\zeta}_j,$$
$$\ldots, \quad (9)$$

results in combinations of the coefficients of Equation (2) (the index of the function $g_{index}(\ldots)$ relates to the specific variable(s) $\tilde{\zeta}_{index}$ which are omitted in the integral $g_{index} = \int \ldots \int g(\tilde{\zeta}) d\tilde{\zeta}_{\{index\}}$). The higher the value of a Sobol coefficient with respect to the others is, the more it is advantageous to reduce the uncertainty of its associated variable in order reduce the uncertainty of the quantity of interest.

## 3. Ion-Solid Interaction Program SDTrimSP

SDTrimSP [5,6] is a parallelized Monte-Carlo code which simulates transport of energetic particles through a target by employing sequentially two-body collision approximation to compute collision-cascades in three dimensions. This approximation has been shown to be valid (i.e., the stochastic fluctuations of the collision processes exceed the approximation error) for impact energies larger than about 50 eV [7]. Versions of the SDTrimSP-code differ in the description of the target composition, e.g., as one-dimensional (c(x) [6]), two-dimensional (c(x,y) [8]) or three-dimensional (c(x,y,z) [9]). Common to all versions (and key to the high code efficiency) is the assumption of amorphous targets, which circumvents the storage of sample atom coordinates. The simulations were performed with standard settings, i.e., considering a static one-dimensional target (the concentral profile c(x) was kept constant) and the scattering integral was computed using the Gauss-Mehler quadrature scheme with eight pivots. The varied parameters were the projectile energy and the impact angle (with zero degrees corresponding to a perpendicular impact, parallel to the surface normal).

## 4. Results and Discussion

The above program is applied to simulate ion-solid interactions for the case of incident deuterium ions with an energy of $E_0 = 200$ eV at an impact angle of $\alpha = 45$ degrees to a surface consisting of iron with a commonly used surface binding energy of $E_{SB} = 4.28$ eV. We assume the parameters to be normally distributed within a standard deviation of roughly 10%, i.e., $\sigma_{E_0} = 20$ eV, $\sigma_{E_{SB}} = 0.4$ degrees and $\sigma_\alpha = 4$ eV.

First, in order to have a calibration standard to compare with we employ random sampling of the model response. For each realization of the random variable $\{\tilde{\zeta}_1, \ldots, \tilde{\zeta}_N\}$ there exists a model response $R_i = R(\tilde{\zeta}_i)$ constituting the sample solution set $\{R_1, \ldots, R_N\}$ from which moments can be computed. The expected mean is and its variance read

$$\langle R \rangle = \frac{1}{N} \sum_{i=1}^{N} R(\tilde{\zeta}_i), \quad \text{with} \quad \text{var}(R) = \langle R^2 \rangle - \langle R \rangle^2. \quad (10)$$

In Figure 1 the results of $20^3 = 8000$ samples are shown for the above parameter settings. The mean value for the sputter yield is $Y_{MC} = 0.052$ with a standard deviation of $\sigma_{MC} = 0.013$. Even more, the full uncertainty distribution may be established with help of a histogram if the sample solution set is sufficiently large ($N \gtrsim 1000$). But, although this procedure is straightforward and automatically contains the full model answer with all correlations, it has the vital drawback of a comparatively low convergence rate. If the computation time of a single model output is not in the order of seconds or becomes more sophisticated with a higher number of variables (curse of dimension), the mere accumulation of sample point densities to infer the complete distribution is futile. Much more promising in this respect is the spectral approach of Section 2 which results will be discussed next.

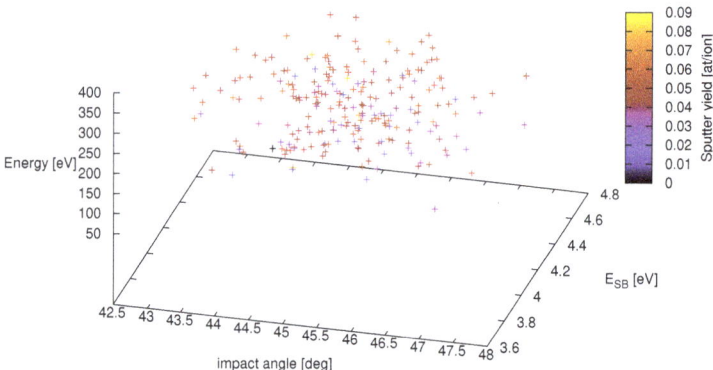

Figure 1. Sputter yield for deuterium on iron from SDTrimSP-simulations with energy, angle and surface binding energy distribution for $20^3$ samples. The respective input variables are $E_0 = 200 \pm 20$ eV, $E_{SB} = 4.28 \pm 0.4$ eV and an incident angle of $45 \pm 4$ degrees, The resulting sputter yield is plotted with a color scheme ranging from dark blue at zero up to light yellow at 0.09 sputtered atoms per incoming ion. The mean value of the sputter yield is $Y_{MC} = 0.052$ with a standard deviation of $\sigma_{MC} = 0.013$.

Extending the formulas of Section 2 to three random variables $\vec{\xi} = (\xi_1, \xi_2, \xi_3)$ with $\hat{E}_0 = E_0 + \xi_1 \sigma_{E_0}$, $\hat{E}_{SB} = E_{SB} + \xi_2 \sigma_{E_{SB}}$ and $\hat{\alpha} = \alpha + \xi_3 \sigma_\alpha$, the summation of the terms in Equation (6) runs over three indices $l_1$, $l_2$ and $l_3$ with an upper boundary of $P + 1 = 5$ in the present setup of fourth order polynomials (for numerical accuracy of the Gaussian quadrature it is expedient to be one order higher than the polynomial order of the spectral expansion). This results in a total of 216 terms (three nested summations, each running from $l_i = 0$ to 5 with $i = 1, 2, 3$) over the collocation points composed of 6 Gaussian quadrature abscissas assigned to $\xi_{l_i}$ and 6 weights $w_{l_i}$. The value for the function $g(\xi_{l_1}, \xi_{l_2}, \xi_{l_3})$ is obtained from a SDTrimSP run, which roughly takes 3 min on a modern CPU. However, the complete run for the 216 terms can be speeded up enormously since the calculations are independent and can be done in parallel. Once calculated, the 35 coefficients of Equation (6) establish a fast surrogate model, which is simply the evaluation of a polynomial. This is shown in Figure 2 as the red mesh. The respective sputter yield, for which the uncertainty quantification was performed, is depicted in the center as the green sphere with $Y_{UQ} = 0.050$ at/ion and its standard deviation of $\sigma_{UQ} = 0.011$ as the green perpendicular line. The comparison with the result of the sampling approach above ($Y_{MC} = 0.052 \pm 0.013$) shows excellent agreement.

Without the need to do any further simulations, various quantities may be inferred from the coefficients, e.g., the variance as in Equation (2), or the Sobol coefficients, which allow to investigate the sensitivity of the result on the uncertainty of the input variables. For the above variables $E_0$, $\alpha$ and $E_{SB}$ we get a relationship of 10:20:70 in the Sobol coefficients (only first order is numerically significant) for $\{E_0, \alpha, E_{SB}\}$, indicating that the improvement of the knowledge of $E_{SB}$ is most rewarding if one wants to reduce the uncertainty of the sputter yield.

Following this trail, we performed a series of experimental measurements of the sputter yield for different impact angles with $\alpha = 0, 45, 60$ and 75 degrees at $E_0 = 2$ keV. Then we applied the uncertainty quantification method discussed above in order to provide quantitative estimates of the sputter yields at a variety of settings for the surface binding energy $E_{SB}$. It turned out that the most probable value for the surface binding energy is $E_{SB}^{new} = 4.8 \pm 0.4$ eV, one and a half standard deviations larger than the value commonly used up to now [1], i.e., $E_{SB}^{old} = 4.2 \pm 0.4$ eV. With the revised setting of $E_{SB}$ we compared (see Figure 3) simulations of the sputter yield for different incident energies of deuterium with results from Rutherford backscattering (RBS) and weight loss (WL) experiments and got an improved agreement (except for $E_0 = 1$ keV). With these results the Bayes factor rules out another competitor to SDTrimSP (i.e., Monte Carlo decision for the occurrences of collisions of incident ions with atoms in the target) being the SRIM-model, which employs a quantum mechanical treatment of ion-atom collisions and seems not to comprise all important effects present..

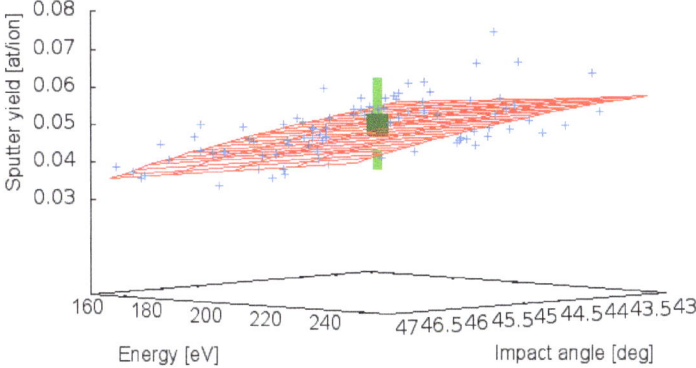

**Figure 2.** Sputter yield reproduced by the surrogate model from the uncertainty quantification of SDTrimSP calculations for $E_0 = 200$ eV, $E_{SB} = 4.28$ eV and an incident angle of 45 degrees. The filled circle in dark green shows a yield of 0.050 for these settings with a standard deviation of 0.011 (light green line). For reason of portrayal the surrogate model (red mesh) was varied in two dimensions only for $E_0$ and $\alpha$, while $E_{SB}$ was kept fixed at 4.28 eV. In addition, the blue plus signs show the scatter data from the sampling approach already shown in Figure 1.

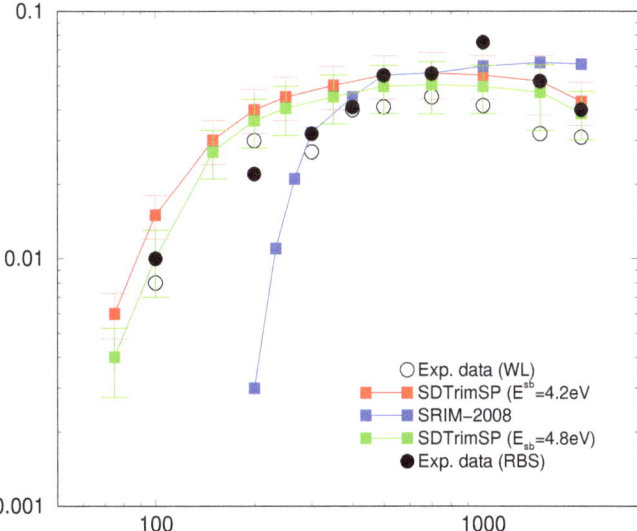

**Figure 3.** Comparison of the sputter yield for the previously set surface binding energy $E_{SB} = 4.2 \pm 0.4$ eV (red squares) and the newly acquired setting of energy $E_{SB} = 4.8 \pm 0.4$ eV (green squares) with data from experiments done in a Rutherford backscattering setup (RBS, filled circles) and a weight loss setup (WL, open circles). A further model, SRIM (blue squares), can almost certainly be ruled out. All lines shown are guide to the eye.

## 5. Summary and Conclusions

The non-intrusive polynomial chaos expansion for quantifying the propagation of uncertainty through the model has been proven to be a valuable tool in describing the reliability of a model outcome. The experience with the employed algorithm revealed that the spectral expansion with moderate settings of employing only up to 4th order polynomials and six Gaussian quadrature abscissa, which requires less than 1000 simulation runs, is well suited for the determination of a medium number of uncertain parameters. We applied the method to SDTrimSP simulations in determining the sputtering yield and its standard deviation for the example of incident deuterium ions on an iron target. Residing on both quantities we could rule out the existing parameter setting for the surface binding energy and assigned a new much more accurate one. With this newly set input parameter it was possible to get a better agreement with available experimental data and eventually put us in the position to rule out another physical model.

**Author Contributions:** R.A. performed the experiments and provided the data; R.P. and U.v.T. analyzed the data, contributed analysis tools and wrote the paper.

**Conflicts of Interest:** The authors declare no conflict of interest.

## References

1. Eckstein, W. *Computer Simulation of Ion-Solid Interactions*; Springer Series in Materials Science; Springer-Verlag: Berlin, Heidelberg; New York, NY, USA, 1991; Volume 10.
2. Wiener, N. The homogeneous chaos. *Am. J. Math.* **1938**, *60*, 897–936.
3. Press, W.H.; Teukolsky, S.A.; Vetterling, W.T.; Flannery, B.P. *Numerical Recipes: The Art of Scientific Computing*, 3rd ed.; Cambridge University Press: Cambridge 2007.
4. Smith, R.C. *Uncertainty Quantification: Theory, Implementation, and Applications*; SIAM: Philadelphia, PA, USA, 2014.
5. Möller, W.; Eckstein, W.; Biersack, J. Tridyn-binary collision simulation of atomic collisions and dynamic composition changes in solids. *Comput. Phys. Commun.* **1988**, *51*, 355–368.
6. Eckstein, W.; Dohmen, R.; Mutzke, A.; Schneider, R. *SDTrimSP: Ein Monte-Carlo Code zur Berechnung von Stossereignissen in Ungeordneten Targets (SDTrimSP: A Monte-Carlo Code for Calculating Collision Phenomena in Randomized Targets)*; Technical Report IPP 12/3; Max-Planck-Institut für Plasmaphysik: Garching, Germany, 2007.
7. Behrisch, R.; Eckstein, W. *Sputtering by Particle Bombardment*; Springer: Berlin/Heidelberg 2007.
8. Mutzke, A.; Schneider, R. *SDTrimSP-2D: Simulation of Particles Bombarding on a Two Dimensional Target Version 1.0*; Technical Report IPP 12/4; Max-Planck-Institut für Plasmaphysik: Garching, Germany, 2009.
9. von Toussaint, U.; Mutzke, A.; Manhard, A. Sputtering of rough surfaces: A 3D simulation study. *Phys. Scr.* **2017**, *T170*, 014056.

© 2019 by the authors. Licensee MDPI, Basel, Switzerland. This article is an open access article distributed under the terms and conditions of the Creative Commons Attribution (CC BY) license (http://creativecommons.org/licenses/by/4.0/).

*Proceedings*

# On the Diagnosis of Aortic Dissection with Impedance Cardiography: A Bayesian Feasibility Study Framework with Multi-Fidelity Simulation Data [†]

Sascha Ranftl [1,*], Gian Marco Melito [2], Vahid Badeli [3], Alice Reinbacher-Köstinger [3], Katrin Ellermann [2] and Wolfgang von der Linden [1]

1. Institute of Theoretical Physics-Computational Physics, Graz University of Technology, 8010 Graz, Austria; vonderlinden@tugraz.at
2. Institute of Mechanics, Graz University of Technology, 8010 Graz, Austria; gmelito@tugraz.at (G.M.M.); ellermann@tugraz.at (K.E.)
3. Institute of Fundamentals and Theory in Electrical Engineering, Graz University of Technology, 8010 Graz, Austria; vahid.badeli@tugraz.at (V.B.); alice.koestinger@tugraz.at (A.R.-K.)
* Correspondence: ranftl@tugraz.at
† Presented at the 39th International Workshop on Bayesian Inference and Maximum Entropy Methods in Science and Engineering, Garching, Germany, 30 June–5 July 2019.

Published: 9 December 2019

**Abstract:** Aortic dissection is a cardiovascular disease with a disconcertingly high mortality. When it comes to diagnosis, medical imaging techniques such as Computed Tomography, Magnetic Resonance Tomography or Ultrasound certainly do the job, but also have their shortcomings. Impedance cardiography is a standard method to monitor a patients heart function and circulatory system by injecting electric currents and measuring voltage drops between electrode pairs attached to the human body. If such measurements distinguished healthy from dissected aortas, one could improve clinical procedures. Experiments are quite difficult, and thus we investigate the feasibility with finite element simulations beforehand. In these simulations, we find uncertain input parameters, e.g., the electrical conductivity of blood. Inference on the state of the aorta from impedance measurements defines an inverse problem in which forward uncertainty propagation through the simulation with vanilla Monte Carlo demands a prohibitively large computational effort. To overcome this limitation, we combine two simulations: one simulation with a high fidelity and another simulation with a low fidelity, and low and high computational costs accordingly. We use the inexpensive low-fidelity simulation to learn about the expensive high-fidelity simulation. It all boils down to a regression problem—and reduces total computational cost after all.

**Keywords:** bayesian probability theory; uncertainty quantification; impedance cardiography; aortic dissection

## 1. Introduction

The largest blood vessel in the human body is the aorta. The wall of the aorta is made of aortic tissue, a layered composition of muscle cells, collagen, elastin fibres, etc. In Aortic Dissection (AD), a tear in the innermost layer of the aortic wall permits blood to flow in between the layers, effectively forcing apart the layers and deforming the geometry of the aorta. Obviously, AD affects blood circulation unfavourably ([1] p. 459). This pathology is illustrated in Figure 1.

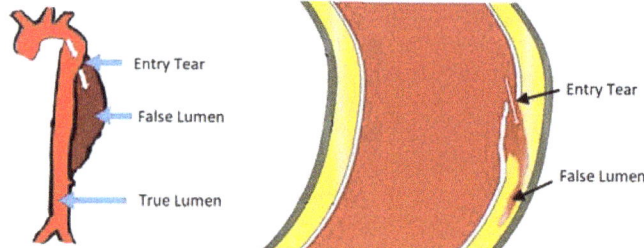

**Figure 1.** Illustration of a dissected aorta. Left: The whole organ. Right: Close-up to the entry tear [2]. Blood pushes from the anatomically correct cavity (medical parlance: true lumen) through a tear into the aortic wall. The tear grows and builds another cavity (medical parlance: false lumen), affecting blood circulation unfavourably.

The condition AD is often acute and requires immediate treatment, but diagnosis is difficult. Physicians use a variety of imaging techniques to diagnose AD, among which are Magnetic Resonance Tomography (MRT), Computed Tomography (CT) and Echocardiography, the latter of which is based on an ultrasound device [3]. Ultrasound devices are comparably cheap, fast and easy to handle. But if wave propagation is obfuscated by, e.g., the rib cage, the technique is not applicable. CT and MRT do not have this limitation due to full radiation penetration of the body, but show a number of drawbacks: long measurement times, radiation exposure, high costs, require specialized personnel (radiologists) and most importantly, MRT/CT is not available on a whim. A fast response, and a fast diagnosis, hence, is key to the treatment of AD patients. In this work, we analyse the proposal of [4] to use impedance cardiography (ICG) [5] for AD diagnosis. In ICG, one places a pair of electrodes on the thorax (upper body), injects a defined low-amplitude, alternating electric current into the body and measures the voltage drop. The generic experimental setup is illustrated in Figure 2. The specific al resistance (impedance) of blood is much lower than that of muscle, fat or bone [6]. Electric current seeks the path of least resistance, and thus the current propagates through the aorta rather than through, e.g., the spine. If blood is redistributed within the body due to AD, the path of least resistance is expected to change, and so the overall resistance of the body. ICG is bad to distinguish between different types of blood redistribution, e.g., AD or lung edema [7]. Still, ICG yields yet another clue in a physician's diagnostic procedure. ICG is fast, cheap, available on a whim and does not require specialized personnel. A new medical detection device based on ICG would thus close a gap left open by existing procedures.

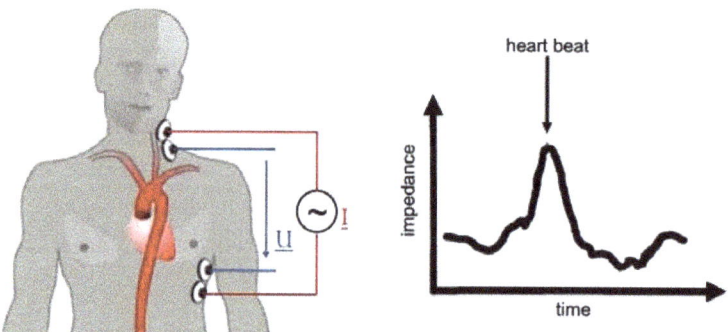

**Figure 2.** Left: Experimental setup of ICG. Right: Generic ICG signal.

To develop such a device, it is necessary to perform experiments which are extraordinarily difficult, both technically as well as ethically. One would need ICG measurements as well as high quality tomography data stemming from the same person, before and after AD happened. That kind of data is not available, and we resort to Finite Element (FE) simulations [8] instead. In these simulations, we find a number of input parameters which are well-defined, but usually neither known precisely nor accessible in the clinical setting. For example, a patient's blood conductivity varies from day to day. The input parameters are thus afflicted with uncertainty, and this uncertainty propagates through the simulation to the output, which here is the measured impedance. If we wanted a meaningful statement on the condition of the aorta, we therefore need to quantify the uncertainty in the measurement. It is an inverse problem, which involves the forward Uncertainty Quantification (UQ) first. UQ has become a term on its own in the engineering community. A rather chunky, but quite comprehensive collection of reviews on the various aspects of UQ can be found in Reference [9]. A Bayesian perspective is discussed in [10–12].

UQ usually requires quite some computational effort, depending on the number of uncertain parameters and the computational cost of a single simulation itself. If this computational effort is prohibitively large, one may use a surrogate model. The two most widely used surrogate models are Polynomial Chaos Expansion (PCE) [13–16] and Gaussian Process Regression (GPR) [17,18]. PCE is particularly widely spread within then engineering community, while GPR has had its renaissance recently within the machine learning community [19,20].

This work is inspired by the article of Kennedy and O'Hagan in 2000 [21]. They performed UQ by making use of a computer simulation with different levels of 'sophistication' or 'fidelity'. In other words, a cheap simplified simulation serves as a surrogate. Koutsourelakis follows this idea later on [22]. While UQ in general has arrived fully in the Biomedical Engineering community [23], the Bayesian approach has not. Biehler et al. [24] were, to the best knowledge of the authors, the first to apply a Bayesian Multi-Fidelity Scheme in the context of computational Bio-mechanics.

In Section 2 we build a physical model of an impedance cardiography measurement applied to the described physiological system. In Section 3, we develop a Bayesian Multi-Fidelity scheme, which is then used for Uncertainty Quantification of the physical model. The results, i.e., the uncertainty bands of the ICG signal, are presented and discussed in Section 4. We draw our conclusions and suggest possible future improvements in Section 5.

## 2. The Physical Model

We start from the Maxwell's equations and recognize that one cardiac cycle, i.e., the time span between two heart beats, is on the order of one second, and the frequency of the injected current is on the order of a hundred kilo-Hertz. Thus we can assume the electric field to be quasi-static [4], and the Maxwell equations boil down to Laplace's equation (in complex notation),

$$\nabla\Big((\sigma + i\omega\varepsilon)\nabla V\Big) = 0, \tag{1}$$

with electric potential $V$, electrical conductivity $\sigma$, angular frequency $\omega$, permittivity $\varepsilon$ and imaginary unit $i$. Equation (1) is then to be solved on the geometry as depicted in Figure 3 and described as follows. The thorax (upper body) is modelled by an elliptic cylinder with a spatially homogeneous conductivity and permittivity. The aorta is an up-side-down umbrella stick. We consider the whole organ to be filled with blood and neglect the vessel walls. In clinical parlance, the blood-filled cavity caused by the aortic dissection is called "false lumen", while the anatomically correct cavity is called "true lumen". The true lumen is modelled as a circular cylinder, and the false lumen is a holed out circular cylinder attached to the true lumen. The dynamics are modelled via a time-dependent true and false lumen radius, which arises from pressure waves in a pulsatile flow. Further, the blood conductivity depends on the blood flow velocity, and is thus time-dependent in the true lumen, but constant in the false lumen due to a negligible flow velocity [4,25]. The boundary conditions are specified by the body surface and two source electrodes. The two source electrodes are modelled by two patches (top and bottom) on the left-hand side of the patient. For the top patch, the injection current is held constant at 4 mA and a frequency of 100 kHz via a constant surface integral of the current density.

The bottom patch is defined as ground, i.e., a constant voltage $V_{bottom} = 0$ V. Considering the relatively low conductivity of air, we assume the rest of the body surface to be perfectly insulating. Equation (1) is then discretized in space and solved with the Finite Elements method [8]. The quality and fidelity of the space discretization, colloquially termed as "the mesh", is crucial to the quality of the solution, but also to the amount of computational effort. We use a rather coarse mesh of low fidelity, and a rather detailed mesh of high fidelity, with two examples illustrated in Figure 4.

We distinguish observable and unobservable (uncertain) parameters (also termed hidden or latent variables). An obvious observable parameter is time $t$. From the plethora of unobservable parameters, we choose the false lumen radius, $r_{fl}$, and perform a number of simulations with sensible values within the physiological and physical range, i.e., 5.0–25.0 mm with a step size of 1.0 mm. The physical lower boundary would be 0 mm, yet below 5 mm meshing problems occur in the LoFi model, i.e., badly shaped elements become frequent, geometry is approximated badly and thus space discretization fails. This is not surprising, since the LoFi model's size of finite elements is on the order of 5.0 mm, and therefore cannot exhibit features of more detail. For any simulation, the voltage drop between any two points can now be measured. In the clinical setting, there would be a number of probe electrodes attached to the patient's chest, back, neck and/or limbs, and voltage drops measured between the many pairs of probe electrodes. Here, we limit ourselves to just one pair of probe electrodes, with one probe electrode right beneath the upper injection electrode, and one probe electrode right above the lower injection electrode. The positions of the probe electrodes, relative to the injection electrodes, are indicated in Figure 3 by $V_{top}$ and $V_{bottom}$.

We used the Comsol Multiphysics software to perform the modelling [26].

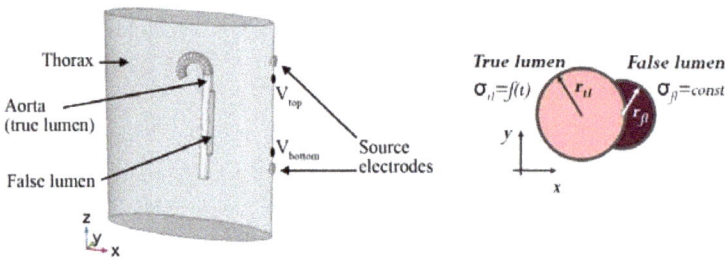

**Figure 3.** Left: CAD Model of Thorax with dissected Aorta. Right: Ground view on True Lumen and False Lumen [4].

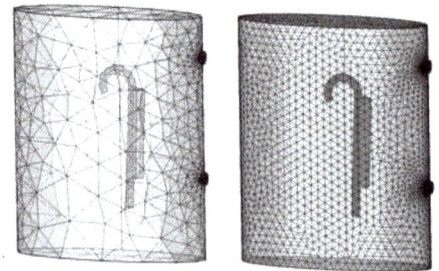

**Figure 4.** Left: Low fidelity mesh, geometry represented by ~10,000 tetrahedral elements. Right: High fidelity mesh, geometry represented by ~300,000 tetrahedral elements.

## 3. Bayesian Multi-Fidelity Scheme

Having computed all the HiFi impedance times series (red) in Figure 5, the uncertainty quantification is pretty much done already. The uncertainty bands of the Brute Force HiFi data in Figure 6 (blue) are inferred from a Gaussian Process model with a squared exponential kernel. In the following paragraphs we will instead use the LoFi impedance time series (black) and only a few of the data points of the HiFi impedance time series (red). In other words; in order to compute the uncertainty bands, instead of doing all the HiFi simulations, we rather do all the LoFi simulations and only a few HiFi simulations.

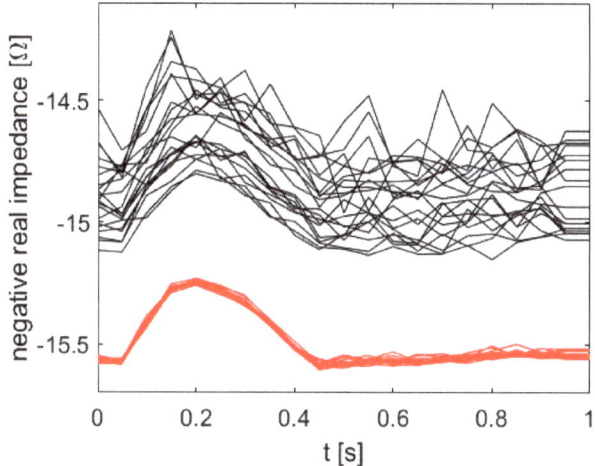

**Figure 5.** Data. Negative real part of the impedance measured with the probe electrodes indicated in Figure 3 (left). Simulations were done for one cardiac cycle with a time step of 50 ms and with 25 values of the false lumen radius (5 mm–25 mm) with the HiFi model (red) and the LoFi model (black).

We distinguish the observable physical parameter time, $t$, from unobservable (uncertain) physical parameters $\zeta$. We consider two disjoint data sets, $D_1$ and $D_2$. $D_1$ is a large set of input parameters, $\vec{\zeta}_1 = \{\zeta_1^{(i)}\}_{i=1}^{N_{\zeta,1}}$, $\vec{t}_1 = \{t_1^{(k)}\}_{k=1}^{N_{t,1}}$, and corresponding (noisy) outputs of the LoFi model, $\vec{V}_{L,1} = \{V_{L,1}^{(i,k)}\}_{i,k=1}^{N_{\zeta,1},N_{t,1}}$, where each individual output $V_{L,1}^{(i,k)}$ is the voltage drop with parameter $\zeta_1^{(i)}$ at time instance $t_1^{(k)}$. $\vec{t}_1$ covers the time series of one whole cardiac cycle. This means $D_1 = \{\vec{t}_1, \vec{\zeta}_1, \vec{V}_{L,1}\} = \{(t_1^{(k)}, \zeta_1^{(i)}, V_{L,1}^{(i,k)})\}_{i,k=1}^{N_{\zeta,1},N_{t,1}}$. In principle, the solver is deterministic, yet the solution depends on the mesh. Since the false lumen radius is treated as a random variable, the geometry is random as well, and each mesh a specific realization of it. We then choose a small subset of the outputs of $D_1$ with size $N_{\zeta,2}$, and $N_{t,2}$ respectively, for which we additionally compute the (noisy) HiFi solution. This subset is chosen such that the support of $\vec{V}_{L,1}$ is appropriately covered. Given input parameters $\vec{\zeta}_2 = \{\zeta_2^{(i)}\}_{i=1}^{N_{\zeta,2}}$, $\vec{t}_2 = \{t_2^{(k)}\}_{k=1}^{N_{t,2}}$, the corresponding output of the HiFi model is $\vec{V}_{H,2} = \{V_{H,2}^{(j,m)}\}_{j,m=1}^{N_{\zeta,2},N_{t,2}}$. We gather these tuples of LoFi-output and corresponding HiFi-output in data set $D_2 = \{\vec{t}_2, \vec{\zeta}_2, \vec{V}_{L,2}, \vec{V}_{H,2}\} = \{(t_2^{(m)}, \zeta_2^{(j)}, V_{L,2}^{(j,m)}, V_{H,2}^{(j,m)})\}_{j,m=1}^{N_{\zeta,2},N_{t,2}}$. In other words, $D_2$ shall be a small subset of $D_1$ which "in hindsight" is augmented with the corresponding HiFi data.

We acknowledge that time $t$ is observable in our experiment, but the latent variables $\zeta$ are not. Let $V_H^*$ and $V_L^*$ be the true values of high fidelity and low fidelity model respectively, corresponding to given input parameters (latent variables) $\zeta$ and a time instance $t$. Note the distinction of the true values $V_H^*$ and $V_L^*$ from the noisy observations in $D_1$ and $D_2$. Let $\mathcal{C}$ be the conditional complex. We want to compute the uncertainty bands of the ICG signal in Figure 5, meaning the posterior pdf of $V_H^*$ given $t$, and introduce $V_L^*$ via marginalisation

$$p(V_H^* \mid t, D_1, D_2, \mathcal{C}) = \int p(V_H^* \mid V_L^*, t, D_1, D_2, \mathcal{C}) \, p(V_L^* \mid t, D_1, D_2, \mathcal{C}) \, dV_L^*, \quad (2)$$

where we constructed the data sets such that we can partially cross them out here. For the sake of easier notation, we will omit the conditional complex $\mathcal{C}$ from here on. Let us first discuss the first term in the integral. It implies a one-dimensional regression problem $V_L^* \mapsto V_H^*$. This is convenient since the original problem, $\xi \mapsto V_H^*$, is usually multi-dimensional, and will come in handy once we scale up the number of uncertain parameters. We need to choose a regression function $f$, acknowledge noise induced by the discretization error with $\sigma$, and marginalize $f$'s hyperparameters $\theta$, i.e.,

$$p(V_H^* \mid V_L^*, t, D_2, f, \sigma) = \int p(V_H^* \mid V_L^*, t, \cancel{D_2}, f, \sigma, \theta) \, p(\theta \mid \cancel{V_L^*}, t, D_2, f, \sigma) d\theta \qquad (3)$$

Note that $f$ is actually included in the conditional complex $\mathcal{C}$, but explicitly written out here. In Equation (3), we find a belated, formal justification for replacing the original regression problem $\xi \mapsto V_H^*$. In the conditional pdf in Equation (3), $\mathcal{C}$ implies that the knowledge of $f$, $\theta$, and $V_L^*$ already determines $V_H^*$ apart from noise $\sigma$, and thus $D_2$ is superfluous.

By looking at the data, we recognize that a linear regression function $f$ will capture the salient features, and is hence sufficient for all time instances. Thus the prior reads

$$p(\theta \mid f) = p(a, b \mid f, a_0, b_0) = a_0(1+a)^{-3/2} \Theta(\mid b \mid \leq b_0), \qquad (4)$$

with inclination $a$, constant offset $b$ and $\Theta$ being the Heaviside function. We find no apparent outliers in the data, and the likelihood shall be Gaussian with constant noise level, which was estimated from the data to $\sigma = 0.01$.

The second term in Equation (2) is approximated by weighted samples ($D_1$), and the integral boils down to a discreet sum over these.

## 4. Results & Discussion

An example of the intermediate result of Equation (3) is shown in Figure 7. We see the predictive probability of the HiFi impedance given the LoFi impedance for the time instance at peak systole ($t = 200$ ms). Naturally, linear regression requires at least two training data points, which was deemed good enough in this experiment. In Figure 6, one can see the final result in comparison to the reference solution, i.e., with all the HiFi simulations. Expectations as well as uncertainties ($2\sigma$) match the reference solution quite well in the time range of 0–500 ms (systolic part), while the uncertainties in the time range 500–1000 ms (diastolic part) are a bit larger. This is reasonable, since the discretisation error in this regime is notably higher, which is obvious from the data, Figure 5. Since changes to the impedance due to aortic dissection are particularly expected to arise in the systolic part, we find the result satisfying nevertheless. The computational effort is documented in Table 1. The Bayesian Multi-fidelity scheme reduces the computational effort roughly by a factor of 3.5. This might seem disappointing at first, but is actually quite close to the theoretical limit of a factor of 4, which is determined by the ratio of computational effort of one LoFi simulation over one HiFi simulation, and specific for any experiment thus. Specifically, the HiFi computational effort is defined by the user's desired fidelity, e.g., mesh convergence, while the LoFi computational effort is determined by the HiFi model's cheapest simplification which still shows statistical correlation with the HiFi model.

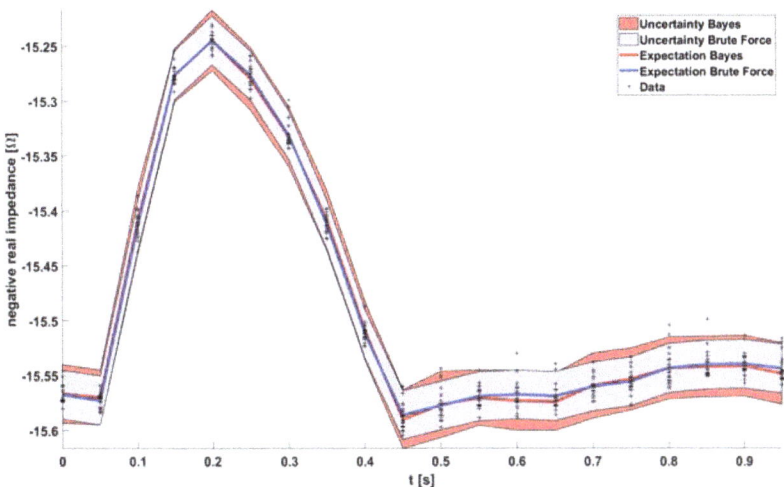

**Figure 6.** Resulting expectations and uncertainty bands ($2\sigma$) for the HiFi impedances. Red (Bayes) and Blue (brute force).

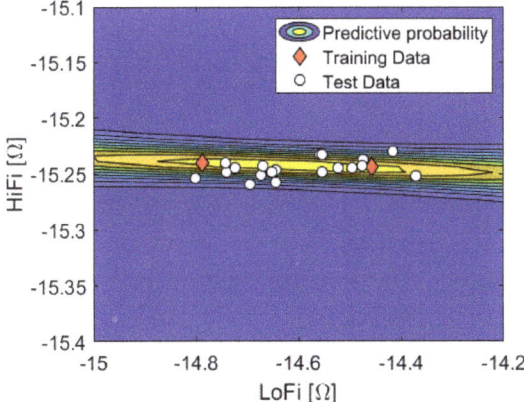

**Figure 7.** Linear regression at peak systole ($t$ = 200 ms). Predictive probability of HiFi impedances given the LoFi impedances, trained with two data points corresponding to the minimum and maximum used false lumen radius.

**Table 1.** Comparison of computational resources. Each experiment was performed on 20 cores in parallel (with trivial parallelization) on Xeon E5-2640 with 8GB RAM/CPU. Degrees of freedom vary from time instance to time instance since the aortic radius is a function of time. In Figure 6, we compare the Brute Force HiFi results (blue) with the Bayesian HiFi results (red). The LoFi results are a mere means to compute the Bayesian HiFi results and thus not shown, yet their computational effort is documented since needed to quantify the reduction of the computational effort.

| Model | Degrees of Freedom | Samples | CPU Time [s] | Wall Clock Time [s] |
|---|---|---|---|---|
| LoFi | 9000–15,000 | 25 | 39,500 | 1975 |
| Brute Force HiFi | 100,000–550,000 | 25 | 192,480 | 9624 |
| Bayesian HiFi | | 25 LoFi, 2 HiFi | 54955 | 2748 |

## 5. Conclusions & Outlook

We have set up a framework to systematically study the uncertainties of theoretical impedance cardiography signals associated with aortic dissection. Since the computational effort is about to skyrocket soon, we employed a Bayesian multi-fidelity scheme rather than using brute force. We did first experiments as a proof of principle, and computed the uncertainty bands of the simulation given an unknown false lumen radius. We achieved a solution quantitatively comparable to the reference solution, while reducing computational effort by roughly a factor of 3.5, which is close to the theoretical limit of 4. With increasing computational effort per simulation, we expect the reduction factor to increase.

The physical model will be improved by adding organs (e.g., lungs, heart) to the model one by one, and dispersion effects investigated by varying the injection current frequency (i.e., the boundary conditions). In terms of data analysis, the next step is to make further use of the pronounced structure of the signal, and sparsify simulation runs of the HiFi model in the time domain. This could be done by, e.g., interpreting each time series as a sample drawn from a Gaussian Process. i.e., we impose a GP prior onto the signal. Ultimately, we want to answer the inverse problem, "Is the aorta healthy or dissected?", and thus need to compute the evidences. We strongly believe that this question can only be answered unambiguously by considering the signals of multiple electrodes at different positions on the human body at once.

**Author Contributions:** conceptualization, S.R.; methodology, S.R.; software, S.R., G.M.M., V.B. and A.R.-K.; formal analysis, S.R.; investigation, S.R. and G.M.M.; resources, S.R. and G.M.M.; data curation, S.R. and G.M.M.; writing—original draft preparation, S.R.; writing—review and editing, W.v.d.L.; visualization, S.R.; supervision, W.v.d.L., K.E. and A.R.-K.; project administration, S.R.; funding acquisition, W.v.d.L. and K.E.

**Funding:** This work was funded by the Lead project "Mechanics, Modeling and Simulation of Aortic Dissection" of the TU Graz (biomechaorta.tugraz.at).

**Acknowledgments:** The authors would like to acknowledge the use of HPC resources provided by the ZID of Graz University of Technology.

**Conflicts of Interest:** The authors declare no conflict of interest.

**Sample Availability:** Data and code are available from the authors upon reasonable request.

## References

1. Humphrey, J.D. *Cardiovascular Solid Mechanics*; Springer: New York, NY, USA: 2002; p. 459, doi:10.1007/978-0-387-21576-1.
2. Heuser, J. *Distributed under a CC-BY-SA-3.0 License*; Tech. Rep.; Wikimedia Commons: Omaha, NE, USA, 2016.
3. Khan, I.A.; Nair, C.K. Clinical, diagnostic, and management perspectives of aortic dissection. *Chest* **2002**, *122*, 311–328, doi:10.1378/chest.122.1.311.
4. Reinbacher-Köstinger, A.; Badeli, V.; Biro, O.; Magele, C. Numerical Simulation of Conductivity Changes in the Human Thorax Caused by Aortic Dissection. *IEEE Trans. Magn.* **2019**, *55*, 1–4, doi:10.1109/tmag.2019.2895418.
5. Miller, J.C.; Horvath, S.M. Impedance Cardiography. *Psychophysiology* **1978**, *15*, 80–91, doi:10.1111/j.1469-8986.1978.tb01340.x.
6. Gabriel, C.; Gabriel, S.; Corthout, E.C. The dielectric properties of biological tissues: I. Literature survey. *Phys. Med. Biol.* **1996**, *41*, 2231–2249.
7. De Sitter, A.; Verdaasdonk, R.M.; Faes, T.J. Do mathematical model studies settle the controversy on the origin of cardiac synchronous trans-thoracic electrical impedance variations? A systematic review. *Physiol. Meas.* **2016**, *37*, R88–R108, doi:10.1088/0967-3334/37/9/R88.
8. Zienkiewicz, O.C.; Taylor, R.L.; Zhu, J.Z. *The Finite Element Method: Its Basis and Fundamentals*; Elsevier: London, UK, 1967.
9. Ghanem, R.G.; Owhadi, H.; Higdon, D. *Handbook of Uncertainty Quantification*; Springer: New York, NY, USA, 2017; doi:10.1007/978-3-319-12385-1.

10. Haylock, R. Bayesian Inference about Outputs of Computationally Expensive Algorithms with Uncertainty on the Inputs. Ph.D. Thesis, Universtity of Nottingham, Nottingham, UK, 1997. Available online: http://eprints.nottingham.ac.uk/13193/1/338522.pdf (accessed on 22 Novemebr 2019).
11. O'Hagan, A.; Kennedy, M.C.; Oakley, J.E. Uncertainty analysis and other inference tools for complex computer codes. In *Bayesian Staistics 6*; Oxford Science Publications: Oxford, UK, 1999; pp. 503–524.
12. O'Hagan, A. Bayesian analysis of computer code outputs: A tutorial. *Reliab. Eng. Syst. Saf.* **2006**, *91*, 1290–1300, doi:10.1016/j.ress.2005.11.025.
13. Wiener, N. The Homogeneous Chaos. *Am. J. Math.* **1938**, *60*, 897–936.
14. Ghanem, R.G.; Spanos, P.D. *Stochastic Finite Elements: A Spectral Approach*; Springer: New York, NY, USA, 1991; doi:10.1007/978-1-4612-3094-6.
15. Xiu, D.; Karniadakis, G.E. The Wiener-Askey polynomial chaos for stochastic differential equations. *SIAM J. Sci. Comput.* **2005**, *27*, 1118–1139.
16. O'Hagan, A. Polynomial Chaos: A Tutorial and Critique from a Statistician's Perspective. Available online: http://tonyohagan.co.uk/academic/pdf/Polynomial-chaos.pdf (accessed 25 June 2019).
17. O'Hagan, A. Curve Fitting and Optimal Design for Prediction. *J. R. Stat. Soc. Ser. B (Methodol.)* **1978**, *40*, 1–42, doi:10.2307/2984861.
18. Rasmussen, C.E.; Williams, C.K. *Gaussian Processes for Machine Learning*; The MIT Press: Cambridge, MA, USA, 2006; doi:10.1142/S0129065704001899.
19. Bishop, C. *Neural Networks for Pattern Recognition*; Oxford University Press: Oxford, UK, 1996.
20. MacKay, D.J. *Information Theory, Inference, and Learning Algorithms*; Cambridge University Press: Cambridge, UK, 2003; doi:10.1166/asl.2012.3830.
21. Kennedy, M.C.; O'Hagan, A. Predicting the output from a complex computer code when fast approximations are available. *Biometrika* **2000**, *87*, 1–13, doi:10.1093/biomet/87.1.1.
22. Koutsourelakis, P.S. Accurate Uncertainty Quantification using inaccurate Computational Models. *SIAM J. Sci. Comput.* **2009**, *31*, 3274–3300, doi:10.1137/080733565.
23. Eck, V.G.; Donders, W.P.; Sturdy, J.; Feinberg, J.; Delhaas, T.; Hellevik, L.R.; Huberts, W. A guide to uncertainty quantification and sensitivity analysis for cardiovascular applications. *Int. J. Numer. Methods Biomed. Eng.* **2016**, *32*, e02755, doi:10.1002/cnm.2755.
24. Biehler, J.; Gee, M.W.; Wall, W.A. Towards efficient uncertainty quantification in complex and large-scale biomechanical problems based on a Bayesian multi-fidelity scheme. *Biomech. Model. Mechanobiol.* **2015**, *14*, 489–513, doi:10.1007/s10237-014-0618-0.
25. Alastruey, J.; Xiao, N.; Fok, H.; Schaeffter, T.; Figueroa, C.A. On the impact of modelling assumptions in multi-scale, subject-specific models of aortic haemodynamics. *J. R. Soc. Interface* **2016**, *13*, doi:10.1098/rsif.2016.0073.
26. Comsol, A.B.; Stockholm, S. Comsol Multiphysics Version 5.4. Available online: http://www.comsol.com (accessed on 4 June 2019).

© 2019 by the authors. Licensee MDPI, Basel, Switzerland. This article is an open access article distributed under the terms and conditions of the Creative Commons Attribution (CC BY) license (http://creativecommons.org/licenses/by/4.0/).

*Proceedings*

# A Complete Classification and Clustering Model to Account for Continuous and Categorical Data in Presence of Missing Values and Outliers [†]

**Guillaume Revillon** [‡] **and Ali Mohammad-Djafari** *

L2S, CentraleSupélec-Univ Paris Saclay, 91192 Gif-sur-Yvette, France; guillaume.revillon@l2s.centralesupelec.fr
* Correspondence: Ali.Mohammad-Djafari@l2s.centralesupelec.fr; Tel.: +33-6-19-97-11-17
† Presented at the 39th International Workshop on Bayesian Inference and Maximum Entropy Methods in Science and Engineering, Garching, Germany, 30 June–5 July 2019.
‡ These authors contributed equally to this work.

Published: 9 December 2019

**Abstract:** Classification and clustering problems are closely connected with pattern recognition where many general algorithms have been developed and used in various fields. Depending on the complexity of patterns in data, classification and clustering procedures should take into consideration both continuous and categorical data which can be partially missing and erroneous due to mismeasurements and human errors. However, most algorithms cannot handle missing data and imputation methods are required to generate data to use them. Hence, the main objective of this work is to define a classification and clustering framework that handles both outliers and missing values. Here, an approach based on mixture models is preferred since mixture models provide a mathematically based, flexible and meaningful framework for the wide variety of classification and clustering requirements. More precisely, a scale mixture of Normal distributions is updated to handle outliers and missing data issues for any types of data. Then a variational Bayesian inference is used to find approximate posterior distributions of parameters and to provide a lower bound on the model log evidence used as a criterion for selecting the number of clusters. Eventually, experiments are carried out to exhibit the effectiveness of the proposed model through an application in Electronic Warfare.

**Keywords:** classification; clustering; mixture models; bayesian framework; outliers; missing data

## 1. Introduction

Classification and clustering problems are closely connected with pattern recognition [1] where many general algorithms [2–4] have been developed and used in various fields [5,6]. Depending on the complexity of patterns in data, classification and clustering procedures should take into consideration both continuous and categorical data which can be partially missing and erroneous due to mismeasurements and human errors. However, most algorithms cannot handle missing data and imputation methods [7] are required to generate data to use them. Hence, the main objective of this work is to define a classification and clustering framework that handles both outliers and missing values. Here, an approach based on mixture models is preferred since mixture models provide a mathematically based, flexible and meaningful framework for the wide variety of classification and clustering requirements [8]. Two families of models emerge from finite mixture models fitting mixed-type data:

- The location mixture model [9] that assumes that continuous variables follow a multivariate Gaussian distribution conditionally on both component and categorical variables.

- The underlying variables mixture model [10] that assumes that each discrete variable arises from a latent continuous variable and that all continuous variables follow a Gaussian mixture model.

In this work, the location mixture model approach is retained since it better models relations between continuous and categorical features when data patterns are mostly designed by first choosing patterns of categorical features to achieve a specific goal and then choosing continuous features that meet constraints related to the chosen patterns and the problem environment. Indeed regarding clustering approach, each cluster groups observations that share same combinations of categorical features where continuous features belong to a peculiar subset. Hence, the location mixture model naturally responds to that dependence structure by assuming that continuous variables are normally distributed conditionally to categorical variables. More precisely, a scale mixture of conditional Gaussian distributions [11] is updated to handle outliers and missing data issues for any types of data. Then a variational Bayesian inference [12] is used to find approximate posterior distributions of parameters and to provide a lower bound on the model log evidence used as a criterion for selecting the number of clusters. An application of the resulting model in Electronic Warfare [13] is proposed to perform Source Emission Identification which is a supreme asset for decision making in military tactical situations. By providing information about the presence of threats, classification and clustering of radar emitters have a significant role ensuring that countermeasures against enemies are well-chosen and enabling detection of unknown radar signals to update databases. As a pulse-to-pulse modulation pattern [14], a radar signal pattern is decomposed into a relevant arrangement of sequences of pulses where each pulse is defined by continuous features and each sequence is characterized by categorical features. However, a radar signal is often partially observed due to the presence of many radar emitters in the electromagnetic environment causing mismeasurements and measurement errors. Therefore the proposed model is suitable for radar emitter classification and clustering. The outline of the paper is as follows. Assumptions on mixed-type data are presented in Section 2. Then, the proposed model and inference procedure are introduced in Section 3. Finally, evaluation of the model is proposed through different experiments on radar emitter datasets in Section 4.

## 2. Mixed-Type Data

In this section, a joint distribution for mixed data is introduced to model the dependence structure between continuous and categorical data. Then, outliers and missing values are tackled by taking advantage of the joint distribution.

### 2.1. Assumptions on Mixed-Type Data

Data $x$ consist of $J$ observations $(x_j)_{j=1}^J$ gathering continuous features $x_q = (x_{qj})_{j=1}^J$ and categorical features $x_c = (x_{cj})_{j=1}^J$. Let $x_j = (x_{qj}, x_{cj})$ the $j^{th}$ observation vector of mixed variables where

- $x_{qj} \in \mathbb{R}^d$ is a vector of $d$ continuous variables,
- $x_{cj} = \left(x_{cj}^0, \ldots, x_{cj}^{q-1}\right) \in \mathcal{C}_q$ is a vector of $q$ categorical variables where $\mathcal{C}_q = \mathcal{C}_0 \times \ldots \times \mathcal{C}_{q-1}$ is the tensor gathering each space $\mathcal{C}_i = \left\{m_1^i, \ldots, m_{|\mathcal{C}_i|}^i\right\}$ of events that $x_{cj}^i$ can take $\forall i \in \{0, \ldots, q-1\}$.

### 2.2. Distribution of Mixed-Type Data

Considering that the retained approach focuses on conditioning continuous data $x_q = (x_{qj})_{j=1}^J$ according to categorical data $x_c = (x_{cj})_{j=1}^J$, the following joint distribution is introduced

$$\forall j \in \{1,\ldots,J\}, \ p(x_{qj}, x_{cj}) = \prod_{c \in \mathcal{C}_q} \left(\pi_c \mathcal{N}\left(x_{qj}|\mu_c, \Sigma\right)\right)^{\delta_{x_{cj}}^c} \tag{1}$$

where continuous variables $x_{qj}$ are normally distributed according to categorical variables $x_{cj}$ with means $(\mu_c)_{c \in \mathcal{C}_q}$ and variance $\Sigma$. As for categorical variables $x_{cj}$, they are jointly distributed according to a multivariate categorical distribution $\mathcal{MC}(x_{cj}|\pi)$ parametrized by weights $\pi = (\pi_c)_{c \in \mathcal{C}_q}$ and defined by

$$\mathcal{MC}(x_{cj}|\pi) = \prod_{c \in \mathcal{C}_q} \pi_c^{\delta_{x_{cj}}^c} \qquad (2)$$

where $\forall c = (c^0, \ldots, c^{q-1}) \in \mathcal{C}_q = \mathcal{C}_0 \times \ldots \times \mathcal{C}_{q-1}$:

$$\sum_{c \in \mathcal{C}_q} \pi_c = 1 \text{ and } \delta_{x_{cj}}^c = \begin{cases} 1 \text{ if } x_{cj}^0 = c^0, \ldots, x_{cj}^{q-1} = c^{q-1} \\ 0 \text{ otherwise} \end{cases}.$$

This multivariate categorical distribution is proposed to tackle issues related to missing data by modelling a dependence structure for $x_{cj}$ that enables inference on missing categorical features.

### 2.3. Outlier Handling

Outliers are only considered for continuous data $x_q = (x_{qj})_{j=1}^J$ since only reliable categorical variables are assumed to be filled in databases and unreliable ones are processed as missing data. Then, continuous outliers are handled by introducing scale latent variables $u = (u_j)_{j=1}^J$ conditionally to categorical data $x_c$ due to the dependence structure established in (1) such that

$$\forall j \in \{1, \ldots, J\}, \ x_{qj}|u_j, x_{cj} \sim \prod_{c \in \mathcal{C}_q} \mathcal{N}\left(x_{qj}|\mu_c, u_j^{-1}\Sigma\right)^{\delta_{x_{cj}}^c} \text{ and } u_j|x_{cj} \sim \prod_{c \in \mathcal{C}_q} \mathcal{G}\left(u_j|\alpha_c, \beta_c\right)^{\delta_{x_{cj}}^c},$$

where each $u_j$ follows conditionally to categorical data $x_{cj}$ a Gamma distribution with rate and shape parameters $(\alpha_c, \beta_c) \in \mathbb{R}^{*+} \times \mathbb{R}^{*+}$.

### 2.4. Missing Data Handling

Both continuous and categorical data $(x_{qj}, x_{cj})_{j=1}^J$ can be partially observed. Hence $(x_{qj}, x_{cj})_{j=1}^J$ are decomposed into observed features $(x_{qj}^{obs}, x_{cj}^{obs})_{j=1}^J$ and missing features $(x_{qj}^{miss}, x_{cj}^{miss})_{j=1}^J$ such that

$$\forall j \in \{1, \ldots, J\}, \quad x_{qj} = \begin{pmatrix} x_{qj}^{miss} \\ x_{qj}^{obs} \end{pmatrix} \text{ with } (x_{qj}^{miss}, x_{qj}^{obs}) \in \mathbb{R}^{d_j^{miss}} \times \mathbb{R}^{d_j^{obs}} \text{ and } d_j^{miss} + d_j^{obs} = d,$$

$$x_{cj} = \begin{pmatrix} x_{cj}^{miss} \\ x_{cj}^{obs} \end{pmatrix} \text{ with } (x_{cj}^{miss}, x_{cj}^{obs}) \in \mathcal{C}_{q_j^{miss}} \times \mathcal{C}_{q_j^{obs}} \text{ and } q_j^{miss} + q_j^{obs} = q.$$

where $(\mathbb{R}^{d_j^{miss}}, \mathcal{C}_{q_j^{miss}})$ and $(\mathbb{R}^{d_j^{obs}}, \mathcal{C}_{q_j^{obs}})$, are disjoint subsets of $(\mathbb{R}^d, \mathcal{C}_q)$ embedding missing features $(x_{qj}^{miss}, x_{cj}^{miss})$ and observed features $(x_{qj}^{obs}, x_{cj}^{obs})$. Missing continuous data $x_q^{miss} = (x_{qj}^{miss})_{j=1}^J$ are handled by taking advantage of properties of the multivariate normal distribution to obtain a distribution for missing values. Due to the dependence structure established in (1), missing continuous data $x_q^{miss} = (x_{qj}^{miss})_{j=1}^J$ are distributed conditionally to observed continuous data $x_q^{obs} = (x_{qj}^{obs})_{j=1}^J$ and categorical data $x_c$ as follows

$$\forall j \in \{1, \ldots, J\}, \ x_{qj}^{miss}|x_{qj}^{obs}, x_{cj} \sim \prod_{c \in \mathcal{C}} \mathcal{N}\left(x_{qj}^{miss}|\mu_{jc}^{x_q^{miss}}, \Sigma^{x_q^{miss}}\right)^{\delta_{x_{cj}}^c}, \ x_{qj}^{obs}|x_{qj}^{obs}, x_{cj} \sim \prod_{c \in \mathcal{C}} \mathcal{N}\left(x_{qj}^{obs}|\mu_{jc}^{x_q^{obs}}, \Sigma^{x_q^{obs}}\right)^{\delta_{x_{cj}}^c},$$

where $\forall j \in \{1, \ldots, J\}$, $\forall c \in C_q$:

$$\mu_{jc}^{x_q^{miss}} = \mu_c^{miss} + \Sigma^{cov} {\Sigma^{obs}}^{-1} \left( x_{qj}^{obs} - \mu_c^{obs} \right), \quad \mu_{jc}^{x_q^{obs}} = \mu_c^{obs},$$

$$\Sigma^{x_q^{miss}} = \Sigma^{miss} - \Sigma^{cov} {\Sigma^{obs}}^{-1} \Sigma^{cov'} \text{ and } \Sigma^{x_q^{obs}} = \left( {\Sigma^{obs}}^{-1} + 2 \times {\Sigma^{obs}}^{-1} \Sigma^{cov'} \left( \Sigma^{x_q^{miss}} \right)^{-1} \Sigma^{cov} {\Sigma^{obs}}^{-1} \right)^{-1}.$$

Noting that the dependence structure between categorical features is modeled through Kronecker symbols $(\delta_{x_{cj}}^c)_{c \in C_q}$, this dependence structure can be exploited to handle missing features such that the missing features $x_{cj}^{miss}$ follow a multivariate categorical distribution conditionally to observed features $x_{cj}^{obs}$ given by

$$p(x_{cj}^{miss} = c^{miss} | x_{cj}^{obs} = c^{obs}) = \frac{\pi_{c^{miss}, c^{obs}}}{\sum_{c^{miss} \in C_{q^{miss}}} \pi_{c^{miss}, c^{obs}}}$$

where $\pi_{c^{miss}, c^{obs}}$ is the joint probability $\pi_c$ defined in (2) for $c = (c^{miss}, c^{obs}) \in C_{q_j^{miss}} \times C_{q_j^{obs}}$.

## 3. Model and Inference

In this section, the proposed model is briefly presented as a hierarchical latent variable model handling missing values and outliers. Then, the inference procedure is developed through a variational Bayesian approximation. At last, classification and clustering algorithms are introduced by using the proposed model.

### 3.1. Model

According to a dataset $x^{obs}$ of i.i.d observations, independent latent variables $h = (x^{miss}, u, z)$, parameters $\Theta = (a, \pi, \alpha, \beta, \mu, \Sigma)$ of the $K$ clusters and assumptions on mixed data defined in Section 2.1, the complete likelihood of the proposed mixture model can be expressed as

$$p(x^{obs}, h | \Theta, K) = \prod_{j=1}^{J} \prod_{k=1}^{K} \left( a_k \prod_{c \in C_q} \left( \pi_{kc} \mathcal{N} \left( \begin{pmatrix} x_{qj}^{miss} \\ x_{qj}^{obs} \end{pmatrix} \bigg| \mu_{kc}, u_j^{-1} \Sigma_k \right) \mathcal{G}(u_j | \alpha_{kc}, \beta_{kc}) \right)^{\delta_c^c \left( x_{cj}^{miss}, x_{cj}^{obs} \right)} \right)^{\delta_{z_j}^k}$$

where

- $x^{obs} = (x_{qj}^{obs}, x_{cj}^{obs})_{j=1}^{J}$ are the observed features,
- $x^{miss} = (x_{qj}^{miss}, x_{cj}^{miss})_{j=1}^{J}$ are the latent variables modelling the missing features,
- $z = (z_j)_{j=1}^{J}$ the independent labels for continuous and categorical observations $x = (x_{qj}, x_{cj})_{j=1}^{J}$
- $u = (u_j)_{j=1}^{J}$ the scale latent variables handling outliers for quantitative data $x_q$ and distributed according to a Gamma distribution with shape and rate parameters $(\alpha, \beta) = (\alpha_{kc}, \beta_{kc})_{(k,c) \in \{1, \ldots, K\} \times C_q}$,
- $a = (a_k)_{k=1}^{K}$ are the weights related to component distributions,
- $(\mu, \Sigma) = ((\mu_{kc})_{c \in C_q}, \Sigma_k)_{k=1}^{K}$ the mean and the variance parameters of quantitative data $x_q$ for each cluster,
- $\pi = (\pi_k)_{k=1}^{K}$ the weights of the multivariate Categorical distribution of categorical data $x_c$ for each cluster.

Eventually, the Bayesian framework imposes to specify a prior distribution $p(\Theta|K)$ for $\Theta$ which is chosen as

$$p(\Theta|K) = p(a|K)p(\pi|K)p(\alpha,\beta|K)p(\mu,\Sigma|K)$$
$$= \mathcal{D}(a|\kappa_0)\prod_{k=1}^{K}\mathcal{D}(\pi_k|\pi_0)\prod_{c\in\mathcal{C}_q}p(\alpha_{kc},\beta_{kc}|p_0,q_0,s_0,r_0)\mathcal{N}\left(\mu_{kc}|\mu_{0_{kc}},\eta_{0_{kc}}^{-1}\Sigma_k\right)\mathcal{IW}(\Sigma_k|\gamma_0,\Sigma_0)$$

where $\mathcal{D}(\cdot|\cdot)$ and $\mathcal{IW}(\cdot|\cdot)$ denote the Dirichlet and Inverse-Wishart distributions and $p(\cdot,\cdot|p,q,s,r)$ is a particular distribution designed to avoid a non-closed-form posterior distribution for $(\alpha,\beta)$ such that $\forall (\alpha,\beta) \in \mathbb{R}^{*+} \times \mathbb{R}^{*+}$, $p(\alpha,\beta|p,q,s,r) \propto p^{\alpha-1}e^{-q\beta}\beta^{s\alpha}\Gamma(\alpha)^{-r}$.

## 3.2. Variational Bayesian Inference

The intractable posterior distribution $P = p(h,\Theta|x^{\text{obs}},K)$ is approximated by a tractable one $Q = q(h,\Theta|K)$ whose parameters are chosen via a variational principle to minimize the Kullback-Leibler (KL) divergence

$$KL\left[Q||P\right] = \int q(h,\Theta|K)\log\left(\frac{q(h,\Theta|K)}{p(h,\Theta|x^{\text{obs}},K)}\right)\partial h\partial\Theta = \log p(x^{\text{obs}}|K) - \mathcal{L}(q|K)$$

with $\mathcal{L}(q|K)$ a lower bound for the log evidence $\log p(x^{\text{obs}}|K)$ given by

$$\mathcal{L}(q|K) = \mathbb{E}_{h,\Theta}\left[\log p(x^{\text{obs}},h,\Theta|K)\right] - \mathbb{E}_{h,\Theta}\left[\log q(h,\Theta|K)\right], \quad (3)$$

where $\mathbb{E}_{h,\Theta}[\cdot]$ denotes the expectation with respect to $q(h,\Theta|K)$. Then, minimizing the KL divergence is equivalent to maximizing $\mathcal{L}(q|K)$. Assuming that $q(h,\Theta|K)$ can be factorized over the latent variables $h$ and the parameters $\Theta$, a free-form maximization with respect to $q(h|K)$ and $q(\Theta|K)$ leads to the following update rules :

$$\textbf{VBE-step}: q(h|K) \propto \exp\left(\mathbb{E}_{\Theta}\left[\log p(x^{\text{obs}},h|\Theta,K)\right]\right),$$
$$\textbf{VBM-step}: q(\Theta|K) \propto \exp\left(\mathbb{E}_{h}\left[\log p(x^{\text{obs}},h,\Theta|K)\right]\right).$$

Thereafter, the algorithm iteratively updates the variational posteriors by increasing the bound $\mathcal{L}(q|K)$. Even if latent variables $h$ and parameters $\Theta$ are assumed to be independent a posteriori, their conditional structures are preserved as follows

$$q(h|K) = q(x_q^{\text{miss}}|u,x_c^{\text{miss}},z,K)q(u|x_c^{\text{miss}},z,K)q(x_c^{\text{miss}}|z,K)q(z|K),$$
$$q(\Theta|K) = q(a|K)q(\pi|K)q(\alpha,\beta|K)q(\mu,\Sigma|K).$$

Eventually, the following conjugate variational posterior distributions are obtained according to the previous assumptions

$$q(h|K) = \prod_{j=1}^{J}\prod_{k=1}^{K}\left(\tilde{r}_{jk}\prod_{c_{\text{miss}}\in\mathcal{C}_{q_j^{\text{miss}}}}\left(\tilde{r}_{jkc_{\text{miss}}}^{x_c^{\text{miss}}}\prod_{c_{\text{obs}}\in\mathcal{C}_{q_j^{\text{obs}}}}\left(\mathcal{N}\left(x_{qj}^{\text{miss}}|\tilde{\mu}_{jkc}^{x_q^{\text{miss}}},u_j^{-1}\tilde{\Sigma}_k^{x_q^{\text{miss}}}\right)\mathcal{G}\left(u_j|\tilde{\alpha}_{jkc},\tilde{\beta}_{jkc}\right)\right)^{\delta_{cj}^{c_{\text{obs}}}}\right)^{\delta_{\text{miss}}^{c_{\text{miss}}}}\right)^{\delta_{z_j}^{k}},$$

$$q(\Theta|K) = \mathcal{D}(a|\tilde{\kappa})\prod_{k=1}^{K}\mathcal{D}(\pi|\tilde{\pi}_k)\prod_{c\in\mathcal{C}_q}p(\alpha_{kc},\beta_{kc}|\tilde{p}_k,\tilde{q}_k,\tilde{s}_k,\tilde{r}_k)\mathcal{N}\left(\mu_{kc}|\tilde{\mu}_{kc},\tilde{\eta}_{kc}^{-1}\Sigma_k\right)\mathcal{IW}(\Sigma_k|\tilde{\gamma}_k,\tilde{\Sigma}_k).$$

Their respective parameters are estimated during the VBE and VBM-steps by developing expectations $\mathbb{E}_{\Theta}\left[\log p(x^{\text{obs}},h|\Theta,K)\right]$ and $\mathbb{E}_{h}\left[\log p(x^{\text{obs}},h,\Theta|K)\right]$.

## 3.3. Classification and Clustering

According to the degree of supervision, three problems can be distinguished: supervised classification, semi-supervised classification and unsupervised classification known as clustering. The supervised classification problem is decomposed into a training step and a prediction step. The training step consists in estimating parameters $\Theta$ given the number of classes $K$ and a set of training data $x$ with known labels $z$. Then, the prediction step results in associating label $z^*$ of a new sample $x^*$ to its class $k^*$ chosen as the Maximum A Posteriori (MAP) solution

$$k^* = \arg\max_{k=1}^{K} q(z^* = k|K)$$

given the previous estimated parameters $\Theta$. In the semi-supervised classification, only the number of classes $K$ is known and both labels $z$ of the dataset $x$ and parameters $\Theta$ have to be determined. As for the prediction step, the MAP criterion is retained for affecting observations to classes such that

$$k^* = \arg\max_{k=1}^{K} q(z = k|K) .$$

Given a set of data $x$, the clustering problem aims to determine the number of clusters $\tilde{K}$, labels $z$ of data and parameters $\Theta$. Selecting the appropriate $\tilde{K}$ seems like a model selection issue and is usually based on a maximized likelihood criterion given by

$$\tilde{K} = \arg\max_{K} \log p(x|K) = \arg\max_{K} \log \int p(x,\Theta|K)d\Theta . \qquad (4)$$

Unfortunately, $\log p(x|K)$ is intractable and the lower bound in (3) is preferred to penalized likelihood criteria [8,15,16] since it does not depend on asymptotical assumptions and does not require Maximum Likelihood estimates. Then according to an a priori range of numbers of clusters $\{K_{\min}, \ldots, K_{\max}\}$, the semi-supervised classification is performed for each $K \in \{K_{\min}, \ldots, K_{\max}\}$ and both $z^K$ and $\Theta^K$ are estimated. Finally, the number of classes $\tilde{K}$ in (4) is chosen as the maximizer of the lower bound $\mathcal{L}(q|K)$ :

$$\tilde{K} = \arg\max_{K} \mathcal{L}(q|K) . \qquad (5)$$

After determining $\tilde{K}$, only $z^{\tilde{K}}$ and $\Theta^{\tilde{K}}$ are kept as estimated labels and parameters.

## 4. Application

In this section, the proposed method is performed on a radar emitter dataset. For comparison, a standard neural network (NN), the k-nearest neighbours (KNN) algorithm, Random Forests (RdF) the k-means algorithm are also evaluated. Two experiments are carried out to evaluate classification and clustering performance with respect to a range of percentages of missing values.

### 4.1. Data

Realistic data are generated from an operational database gathering 55 radar emitters presenting various patterns. Each pattern consists of a sequence of pulses which are defined by a triplet of continuous features (pulse features) and a quartet of categorical features (pulse modulations) listed among 42 combinations of the categorical features. For each radar emitter, 100 observations $(x_j)_{j=1}^{100}$ are simulated from its pattern of pulses such that an observation $x_j = (x_{qj}, x_{cj})$ is made up of continuous features $x_{qj}$ and categorical features $x_{cj}$ related to one of the pulses. Extra missing values are added to evaluate limits of the proposed approach by randomly deleting coordinates of $(x_{qj})_{j=1}^{100}$ and $(x_{cj})_{j=1}^{100}$ for each of the 55 radar emitters. Therefore, imputation methods [17] are used to handle missing data for comparison algorithms. As for continuous missing data, they are handled through the Mean and k-nearest neighbours imputation methods whereas missing categorical data are handled through the k-nearest neighbours and mode imputation methods. These imputation methods are compared with the proposed approach where missing continuous data are reconstructed through the variational posterior marginal mean of missing continuous data given by $\forall j \in \{1, \ldots, J\}$,

$$\tilde{x}_{qj}^{\text{miss}} = \mathbb{E}_{x_{qj}^{\text{miss}}}\left[\int q(x_{qj}^{\text{miss}}, u_j, x_{cj} z_j) \partial u_j \partial x_{cj} \partial z_j\right] = \sum_{k=1}^{K} \tilde{r}_{jk} \sum_{c^{\text{obs}} \in \mathcal{C}_{q_j^{\text{obs}}}} \delta_{x_{cj}^{\text{obs}}}^{c^{\text{obs}}} \sum_{c^{\text{miss}} \in \mathcal{C}_{q_j^{\text{miss}}}} \tilde{r}_{jkc^{\text{miss}}}^{x_c^{\text{miss}}} \tilde{\mu}_{jkc^{\text{obs}}c^{\text{miss}}}^{x_q^{\text{miss}}} \quad (6)$$

and missing categorical data are reconstructed through the variational posterior marginal mode of missing categorical data given by $\forall j \in \{1, \ldots, J\}$,

$$\tilde{x}_{cj}^{\text{miss}} = \arg\max_{c^{\text{miss}} \in \mathcal{C}_{q_j^{\text{miss}}}} \int q(x_{cj}^{\text{miss}}, z_j) dz_j = \arg\max_{c^{\text{miss}} \in \mathcal{C}_{q_j^{\text{miss}}}} \sum_{k=1}^{K} \tilde{r}_{jk} \tilde{r}_{jkc^{\text{miss}}}^{x_c^{\text{miss}}} . \quad (7)$$

### 4.2. Classification Experiment

The classification experiment evaluates the ability of each algorithm to assign unlabeled data to one of the $K$ classes trained by a set of labeled data. Since comparison algorithms do not handle datasets including missing values, a complete dataset is used to enable their training. During the prediction step, incomplete observations are completed thanks to the mean and KNN imputation methods and the posterior reconstructions defined in (6) and (7). For the classification experiment, results are shown in Figure 1. Without missing data, both algorithms cannot perfectly classify the 55 radar emitters for the 2 datasets. Indeed, both algorithms reach accuracies of 90% for the continuous dataset and 98% for the mixed dataset. These performance can be explained by the non total separability of continuous and categorical datasets since the 55 emitters share 42 combinations of categorical features and some intervals of continuous features. Nonetheless when mixed data are taken into consideration, the dataset becomes more separable leading to higher performance of both algorithms. When the proportion of missing values increases, the proposed model outperforms comparisons algorithms for each dataset. It achieves accuracies of 80% and 95% for 90% of deleted continuous and mixed values whereas accuracies of comparison algorithms are lower than 65% and 75% with missing data imputation from standard methods. These higher performance of the proposed model reveal that the proposed method embeds a more efficient inference method than other imputation methods. That result is confirmed on Figure 1 when comparison algorithms are applied on data reconstructed by the proposed model. Indeed when the proposed inference is chosen, comparison algorithms share the same performance than the proposed model and manage to handle missing data even for 90% of deleted values.

Then, effectiveness of the proposed model can be explained by the fact that missing data imputation methods can create outliers that deteriorate performance of classification algorithms whereas the inference on missing data and labels prediction are jointly estimated in the proposed model. Indeed, embedding the inference procedure into the model framework allows properties of the model, such as outliers handling, to counterbalance drawbacks of imputation methods such as outlier creation.

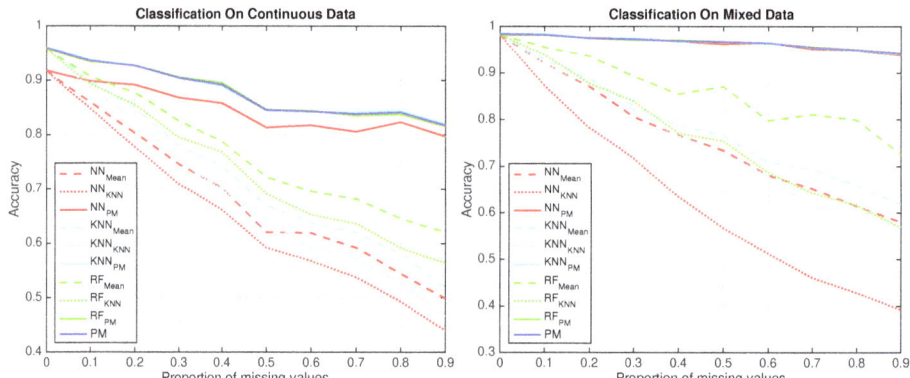

**Figure 1.** Classification performance are presented for the proposed model (PM) in blue, the NN in red, the RnF in green and the KNN in cyan. For each figure, solid lines represent accuracies with a posteriori reconstructed missing data, doted dashed lines stand for accuracies with mean/mode imputation whereas dashed lines show accuracies with KNN imputation for the comparison algorithms.

### 4.3. Clustering Experiment

The clustering experiment tests the ability of each algorithm to find the true number of clusters $\tilde{K}$ among $\{35, \ldots, 85\}$. The lower bound (3) and the average Silhouette score [18] are criteria used to select the optimal number of clusters for the proposed model and the k-means algorithm. Results of the clustering experiment are visible on Figure 2 which presents numbers of clusters selected by the lower bound and average Silhouette scores for the proposed model and k-means algorithm according to different proportions of missing values and imputation methods. Without missing data, the correct number of clusters ($K = 55$) is selected by the two criteria for the k-means algorithm and the proposed model when continuous and mixed data are clustered. In presence of missing values, the average Silhouette score mainly selects $K = 65$ when the k-means algorithm is run on the 2 datasets completed by standard imputation methods. When, the k-means algorithm performs clustering on the posterior reconstructions, the average Silhouette score correctly selects $K = 55$ until 60% of missing values for continuous data and 40% of missing values for mixed data. Eventually when the proposed model does clustering, the two criteria select the correct number of clusters $K = 55$ until 70% of missing values for continuous and mixed data. These results show two main advantages of the proposed model. As previously, the proposed model provides a more robust inference on missing data since the average Silhouette score chooses more representative number of clusters when the k-means algorithm is run on the posterior reconstructions than on data completed by standard imputation methods. Furthermore, since the lower bound criterion also selects the correct number of clusters as the average Silhouette score, it can be used as a valid criterion for selecting the optimal number of clusters and does not require extra computational costs as the Silhouette score since it is computed during the model parameter estimation. Finally, the proposed approach provides a more robust inference on missing data and a criterion for selecting the optimal number of clusters without extra computations.

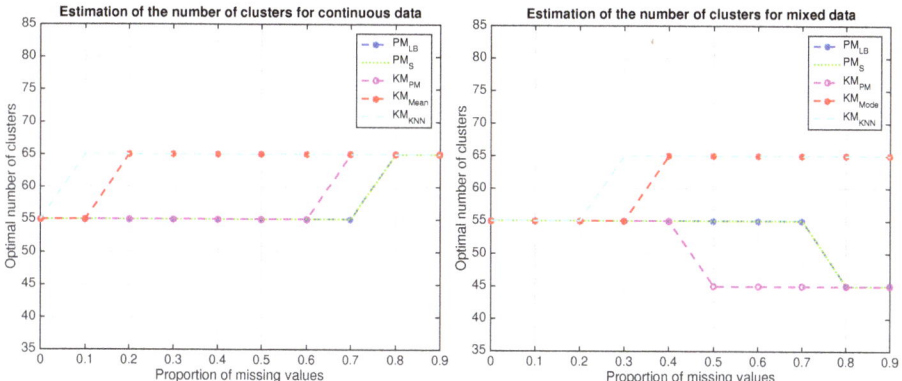

**Figure 2.** Estimation of the number of clusters using the lower bound (LB) and the silhouette score (S) for the proposed model and only the silhouette score (S) for the k-means algorithm.

## 5. Conclusions

In this paper, a mixture model handling both continuous data and categorical data is developed. More precisely, an approach based on the conditional Gaussian mixture model is investigated by establishing conditional relations between continuous and categorical data. Benefiting from a dependence structure designed for mixed-type data, the proposed model shows its efficiency for inferring on missing data, performing classification and clustering tasks and selecting the correct number of clusters. Since the posterior distribution is intractable, model learning is processed through a variational Bayesian approximation where variational posterior distributions are proposed for continuous and categorical missing data. Experiments point out that the proposed approach can handle mixed-type data even in presence of missing values and can outperform standard algorithms in classification and clustering tasks. Indeed the main advantage of our approach is that it enables the counterbalance of imputation methods drawbacks by embedding the inference procedure into the model framework.

## References

1. Bishop, C.M. *Pattern Recognition and Machine Learning (Information Science and Statistics)*; Springer: Berlin/Heidelberg, Germany, 2006.
2. Hartigan, J.A.; Wong, M.A. Algorithm AS 136: A k-means clustering algorithm. *J. R. Stat. Soc. Ser. C (Appl. Stat.)* **1979**, *28*, 100–108.
3. Ester, M.; Kriegel, H.P.; Sander, J.; Xu, X. *A Density-Based Algorithm for Discovering Clusters in Large Spatial Databases with Noise*; Kdd: Portland, OR, USA, 1996; Volume 96, pp. 226–231.
4. Breiman, L. Random forests. *Mach. Learn.* **2001**, *45*, 5–32.
5. Sander, J.; Ester, M.; Kriegel, H.P.; Xu, X. Density-Based clustering in spatial databases: The algorithm GDBSCAN and its applications. *Data Min. Knowl. Discov.* **1998**, *2*, 169–194.
6. Jain, A.K. Data clustering: 50 years beyond K-means. *Pattern Recognit. Lett.* **2010**, *31*, 651–666.
7. Troyanskaya, O.; Cantor, M.; Sherlock, G.; Brown, P.; Hastie, T.; Tibshirani, R.; Botstein, D.; Altman, R.B. Missing value estimation methods for DNA microarrays. *Bioinformatics* **2001**, *17*, 520–525.
8. Biernacki, C.; Celeux, G.; Govaert, G. Assessing a mixture model for clustering with the integrated completed likelihood. *IEEE Trans. Pattern Anal. Mach. Intell.* **2000**, *22*, 719–725.
9. Lawrence, C.J.; Krzanowski, W.J. Mixture separation for mixed-mode data. *Stat. Comput.* **1996**, *6*, 85–92, doi:10.1007/BF00161577.
10. Everitt, B. A finite mixture model for the clustering of mixed-mode data. *Stat. Probab. Lett.* **1988**, *6*, 305–309. doi:10.1016/0167-7152(88)90004-1.
11. Andrews, D.F.; Mallows, C.L. Scale Mixtures of Normal Distributions. *J. R. Stat. Soc. Ser. B (Methodol.)* **1974**, *36*, 99–102.

12. Waterhouse, S.; MacKay, D.; Robinson, T. Bayesian methods for mixtures of experts. In *Advances in Neural Information Processing Systems*; MIT Press: Cambridge, MA, USA, 1996; pp. 351–357.
13. Schleher, D.C. *Introduction to Electronic Warfare*; Technical report; Eaton Corp., AIL Div.: Deer Park, NY, USA, 1986.
14. Richards, M.A. *Fundamentals of Radar Signal Processing*; McGraw-Hill Education: New York, NY, USA, 2005.
15. Akaike, H. Information theory and an extension of the maximum likelihood principle. In *Selected Papers of Hirotugu Akaike*; Springer: New York, NY, USA, 1998; pp. 199–213.
16. Schwarz, G. Estimating the dimension of a model. *Ann. Stat.* **1978**, *6*, 461–464.
17. García-Laencina, P.J.; Sancho-Gómez, J.L.; Figueiras-Vidal, A.R. Pattern classification with missing data: A review. *Neural Comput. Appl.* **2010**, *19*, 263–282.
18. Kaufman, L.; Rousseeuw, P.J. *Finding Groups in Data: An Introduction to Cluster Analysis*; John Wiley & Sons: Hoboken, NJ, USA, 2009; Volume 344.

© 2019 by the authors. Licensee MDPI, Basel, Switzerland. This article is an open access article distributed under the terms and conditions of the Creative Commons Attribution (CC BY) license (http://creativecommons.org/licenses/by/4.0/).

*Proceedings*

# Auditable Blockchain Randomization Tool [†]

**Olivia Saa [‡] and Julio Michael Stern *,[‡]**

IME-USP—Institute of Mathematics and Statistics of the University of São Paulo, Rua do Matão 1010, 05508-090 São Paulo, Brazil; olivia@ime.usp.br or olivia.saa@iota.org

* Correspondence: jstern@ime.usp.br
† Presented at the 39th International Workshop on Bayesian Inference and Maximum Entropy Methods in Science and Engineering, Garching, Germany, 30 June–5 July 2019.
‡ These authors contributed equally to this work.

Published: 2 December 2019

**Abstract:** Randomization is an integral part of well-designed statistical trials, and is also a required procedure in legal systems. Implementation of honest, unbiased, understandable, secure, traceable, auditable and collusion resistant randomization procedures is a mater of great legal, social and political importance. Given the juridical and social importance of randomization, it is important to develop procedures in full compliance with the following desiderata: (a) Statistical soundness and computational efficiency; (b) Procedural, cryptographical and computational security; (c) Complete auditability and traceability; (d) Any attempt by participating parties or coalitions to spuriously influence the procedure should be either unsuccessful or be detected; (e) Open-source programming; (f) Multiple hardware platform and operating system implementation; (g) User friendliness and transparency; (h) Flexibility and adaptability for the needs and requirements of multiple application areas (like, for example, clinical trials, selection of jury or judges in legal proceedings, and draft lotteries). This paper presents a simple and easy to implement randomization protocol that assures, in a formal mathematical setting, full compliance to the aforementioned desiderata for randomization procedures.

**Keywords:** blockchain entropy; statistical randomization; judicial sortition

---

> *Meos tam suspicione quam crimine judico carere oportere.*
> My people should be free from either crime or suspicion.
> Julius Caesar (62BC), in Suetonius (119CE, Sec.I.74.2).

## 1. Introduction: Bad and Good Practices in Randomization

Randomization is a technique used in the design of statistical experiments: in a clinical trial, for example, patients are randomly assigned to distinct groups receiving different treatments with the goal of studding and contrasting their effects. Randomization is nowadays considered a golden standard in statistical practice; its motivation is to prevent systematic biases (like an unfair or tendentious assignment process) that could distort (unintentionally or purposely) the conclusions of the study. For further comments on randomization see [1–3], for Bayesian perspectives see [4,5]. In the legal context, randomization (also known as sortition or allotment) is routinely used for the selection of jurors or judges assigned to a given judicial case; see [6]. For these applications, our initial quotation, from the Roman emperor Julius Caesar, suggests the highest standards of technical quality, and auditability, see [7].

Rerandomization is the practice of rejecting and discarding (for whatever reason) a given randomized outcome, that is subsequently replaced by a new randomization. Repeated rerandomization can be used to completely circumvent the haphazard, unpredictable or aimless nature of randomization, allowing a premeditated selection of a final outcome of choice. There are advanced statistical techniques capable of blending the best characteristics of random and intentional sampling, see for example [8–12]. Nevertheless, rerandomization is often naively used, or abused, with the excuse of (subjectively) "avoiding outcomes that do not look random enough", see for example [13,14]. In the legal context, spurious manipulations of the randomization process are often linked to fraud, corruption and similar maladies, see [6] and references therein.

In order to comply with the best practices for randomization processes, the authors of [6] recommend the use of computer software having a long list of characteristics, for example, being efficient and fully auditable, well-defined and understandable, sound and flexible, secure and transparent. Such requirements are expressed by the following (revised) *desiderata* for randomization procedures:

> *Given the juridical and social importance of the themata under scrutiny, we believe that it is important to develop randomization procedures in full compliance with the following desiderata: (a) Statistical soundness and computational efficiency, see [15–18]; (b) Procedural, cryptographical and computational security, see [19–22]; (c) Complete auditability and traceability, see [23–25]; (d) Any attempt by participating parties or coalitions to spuriously influence the procedure should be either unsuccessful or be detected, see [26–28]; (e) Open-source programming; (f) Multiple hardware platform and operating system implementation; (g) User friendliness and transparency, see [29,30]; (h) Flexibility and adaptability for the needs and requirements of multiple application areas (like, for example, clinical trials, selection of jury or judges in legal proceedings, and draft lotteries), see [6].*

Such requirements conflate several complementary characteristics that may seem, at first glance, incompatible. For example, strong security is often (but wrongly) associated with excessive secrecy, a doctrine known as "security by obscurity", computer routines may be efficient but are often tough as hard to audit, and mathematically well-defined algorithms may be perceived as hard to understand. The bibliographical references given in the formerly stated *desiderata for randomization procedures* already hint at technologies that can be used to achieve a fully compliant randomization procedure, most preeminently, the blockchain. This is the key technology supporting modern public ledgers, cryptocurrencies, and a host of related applications.

A technical challenge for the application under scrutiny is the generation of pseudo-random number sequences that reconcile complementary properties related to computational efficiency, statistical soundness, and cryptographic security. In this respect, the excellent statistical and computational characteristics of linear recurrence pseudo-random number generators (or their modern descendants and relatives), like [16], can be reconciled with the needs concerning unpredictability and cryptographic security by appropriate starts and restarts of the linear recurrence generator. A sequence start for a linear recurrence generator is defined by a *seed* specified by a vector of (typically 1 to 64) integers, while a restart is defined by a *jump-ahead* or *skip-ahead* specified by a single integer (kept small relative to the generator's full period), see [22].

Unpredictable and cryptographically secure *seeds* and *jump-aheads* can be provided by high entropy bit streams extracted from blockchain transactions, an idea that has already been explored in the works of [31–34].

The next section develops a possible implementation of a fully compliant core randomization protocol based on blockchain technology, and also makes a simple prototype available for study and further research. Moreover, in order to make it simple and easy to use, we develop the prototype on top of a readily available crypto-currency platform. We use Bitcoin for this example, but other alternatives like Ethereum or other cryptocurrencies whose miners work under the same incentives model can be used with minor adaptations.

## 2. Results: Core Randomization Protocol in Blockchain

We intend to establish a protocol able to deliver on demand pseudo random numbers, from an auditable and immutable ledger. The procedure will start as follows: the user (the part that wants to receive a random number) shall send a Bitcoin transaction with a register of its purpose embedded in it. (One way to embed a message in a transaction is using the OP_RETURN script, which allows to store up to 40 bytes in a transaction.) The recipient of this transaction may be a proxy representing a competent authority, a pertinent regulatory agency, an agreed custodian, etc. When this transaction is first attached to the blockchain, we concatenate the transaction ID (a 32 bytes, hexadecimal number) and the block header (a 80 bytes, hexadecimal number). In case someone tries to generate more than one transaction for a same purpose, just take the one that was attached first. The resulting 112 bytes hexadecimal number will be the input for some known Verifiable Delay Function (VDF), that should be calibrated accordingly to the purpose of the random number. For instance, a less critical purpose should have a VDF that delays the result in just a few seconds, or even skip completely the VDF step. A critical purpose, with significant interests involved, should have a more complex VDF, with a delay of minutes or even hours. The final result, after the VDF, will be the source for our seeds and jump-aheads.

With the aid of this protocol, one is able to find a different pseudo-random number for each user that demands it. Note that the user does not have any incentive to try to modify its transaction ID, because he does not have any control of the block header. We assume that the user and the miner are not the same person, so a miner will only be interested in trying to control his block header if he is paid to do so. Since the last stage of our protocol involves the calculation of a VDF, it will take a certain amount of time to the miner to decide if the the block he has found will be of interest of the user. Thus, he might even lose his block, if some other miner broadcasts a block of his own before he finishes calculating the VDF.

In the following subsection, the miner's payoff and the necessary delay $T$ for the Verifiable Delay Functions will be explicitly calculated.

### 2.0.1. Preventing Collusion for Spurious Manipulation

Suppose a malicious user tries to bribe a miner that controls a fraction $p$ of the network's computational power. A prize $P = nB$, where $B$ is the Bitcoin block reward, will be paid to the miner if he successfully mines what we call a "desirable block": a block that will deliver a random number in a set $A$, chosen by the malicious user. Let also $\lambda$ be the average rate of incoming blocks and $q$ the probability of a randomly generated number being an element of $A$, i.e., the measure of the set of desirable results for the malicious user. Finally, let $T$ be the expected amount of time needed for the VDF calculations. The moment a miner finds a block that can be accepted by the network, he faces the decision of broadcasting it before checking the VDF, or calculating the VDF before broadcasting. If he decides to check the VDF before broadcasting, he might start another attempt to find a block rightaway.

First, we calculate the expected absolute payoff for the first and second options, called $\mathbb{E}_1$ and $\mathbb{E}_2$, respectively. $\mathbb{E}_1$ will be larger than $B$, since the miner might issue a desirable block by chance:

$$\mathbb{E}_1 = B + qP = B(1 + nq) \tag{1}$$

On the other hand, if the miner chooses to calculate the VDF, he will receive the block reward and the prize $P$, but with a probability given by

$$\begin{aligned}\mathbb{E}_2 =& (B+P)q\mathbb{P}\{\text{no other node finding a block before } t = T\} \\ &+ (B+P)(1-q)\mathbb{P}\{\text{successfully mining a desirable block in another attempt}\} \\ =& B(1+n)q\exp(-(1-p)\lambda T) \\ &+ B(1+n)(1-q)\sum_{i=1}^{\infty}\mathbb{P}\{\text{successfully mining a desirable block after } i \text{ attempts}\}\end{aligned} \quad (2)$$

The probabilities inside the summation, in the last equation, can be calculated as the product of the probability of finding a desirable block after $i$ attempts (that will be a geometric distribution with probability of success $q$) and the probability of finding and checking $i$ blocks before the rest of the network mines one.

Considering

$$P\{\text{attacker finding and analyzing } i \text{ blocks before another node mining one}\}$$

$$= \int_{t=0}^{\infty} p\lambda \exp(-p\lambda t) \frac{(p\lambda t)^{i-1}}{(i-1)!} \exp(-(1-p)\lambda(t+T))dt$$

$$= \frac{(p\lambda)^i \exp(-(1-p)\lambda T)}{(i-1)!} \int_{t=0}^{\infty} \exp(-\lambda t)t^{i-1}$$

$$= \frac{(p\lambda)^n \exp(-(1-p)\lambda T)}{(i-1)!} \lambda^{-i}(i-1)!$$

$$= p^i \exp(-(1-p)\lambda T)$$

it follows that

$$\begin{aligned}\mathbb{E}_2 =& B(1+n)\left[q\exp(-(1-p)\lambda T) + (1-q)\sum_{i=1}^{\infty}q(1-q)^{i-1}p^i\exp(-(1-p)\lambda T)\right] \\ =& B(1+n)\exp(-(1-p)\lambda T)\left(q + \frac{(1-q)pq}{1-p+pq}\right)\end{aligned} \quad (3)$$

Finally, in order to make accepting the bribe not lucrative, we must have $\mathbb{E}_1 > \mathbb{E}_2$, i.e.:

$$\lambda T > \frac{1}{1-p}\log\left(\frac{1+n}{1+nq}\frac{q}{1-p+pq}\right) \quad (4)$$

Since for every $n > 0$ we have $\frac{1+n}{1+nq} < \frac{1}{q}$, if we choose $\lambda T^* = \frac{1}{1-p}\log\left(\frac{1}{q}\frac{q}{1-p+pq}\right)$, we guarantee that the attack will not be lucrative for any bribe $P = nB$. Also, since it can be assumed that $p < 1/2$, a value $\lambda T^* = 2\log\left(\frac{2}{1+q}\right) < 2\log(2)$ will be high enough to prevent an attack for any bribe and any acceptable value of $p$.

## 3. Conclusions and Final Remarks

We formalized a simple and effective protocol to generate on demand pseudo random numbers, in a fully auditable way. We have demonstrated that none of the involved parts has enough financial incentives to try to affect the random number outcome: the part that issues the transaction lacks this power, since it does not have any control on the block header; and the miners do not have enough financial incentives to collude with an attacker, provided a suitable Verifiable Delay Function is applied.

The essentially decentralized, yet completely traceable and auditable nature of the protocol presented in this article, makes the resulting randomization process eminently reliable without recourse of blind trust in any central authority. The authors believe the adoption of such a protocol by the the Brazilian Supreme Court (STF), as recommended in [6], would significantly increase public confidence in the judicial system and be a contributing factor for political and social stability. A simple prototype of the randomization tool described in this article is available in the supplementary materials; it is not intended to be used in a full-fledged application, but only to provide a working example of the key procedures.

**Supplementary Materials:** A simple prototype of the randomization tool described in this article is available online at https://github.com/oliviasaa/random_generator/blob/master/random_generator.py.

**Funding:** This research was funded by FAPESP – the State of São Paulo Research Foundation (grants CEPID-CeMEAI 2013/07375-0 and CEPID-Shell-RCGI 2014/50279-4) and CNPq – the Brazilian National Counsel of Technological and Scientific Development (grant PQ 301206/2011-2, 307648/2018-4 and GD 140490/2016-7).

**Acknowledgments:** The authors are grateful for the support received from IME-USP—the Institute of Mathematics and Statistics of the University of São Paulo and for the advice of Prof. Serguei Popov. The authors also received support from ABJ—the Brazilian Jurimetrics Association; STF—Supremo Tribunal Federal (the Brazilian Supreme Court), which motivated the study and provided the data analysed in Marcondes et al. [6]; the IOTA Foundation; and received helpful comments and advice from Adilson Simonis, Álvaro Machado Dias, Julio Trecenti, Rafael Bassi Stern and Marcelo Guedes Nunes.

**Conflicts of Interest:** The authors declare no conflict of interest. The funders had no role in the design of the study; in the collection, analyses, or interpretation of data; in the writing of the manuscript, or in the decision to publish the results.

## Abbreviations

The following abbreviations are used in this manuscript:

STF    Superior Tribunal Federal—Brazilian Supreme Court
VDF    Verifiable Delay Function

## References

1. Pearl, J. *Causality: Models, Reasoning, and Inference*; Cambridge University Press: Cambridge, UK, 2000.
2. Pearl, J. *Simpson's Paradox: An Anatomy*; Tech. Rep.; University of California: Los Angeles, CA, USA, 1983.
3. Stern, J.M. Decoupling, Sparsity, Randomization, and Objective Bayesian Inference. *Cybern. Hum. Know.* **2008**, *15*, 49–68.
4. Basu, D.; Ghosh, J.K. *Statistical Information and Likelihood, A Collection of Essays by Dr.Debabrata Basu*; Springer: Berlin, Germany, 1988.
5. Gelman, A.; Carlin, J.B.; Stern, H.S.; Rubin, D.B. *Bayesian Data Analysis*; Chapman and Hall/CRC: Boca Raton, FL, USA, 2003.
6. Marcondes, D.; Peixoto, C.; Stern, J.M. Assessing Randomness in Case Assignment: The Case Study of the Brazilian Supreme Court. *Law Probab. Risk* **2019**, doi:10.1093/lpr/mgz006.
7. Tranquillus, S. *The Lives of the Caesars*; Harvard University Press: Cambridge, MA, USA, 1979; Volume 1.
8. Fossaluza, V.; Lauretto, M.S.; Pereira, C.A.B.; Stern, J.M. Combining Optimization and Randomization Approaches for the Design of Clinical Trials. *Springer Proc. Math. Stat.* **2015**, *118*, 173–184.
9. Lauretto, M.S.; Nakano, F.; Pereira, C.A.B.; Stern, J.M. Intentional Sampling by Goal Optimization with Decoupling by Stochastic Perturbation. *Am. Inst. Phys. Conf. Proc.* **2012**, *1490*, 189–201.
10. Lauretto, M.S.; Stern, R.B.; Morgan, K.L.; Clark, M.H.; Stern, J.M. Haphazard Intentional Sampling and Censored Random Sampling to Improve Covariate Balance in Experiments. *Am. Inst. Phys. Conf. Proc.* **2017**, doi:10.1063/1.4985356.
11. Morgan, K.L.; Rubin, D.B. Rerandomization to improve covariate balance in experiments. *Ann. Stat.* **2012**, *40*, 1263–1282.
12. Morgan, K.L.; Rubin, D.B. Rerandomization to balance tiers of covariates. *J. Am. Assoc.* **2015**, *110*, 1412–1421.
13. Bruhn, M.; McKenzie, D. In Pursuit of Balance: Randomization in Practice in Development Field Experiments. *Am. Econ. J. Appl. Econ.* **2009**, *1* 200–232.
14. Ruxton, G.D.; Colegrave, N. *Experimental Design for the Life Sciences*, 2nd ed.; Oxford University Press: Oxford, UK, 2006.

15. Hammersley, J.M.; Handscomb, D.C. *Monte Carlo Methods*; Chapman and Hall: London, UK, 1964.
16. Haramoto, H.; Matsumoto, M.; Nishimura, T.; Panneton, F.; L'Ecuyer, P. Efficient Jump Ahead for F2-Linear Random Number Generators. *INFORMS J. Comput.* **2008**, *20*, 290–298.
17. Knuth, D.E. *The Art of Computer Programming, Volume 2: Seminumerical Algorithms*, 3rd ed.; Addison-Wesley Longman Publ. Co., Inc.: Reading, MA, USA, 1997.
18. Ripley, B.D. *Stochastic Simulation*; Wiley: Hoboken, NJ, USA, 1987.
19. Aumasson, J.-P. *Serious Cryptography: A Practical Introduction to Modern Encryption*; No Starch Press: San Francisco, CA, USA, 2017.
20. Boyar, J. Inferring Sequences Produced by Pseudo-Random Number Generators. *J. ACM* **1989**, *36*, 129–141.
21. Katz, J.; Lindell, Y. *Introduction to Modern Cryptography*; Chapman and Hall: London, UK, 2014.
22. L'Ecuyer, P. Random number generation. In *Handbook of Computational Statistics*; Springer: Berlin/Heidelberg, Germany, 2012; pp. 35–71.
23. Haber, S.; Stornetta, W. How to time-stamp a digital document. *J. Cryptol.* **1991**, *3*, 99–111.
24. Nakamoto, S. Bitcoin: A Peer-to-Peer Electronic Cash System. Unaffiliated Technical Report. 2008. Available online: https://bitcoin.org/bitcoin.pdf (accessed on 30 June 2019).
25. Wattenhofer, R. *Distributed Ledger Technology: The Science of Blockchain*; Inverted Forest: Scotts Valley, CA, USA, 2017.
26. Boneh, D.; Bonneau, J.; Bünz, B.; Fisch, B. Verifiable Delay Functions. Cryptology ePrint Archive, Report 2018/601. 2018. Available online: https://eprint.iacr.org/2018/601 (accessed on 30 June 2019).
27. Goldschlag, D.M.; Stubblebine, S.G. Publicly Verifiable Lotteries: Applications of Delaying Functions. In *International Conference on Financial Cryptography*; Springer: Berlin, Germany, 1998; pp. 214–226.
28. Rabin, M.O. Transaction protection by beacons. *J. Comput. Syst. Sci.* **1983**, *27*, 256–267.
29. Parikh, R.; Pauly, M. What Is Social Software? In *Lecture Notes in Computer Science*; Springer: Berlin/Heidelberg, Germany, 1973; pp. 3–13.
30. Stern, J.M. Verstehen (causal/interpretative understanding), Erklären (law-governed description/prediction), and Empirical Legal Studies. *J. Inst. Theor. Econ.* **2018**, *174*, 105–114.
31. Bonneau, J.; Clark, J.; Goldfeder, S. On Bitcoin as a public randomness source. *IACR Cryptol. ePrint Arch.* **2015**, *2015*, 1015.
32. Kelsey, J.; Schneier, B.; Hall, C.; Wagner, D. Secure applications of low-entropy keys. In *International Workshop on Information Security*; Springer: Berlin, Germany, 1997.
33. Pierrot, C.; Wesolowski, B. Malleability of the Blockchain's Entropy. 2016. Available online: https://eprint.iacr.org/2016/370.pdf (accessed on 30 June 2019).
34. Popov, S. On a decentralized trustless pseudo-random number generation algorithm. *J. Math. Cryptol.* **2017**, *11*, 37–43.

© 2019 by the authors. Licensee MDPI, Basel, Switzerland. This article is an open access article distributed under the terms and conditions of the Creative Commons Attribution (CC BY) license (http://creativecommons.org/licenses/by/4.0/).

*Proceedings*

# Galilean and Hamiltonian Monte Carlo [†]

**John Skilling *** [ID]

Maximum Entropy Data Consultants Ltd., CB4 1XE Kenmare, Ireland
* Correspondence: skilling@eircom.net
† Presented at the 39th International Workshop on Bayesian Inference and Maximum Entropy Methods in Science and Engineering, Garching, Germany, 30 June–5 July 2019.

Published: 5 December 2019

**Abstract:** Galilean Monte Carlo (GMC) allows exploration in a big space along systematic trajectories, thus evading the square-root inefficiency of independent steps. Galilean Monte Carlo has greater generality and power than its historical precursor Hamiltonian Monte Carlo because it discards second-order propagation under forces in favour of elementary force-free motion. Nested sampling (for which GMC was originally designed) has similar dominance over simulated annealing, which loses power by imposing an unnecessary thermal blurring over energy.

**Keywords:** Nested sampling; simulated annealing; Hamiltonian Monte Carlo; Galilean Monte Carlo

**PACS:** 02.50.Ng

---

## 1. Introduction

> *Question*: How does a mathematician find a needle in a haystack?
> *Answer*: Keep halving the haystack and discarding the "wrong" half.

This trick relies on having a test for whether the needle is or is not in the chosen half. With that test in hand, the mathematician expects to find the needle in $\log_2 N$ steps instead of the $O(\frac{1}{2}N)$ trials of direct point-by-point search.

The programmer is faced with a similar problem when trying to locate a small target in a large space. We do not generally have volume-wide global tests available to us, being instead restricted to point-wise evaluations of some quality function $Q(\mathbf{x})$ at selected locations $\mathbf{x}$. A successful algorithm should have two parts.

One part uses quality differences to drive successive trial locations towards better (larger) quality values. This iteration reduces the available possibilities by progressively eliminating bad (low quality) locations. *Nested sampling* [1] accomplishes this without needing to interpret $Q$ as energy or anything else. By relying only on comparisons ($>$ or $=$ or $<$) it's invariant to any monotonic regrade, thereby preserving generality. Its historical precursor was simulated annealing [2], in which $\log Q$ was restrictively interpreted as energy in a thermal equilibrium.

The other part of a successful algorithm concerns how to move location without decreasing the quality attained so far. Here, it will often be more efficient to move systematically for several steps in a chosen direction, rather than diffuse slowly around with randomly directed individual steps. After $n$ steps, the aim is to have moved $\Delta x \propto n$, not just $\sqrt{n}$. *Galilean Monte Carlo* (GMC) accomplishes this with steady ("Galilean") motion controlled by quality value. Its historical precursor was Hamiltonian Monte Carlo (HMC) [3], in which motion was controlled by "Hamiltonian" forces restrictively defined by a quality gradient which sometimes doesn't exist.

In both parts, nested sampling *compression* and GMC *exploration*, generality and power are retained by avoiding presentation in terms of physics. After all, elementary ideas underlie our understanding of physics, not the other way round, and discarding what isn't needed ought to be helpful.

## 2. Compression by Nested Sampling

> *Question:* How does a programmer find a small target in a big space?
> *Answer:* ......

There may be very many ($N$) possible "locations" $\mathbf{x}$ to choose from. For clarity, assume these have equal status *a priori*—there is no loss of generality because unequal status can always be modelled by retreating to an appropriate substratum of equivalent points. For the avoidance of doubt, we are investigating practical computation so $N$ is finite, though with no specific limit.

As the first step, the programmer with no initial guidance available can at best select a random location $\mathbf{x}_1$ for the first evaluation $Q_1 = Q(\mathbf{x}_1)$. The task of locating larger values of $Q$ carries no assumption of geometry or even topology. Locations could be shuffled arbitrarily and the task would remain just the same. Accordingly, we are allowed to shuffle the locations into decreasing $Q$-order without changing the task (Figure 1). If ordering is ambiguous because different locations have equal quality, the ambiguity can be resolved by assigning each location its own (random) key-value to break the degeneracy.

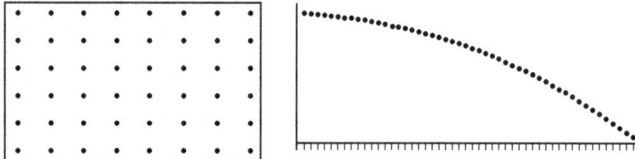

**Figure 1.** $N$ locations (left) ordered (right) by quality $Q$.

Being chosen at random, $\mathbf{x}_1$'s shuffled rank $Nu_1$ marks an equally random fraction of the ordered $N$ locations. Our knowledge of $u_1$ is uniform: $u_1 \sim \text{Uniform}(0,1)$. We can encode this knowledge as one or (better) more samples that simulate what the position might actually have been. If the programmer deems a single simulation too crude and many too bothersome, the mean and standard deviation

$$\log u_1 = -1 \pm 1 \qquad (1)$$

often suffice.

The next step is to discard the "wrong" points with $Q < Q_1$ and select a second location $\mathbf{x}_2$ randomly from the surviving $Nu_1$ possibilities. Being similarly random, $\mathbf{x}_2$'s rank $Nu_1u_2$ marks a random fraction of those $Nu_1$, with $u_2 \sim \text{Uniform}(0,1)$ (Figure 2).

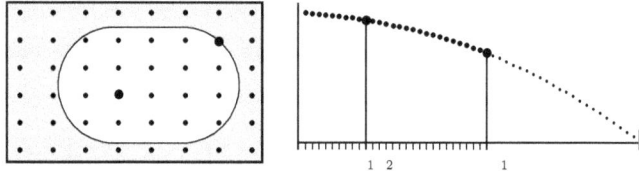

**Figure 2.** Selection of second location $B$ after discarding domain outside $A$.

And so on. After $k$ steps of increasing quality $Q_1 < Q_2 < \cdots < Q_k$, the net compression ratio $X_k = u_1 u_2 \ldots u_k$ can be simulated as several samples from

$$X_k \sim \underbrace{\text{Uniform}(0,1) \cdot \text{Uniform}(0,1) \cdot \ldots \cdot \text{Uniform}(0,1)}_{k} \qquad (2)$$

to get results fully faithful to our knowledge or simply abbreviated as mean and standard deviation

$$\log X_k = \underbrace{(-1 \pm 1) + (-1 \pm 1) + \ldots (-1 \pm 1)}_{k} = -k \pm \sqrt{k} \quad (3)$$

Compression proceeds exponentially until the user decides that $Q$ has been adequately maximised. At that stage, the evaluated sequence $Q_1 < Q_2 < \cdots < Q_k$ of qualities $Q$ has been paired with the corresponding sequence $X_1 > X_2 > \ldots X_k$ of compressions $X$ (Figure 3), either severally simulated or abbreviated in mean.

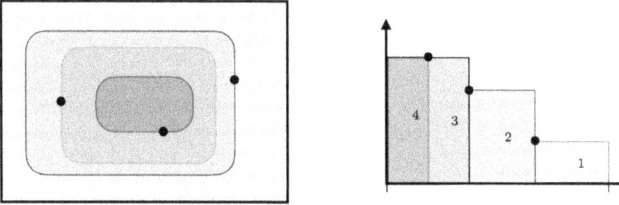

**Figure 3.** Nested sampling produces the relationship $Q(X)$.

$Q$ can then be integrated as

$$Z = \int Q(\mathbf{x})\,d\mathbf{x} = \int_0^1 Q(X)\,dX \approx \sum_{i=1}^k w_i, \quad w_i = Q_i \Delta X_i \quad (4)$$

so that quantification is available for Bayesian or other purposes. Any function of $Q$ can be integrated too from the same run, changing $Q_i$ to some related $Q'_i$ while leaving the $X_i$ fixed. And the statistical uncertainty in any integral $Z$ is trivially acquired from the repeated simulations (2) of what the compressions $X$ might have been according to their known distributions.

That's nested sampling. It requires two user procedures additional to the $Q(\mathbf{x})$ function. The first is to sample an initially random location to start the procedure. The second—which we next address—is to move to a new random location obeying a lower bound on $Q$. Note that there is no modulation within a constrained domain. Locations are either acceptable, $Q(\mathbf{x}) \geq Q^*$, or not, $Q(\mathbf{x}) < Q^*$.

## 3. Exploration by Galilean Monte Carlo

The obvious beginners' MCMC procedure for moving from one acceptable location to another, while obeying detailed balance but not moving so far that $Q$ always disobeys the lower bound $Q^*$, is:

```
Start at x with acceptable quality Q(x) ≥ Q*
Repeat for length of trajectory
    Set v = isotropically random velocity
        x' = x + v = trial location
    if( Q(x) ≥ Q* )   accept new x = x'
    else              reject x' by keeping x
```
(5)

However, randomising $\mathbf{v}$ every step is diffusive and slow, with net distance travelled increasing only as the square root of the number of steps.

All locations within the constrained domain are equally acceptable, so the program might better try to proceed in a straight line, changing velocity only when necessary in an attempt to reflect specularly off the boundary (Figure 4, left). The user is asked to ensure that the imposed geometry makes sense in the context of the particular application, otherwise there will be no advantage.

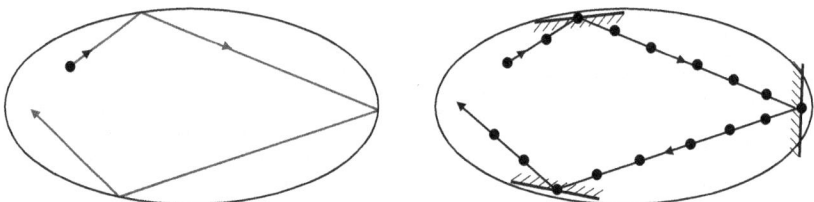

**Figure 4.** The motivation behind Galilean Monte Carlo (GMC).

With finite step lengths, it's not generally possible to hit the boundary exactly whilst simultaneously being sure that it had not already been encountered earlier, so the ideal path is impractical. Instead, we take a current location x and suppose that a corresponding unit vector **n** can be defined there as a proxy for where the surface normal would be if that surface was close at hand (Figure 4, right). Again, it is the user's responsibility to ensure that **n** makes sense in the context of the particular application: exploration procedures cannot anticipate the quirks of individual applications.

Reflection from a plane orthogonal to **n** (drawn horizontally in Figure 5) modifies an incoming velocity **v** (in northward direction from the South) to $\mathbf{v}' = \mathbf{v} - 2\mathbf{n}(\mathbf{n}^T\mathbf{v})$. Depending on the circumstances, the incoming particle may proceed straight on to the North ($+\mathbf{v}$), or be reflected to the East ($+\mathbf{v}'$), or back-reflected to the West ($-\mathbf{v}'$), or reversed back to the South ($-\mathbf{v}$).

*Mistakenly, the author's earlier introduction of GMC in 2011 [4] reduced the possibilities by eliminating West, but at the cost of allowing the particle to escape the constraint temporarily, which damaged the performance and cancelled its potential superiority.*

If the potential destination North is acceptable (bottom left in Figure 5), the particle should move there and not change its direction (so **n** need not be computed). Otherwise, the particle needs to change its direction but not its position.

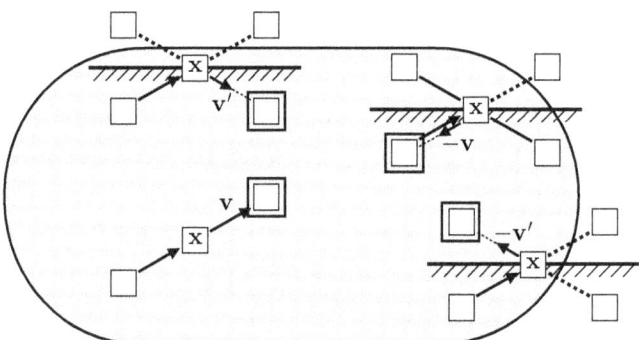

**Figure 5.** North – East – West – South, the four Galilean outcomes.

For a North-South oriented velocity to divert into East-West, either East or West must be acceptable, but not both because East-West particles would then pass straight through without interacting with North-South, so the proposed diversion would break detailed balance. Likewise, for an East-West velocity to divert North-South, either North or South must be acceptable but not both. These conditions yield the following procedure:

```
Start at x with acceptable quality Q(x) ≥ Q*
Set v = isotropically random velocity
Repeat for length of trajectory
```

```
Set N = "Q(x + v) ≥ Q*"
if ( N )                        exit with x = x + v      [go North]
Set v' = Rv = reflection velocity
Set E = "Q(x + v') ≥ Q*",  W = "Q(x − v') ≥ Q*",  S = "Q(x − v) ≥ Q*"
if( S & (E but not W) )         exit with v = v'         [aim East]
if( S & (W but not E) )         exit with v = −v'        [aim West]
                 otherwise      exit with v = −v         [aim South]
```
(6)

Any self-inverse reflection operator **R** will do, though the reflection idea suggests $\mathbf{R} = \mathbf{I} - 2\mathbf{n}\,\mathbf{n}^T$.
That's Galilean Monte Carlo. The trajectory is explored uniformly, with each step yielding an acceptable (though correlated) sample.

## 4. Compression and Exploration

GMC was originally designed for nested-sampling compression, from which probability distributions can be built up after a run by identifying quality as likelihood $L$ in the weighted sequence (4) of successively compressed locations. However, GMC can also be used when exploring a weighted distribution directly.

For *compression* (standard nested sampling, Figure 6, left), only the domain size $X$ is iterated, albeit under likelihood control.

*Compression:*
```
Enter with X and L
   Set constraint L* = L defining X*
   Sample within L* to get X' = uX*
Exit with X' and L'
```
(7)

For *exploration* (standard reversible MCMC, Figure 6, right), the likelihood is relaxed as well through a preliminary random number $u' \sim \mathtt{Uniform}(0,1)$.

*Exploration:*
```
Enter with X and L
   Set constraint L* = u'L defining X*
   Sample within L* to get X' = uX*
Exit with X' and L'
```
(8)

This is equivalent to standard Metropolis balancing "Accept $\mathbf{x}'$ if $L(\mathbf{x}') \geq u'L(\mathbf{x})$", the only difference being that the lower bound $u'L$ is set beforehand instead of checked afterwards.

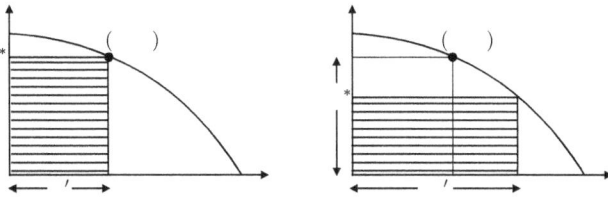

**Figure 6.** GMC for compression (**left**) and exploration (**right**).

## 5. Exploration by Hamiltonian Monte Carlo

Hamiltonian Monte Carlo (HMC) [3] uses a physical analogy with kinetic theory of gases in which a thermal distribution of moving particles, whose position/velocity probability distribution factorises into space and velocity parts

$$\Pr(\mathbf{x},\mathbf{v}) \propto e^{-E(\mathbf{x},\mathbf{v})}, \quad E(\mathbf{x},\mathbf{v}) = V(\mathbf{x}) + T(\mathbf{v}) \qquad (9)$$

with potential energy $V$ defining the spatial target distribution $\Pr(\mathbf{x}) \propto e^{-V(\mathbf{x})}$ and kinetic energy $T = \frac{1}{2}|\mathbf{v}|^2$ distributed as the Boltzmann thermal equilibrium $\Pr(\mathbf{v}) \propto e^{-T(\mathbf{v})}$.
The usual dynamics (Figure 7)

$$\frac{d\mathbf{x}}{dt} = \mathbf{v}, \quad \frac{d\mathbf{v}}{dt} = -\nabla V(\mathbf{x}) \qquad (10)$$

relaxes an initial setting towards the joint equilibrium (9) under occasional collisions which reset $\mathbf{v}$ according to $\Pr(\mathbf{v})$, leaving $\mathbf{x}$ as a sample from the target $\Pr(\mathbf{x})$.

Between collisions, the force field is necessarily digitised into impulses at discrete time intervals $\delta t$, so the computation obeys

$$\delta \mathbf{x} = \mathbf{v}\,\delta t, \quad \delta \mathbf{v} = -\nabla V(\mathbf{x})\,\delta t \qquad (11)$$

To make the trajectory reversible and increase the accuracy order, the impulses are halved at the start $\mathbf{x}$ and end $\mathbf{x}'$, but even this does not ensure full accuracy because the dynamics has been approximated (Figure 8).

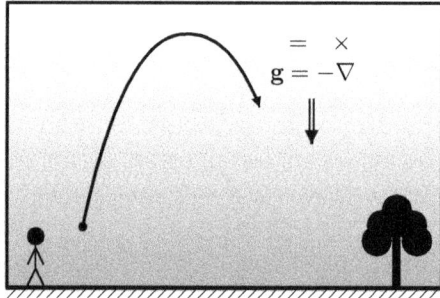

**Figure 7.** The Hamiltonian Monte Carlo idea.

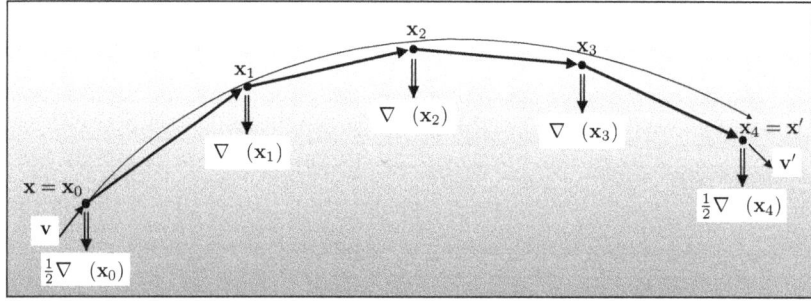

**Figure 8.** Hamiltonian Monte Carlo path approximates the ideal continuous path.

To correct this, the destination $\mathbf{x}'$, whose total energy $E' = V' + T'$ ought to agree with the initial $E = V + T$, is subjected to the usual Metropolis balancing.

$$\text{Accept } \mathbf{x}' \quad \text{iff} \quad e^{-E'} \geq e^{-E} \cdot \texttt{Uniform}(0,1). \qquad (12)$$

In practice, the correction is often ignored because the (reversible) algorithm explores "level sets" whose contours are often an adequately good approximation to the true Hamiltonian provided the fixed timestep is not too large.

That's Hamiltonian Monte Carlo. The trajectory is explored non-uniformly, with successive steps being closer at altitude where the particles are moving slower, so that sampling is closer where the density is smaller—a mismatch which needs to be overcome by the equilibrating collisions. And, of course, HMC requires the potential $V(\mathbf{x})$ (a.k.a. log-likelihood) to be differentiable and generally smooth.

## 6. Compression versus Simulated Annealing

Simulated annealing uses a physical analogy to thermal equilibrium to compress from a prior probability distribution to a posterior. As in HMC, though without the complication of kinetic energy, the likelihood (or quality) is identified as the exponential $L = e^{-E}$ of an energy. In annealing, the energy is scaled by a parameter $\beta$ so that the quality becomes $Q = L^\beta = e^{-\beta E}$ of this scaled energy, with $\beta$ used to connect posterior (where $\beta = 1$) with prior (where $\beta = 0$).

This "simulates" thermal equilibrium at coolness (inverse temperature) $\beta$, and "annealing" refers to sufficiently gradual cooling from prior to posterior that equilibrium is locally preserved. A few lines of algebra, familiar in statistical mechanics, show that the evidence (or partition function) can be accumulated from successive thermal equilibria as

$$\log Z = \int_0^1 \langle \log L \rangle_\beta \, d\beta \tag{13}$$

where $\langle \log L \rangle_\beta$ is the average log-likelihood as determined by sampling the equilibrium appropriate to coolness $\beta$. Equilibrium is defined by weights $L^\beta$ and can be explored either by GMC or (traditionally) by HMC. There is seldom any attempt to evaluate the statistical uncertainty in $\log Z$, the necessary fluctuations being poorly defined in the simulations.

At coolness $\beta$, the equilibrium distribution of locations $\mathbf{x}$, initially uniform over the prior, is modulated by $L^\beta$ so that the samples have probability distribution $\Pr(\mathbf{x}) \propto L(\mathbf{x})^\beta$ which corresponds to

$$\Pr(X) \propto L(X)^\beta \tag{14}$$

in terms of compression. Consequently, samples cluster around the maximum of $\beta \log L + \log X$, where the $\log L(\log X)$ curve has slope $-1/\beta$ (Figure 9, left). Clearly this only works properly if $\log L(\log X)$ is concave ($\frown$). Any zone of convexity ($\smile$) is unstable, with samples heading toward either larger $L$ at little cost to $X$ or toward larger $X$ at little cost to $L$. A simulated-annealing program cannot enter a convex region, and the steady cooling assumed in (13) cannot occur.

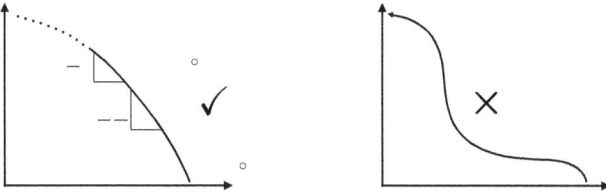

**Figure 9.** Simulated annealing without (**left**) and with (**right**) phase change.

In the physics analogy, this behaviour is a *phase change* and can be exemplified by the transition from steam to water at 100 °C (Figure 9, right). Because of the different volumes (*exponentially* different in large problems), a simulated-annealing program will be unable to place the two phases in algorithmic contact, so will be unable to model the latent heat of the transition. Correspondingly, computation of evidence Z will fail. *Yet our world is full of interesting phase changes, and a method that cannot cope with them cannot be recommended for general use.*

Nested sampling, on the other hand, compresses steadily with respect to the abscissa $\log X$ regardless of the (monotonic) behaviour of $\log L$, so is impervious to this sort of phase change. Contrariwise, simulated annealing cools through the slope, which need not change monotonically. By using thermal equilibria which average over $e^{-\beta E}$, simulated annealing views the system through the lens of a Laplace transform, which is notorious for its ill-conditioning. Far better to deal with the direct situation.

## 7. Conclusions

The author suggests that, just as nested sampling dominates simulated annealing, ...

| Nested sampling | Simulated Annealing |
|---|---|
| Steady compression | Arbitrary cooling schedule for $\beta$ |
| Invariant to relabelling $Q$ | $Q$ has fixed form $L^\beta$ |
| Can deal with phase changes | Cannot deal with phase changes |
| Evidence $Z = \int L\, dX$ with uncertainty | Evidence $Z = \exp \int_0^1 \langle \log L \rangle_\beta\, d\beta$ |

... so does Galilean Monte Carlo dominate Hamiltonian.

| Galilean Monte Carlo | Hamiltonian Monte Carlo |
|---|---|
| No rejection | Trajectories can be rejected |
| Any metric is OK | Riemannian metric required |
| Invariant to relabelling $Q$ | Trajectory explores nonuniformly |
| Quality function $Q(\mathbf{x})$ is arbitrary | Quality $Q(\mathbf{x})$ must be differentiable |
| Step functions OK (nested sampling) | Can not use step functions |
| Can sample any probability distribution | Probability distribution must be smooth |
| Needs 2 work vectors | Needs 3 work vectors |

In each case, reverting to elementary principles by discarding physical analogies enhances generality and power.

**Funding:** This research received no external funding.

**Conflicts of Interest:** The author declares no conflict of interest.

## References

1. Skilling, J. Nested Sampling for general Bayesian computation. *J. Bayesian Anal.* **2006**, *1*, 833–860.
2. Kirkpatrick, S.; Gelatt, C.D.; Vecchi, M.P. Optimization by Simulated Annealing. *Science* **1983**, *220*, 671–680.
3. Duane, S.; Kennedy, A.D.; Pendleton, B.J.; Roweth, D. Hybrid Monte Carlo. *Phys. Lett. B* **1987**, *195*, 216–222.
4. Skilling, J. Bayesian computation in big spaces—Nested sampling and Galilean Monte Carlo. *AIP Conf. Proc.* **2012**, *1443*, 145–156.

© 2019 by the authors. Licensee MDPI, Basel, Switzerland. This article is an open access article distributed under the terms and conditions of the Creative Commons Attribution (CC BY) license (http://creativecommons.org/licenses/by/4.0/).

*Proceedings*

# Information Geometry Conflicts With Independence [†]

## John Skilling

Maximum Entropy Data Consultants Ltd., Kenmare, Ireland; skilling@eircom.net

† Presented at the 39th International Workshop on Bayesian Inference and Maximum Entropy Methods in Science and Engineering, Garching, Germany, 30 June–5 July 2019.

Published: 5 December 2019

**Abstract:** Information Geometry conflicts with the independence that is required for science and for rational inference generally.

**Keywords:** Information Geometry; metric space; probability distribution

**PACS:** 02.50Cw; 02.70Rr

---

## 1. Introduction

Information Geometry [1] assigns a geometrical relationship between probability distributions, using the local curvature (Hessian) of the Kullback-Leibler formula

$$H(\mathbf{p}; \mathbf{q}) = \sum_i p(i) \log \frac{p(i)}{q(i)} \tag{1}$$

as the covariant geometrical metric tensor [2,3] between $\mathbf{q}$ and $\mathbf{p}$. On a $n$-dimensional manifold $\mathbf{p}(\theta)$ specified by parameters $\theta^1, \ldots, \theta^n$, this $n \times n$ Riemannian metric $g$ is

$$g_{jk}(\theta) = \sum_i p(i \mid \theta) \frac{\partial \log p(i \mid \theta)}{\partial \theta^j} \frac{\partial \log p(i \mid \theta)}{\partial \theta^k} \quad \left(\text{or } \int dt\, p(t \mid \theta) \ldots \text{ in continuum form}\right) \tag{2}$$

Geodesic lengths $\ell$ and invariant volumes $V$ follow from $(d\ell)^2 = \sum g_{jk} d\theta^j d\theta^k$ and $dV = \sqrt{\det g}\, d^n\theta$.

Necessarily, lengths are symmetric $\ell(\mathbf{p}, \mathbf{q}) = \ell(\mathbf{q}, \mathbf{p})$ between source and destination, so cannot be isomorphic to $H$ which is from-to asymmetric. Yet (1) is the only connection which preserves independence of separate distributions, $H(\mathbf{x} \times \mathbf{p}; \mathbf{y} \times \mathbf{q}) = H(\mathbf{x}; \mathbf{y}) + H(\mathbf{p}; \mathbf{q})$. Specifically, when $H$ is used to assign an optimal $\mathbf{p}$ (meaning minimally distorted from $\mathbf{q}$) under constraints, that "maximum entropy" selection also depends on separate optimisation $\mathbf{x}$-from-$\mathbf{y}$ unless $H$ has the form (1) [4,5].

It follows that any imposed geometrical connection must introduce interference between supposedly separate distributions. That behaviour is incompatible with the practice of scientific inference, and is confirmed by a counter-example.

## 2. Counter-Example

Consider the 2-parameter family of probability distributions [6]

$$\mathbf{p}_{vw}(t \text{ (mod 1)}) = \begin{cases} f\left(\dfrac{t-v}{w}\right) & \text{for } v < t < v+w, \\ f\left(\dfrac{1+v-t}{1-w}\right) & \text{for } v+w < t < v+1, \end{cases} \tag{3}$$

Parameters $v$ (location) and $w$ (width) lie between 0 and 1. The function $f$ (Figure 1) is monotonically increasing so that it rises from $f(0)$ at $t = v$ to $f(1)$ at $t = v+w$ (mod 1) before falling back to $f(0)$ at

$t = v+1 \pmod 1$. It is positive and normalised to $\int_0^1 f(u)du = 1$ so that the $p_{vw}(\cdot)$'s can be probability distributions on the interval (0,1) — which could model growth and decay in a periodic system.

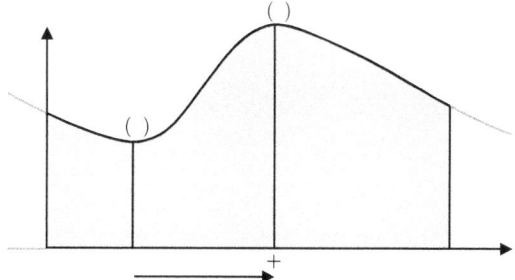

**Figure 1.** Does $v$ affect $w$?

## 2.1. Two Parameters $v$ and $w$

The $2 \times 2$ information-geometry metric evaluates to

$$\begin{bmatrix} g_{vv} & g_{vw} \\ g_{wv} & g_{ww} \end{bmatrix} = \frac{1}{w(1-w)} \int_0^1 \begin{bmatrix} 1 & u \\ u & u^2 \end{bmatrix} \frac{f'(u)^2}{f(u)} du = \begin{bmatrix} A & B \\ B & C \end{bmatrix} / w(1-w) \quad (4)$$

where $A, B, C$ are constants. The table shows their values for two example functions. The first is easy to integrate while the second has vanishing slope $f'(0) = f'(1) = 0$ at the joins (as in Figure 1).

| $f(u)$ | $e^u/(e-1)$ | $(8+6u^2-4u^3)/9$ | |
|---|---|---|---|
| A | $e-1 = 1.71828$ | $\frac{11}{6}\log 2 + \frac{5}{6}\log 5 - \sqrt{15}(\arctan\frac{5}{\sqrt{15}} - \arctan\frac{1}{\sqrt{15}})$ | $= 0.05945$ |
| B | $1 = 1.00000$ | $\frac{89}{12}\log 2 - \frac{25}{12}\log 5 - \frac{\sqrt{15}}{6}(\arctan\frac{5}{\sqrt{15}} - \arctan\frac{1}{\sqrt{15}}) - \frac{4}{3}$ | $= 0.02909$ |
| C | $e-2 = 0.71828$ | $\frac{251}{24}\log 2 + \frac{5}{24}\log 5 + \frac{13\sqrt{15}}{12}(\arctan\frac{5}{\sqrt{15}} - \arctan\frac{1}{\sqrt{15}}) - \frac{31}{3}$ | $= 0.01636$ |

The invariant volume element follows as

$$dV = \sqrt{\det g}\, dv\, dw = \frac{\sqrt{AC - B^2}}{w(1-w)}\, dv\, dw \quad (5)$$

where, by construction, $AC - B^2 > 0$. The total invariant volume is infinite.

$$V = \int_0^1 dv \int_0^1 dw\, \sqrt{\det g} = \infty \quad (6)$$

## 2.2. One Parameter $w$

If $v$ had been fixed, $\mathbf{p}$ would have been confined to a submanifold $\mathbf{p}_w(\cdot)$ parameterised by $w$ alone. The information-geometry metric reduces to

$$g_{ww} = \frac{1}{w(1-w)} \int_0^1 u^2 \frac{f'(u)^2}{f(u)} du = \frac{C}{w(1-w)} \quad (7)$$

The invariant volume element follows as

$$dV = \sqrt{g_{ww}}\, dw = \left(\frac{C}{w(1-w)}\right)^{1/2} dw \quad (8)$$

where, by construction, $C > 0$. The total invariant volume is finite.

$$V = \int_0^1 \sqrt{g_{ww}}\, dw = \pi C^{1/2} \qquad (9)$$

*2.3. Comparison of One and Two Parameters*

Both shape ((5) versus (8)) and integral ((6) versus (9)) over $w$ differ qualitatively according to whether or not $v$ is held fixed.

| Treatment of $v$ influences invariant volumes over $w$ | [Geometry] |

That is a **mathematical fact** of information geometry.

*2.4. Science*

For scientific application, (3) defines a wraparound translation-invariant model in which $v$ does not affect $w$.

| Treatment of $v$ should not influence inference about $w$ | [Science] |

That is a **science requirement**. Any observational consequence of information-geometry's invariant volumes would be rejected by the informed scientist. If there were such consequence, then observation of width $w$ could be used to infer something about location $v$, contrary to the intention of the formulation.

## 3. Conclusions

**Information geometry is not science**. It denies the independence of separate parameters even though such independence is a fundamental requirement of scientific inquiry. The assumption of a geometrical connection between distributions is unnecessary for science and it fails under test.

Information geometry is a self-consistent mathematical structure which (like any other piece of mathematics) may find specialised application within science, but is not fundamental to it. The only fundamental connection is the Kullback-Leibler, which is from-to asymmetric hence not geometric.

**Funding:** This research received no external funding.

**Acknowledgments:** This investigation has been refined by many conversations, in particular with Ariel Caticha.

**Conflicts of Interest:** The author declares no conflict of interest.

## References

1. Amari, S. Differential-geometrical methods in statistics. In *Lecture Notes in Statistics*; Springer-Verlag: Berlin, Germany, 1985.
2. Fisher, R. A. Theory of statistical estimation *Proc. Camb. Philos. Soc.* **1925**, *122*, 700–725.
3. Rao, C.R. Information and the accuracy attainable in the estimation of statistical parameters. *Bull. Calcutta Math. Soc.* **1945**, *37*, 81–89.
4. Shannon, C.F. A Mathematical theory of Communication. *Bell Syst. Tech. J.* **1948**, *27*, 379–423, 623–656.
5. Knuth, K.H.; Skilling, J. Foundations of Inference. *Axioms* **2012**, *1*, 38–73.
6. Skilling, J. Critique of Information Geometry. *AIP Conf. Proc.* **2013**, *1636*, 24–29.

© 2019 by the authors. Licensee MDPI, Basel, Switzerland. This article is an open access article distributed under the terms and conditions of the Creative Commons Attribution (CC BY) license (http://creativecommons.org/licenses/by/4.0/).

*Proceedings*

# Using Entropy to Forecast Bitcoin's Daily Conditional Value at Risk [†]

Hellinton H. Takada [1,*], Sylvio X. Azevedo [2], Julio M. Stern [2] and Celma O. Ribeiro [1]

1. Polytechnic School, University of São Paulo, São Paulo 05508-010, Brazil; celma@usp.br
2. Institute of Mathematics and Statistics, University of São Paulo, São Paulo 05508-090, Brazil; sylvioazevedo@gmail.com (S.X.A.); jstern@ime.usp.br (J.M.S.)
* Correspondence: hellinton@gmail.com
† Presented at the 39th International Workshop on Bayesian Inference and Maximum Entropy Methods in Science and Engineering, Garching, Germany, 30 June–5 July 2019.

Published: 21 November 2019

**Abstract:** Conditional value at risk (CVaR), or expected shortfall, is a risk measure for investments according to Rockafellar and Uryasev. Yamai and Yoshiba define CVaR as the conditional expectation of loss given that the loss is beyond the value at risk (VaR) level. The VaR is a risk measure that represents how much an investment might lose during usual market conditions with a given probability in a time interval. In particular, Rockafellar and Uryasev show that CVaR is superior to VaR in applications related to investment portfolio optimization. On the other hand, the Shannon entropy has been used as an uncertainty measure in investments and, in particular, to forecast the Bitcoin's daily VaR. In this paper, we estimate the entropy of intraday distribution of Bitcoin's logreturns through the symbolic time series analysis (STSA) and we forecast Bitcoin's daily CVaR using the estimated entropy. We find that the entropy is positively correlated to the likelihood of extreme values of Bitcoin's daily logreturns using a logistic regression model based on CVaR and the use of entropy to forecast the Bitcoin's daily CVaR of the next day performs better than the naive use of the historical CVaR.

**Keywords:** entropy; conditional value at risk; cryptocurrency

## 1. Introduction

In finance, risk management is the activity of identifying, analyzing, estimating and controlling the risk of losing money. For our purposes, risk management is a procedure for shaping a loss distribution of an investment. The value at risk (VaR) is the most popular risk measure and it represents how much an investment might lose during usual market conditions with a given probability in a time interval. In other words, VaR is a percentile of a loss distribution. Another very popular risk measure is the conditional value at risk (CVaR), or the expected shortfall. CVaR is a risk measure for investments reintroduced in the literature by Rockafellar and Uryasev [1], for a former reference see Love et al. [2]. According to Sarykalin et al. [3], it approximately (or exactly, under certain conditions) equals the average of some percentage of the worst-case loss scenarios.

Relative to the definitions, there is a near correspondence between VaR and CVaR. For instance, Yamai and Yoshiba [4] defined CVaR as the conditional expectation of loss given that the loss is beyond the VaR level. Consequently, considering the same confidence level, VaR is a lower bound for CVaR. In particular, Rockafellar and Uryasev [1,5] showed that CVaR is superior to VaR in applications related to investment portfolio optimization. In practice, the choice between VaR and CVaR rests on the differences in mathematical properties, stability of statistical estimation, simplicity of optimization procedures, acceptance by regulators, and so on [3]. For instance, in terms of mathematical properties, the CVaR of a portfolio is a continuous and convex function with respect to positions in instruments, whereas the VaR may be even a discontinuous function.

The volatility is the standard deviation of the distribution of logreturns and a very simple and earlier measure of financial risk. The corresponding variance is a natural measure of the statistical uncertainty but it just captures a small portion of the informational content of the distribution of the logreturns. On the other hand, the entropy is a more general measure of uncertainty than the variance because it may be related to higher-order moments of a distribution [6–8]. According to Dionisio et al. [8], the variance measures the concentration around the mean while the entropy measures the dispersion of the density irrespective of the location of the concentration. Finally, for Pele et al. [9], the entropy of a distribution function is strongly related to its tails and this feature is more important for distributions with heavy tails or with an infinite second-order moment for which the variance does not make sense.

In the literature, there are empirical papers showing that entropy has good predictive power for risk. For instance, Billio et al. [10] showed that entropy has the ability to forecast and predict banking crises using directly the entropy of systemic risk measures. In addition, Pele et al. [9] showed that entropy of the intraday distribution of logreturns is a strong predictor of daily VaR, performing better than the classical GARCH models, for a time series of EUR/JPY exchange rates. Similarly, Pele and Mazurencu-Marinescu-Pele [11], instead of using the entropy of the intraday distribution of logreturns, defined the entropy using symbolic time series analysis (STSA) showing that their entropy is a strong predictor of daily VaR, performing better than the classical GARCH models, using high-frequency data for Bitcoin.

There is a recent interest in the statistical properties and risk behavior of cryptocurrencies [12–14] and, in particular, Bitcoin [15]. Consequently, in this paper, we estimate the entropy of the symbolic intraday distribution of Bitcoin's logreturns through the STSA [11] and we model and forecast the Bitcoin's daily CVaR using the estimated entropy. The main contribution of this paper is the extension of the study performed by Pele and Mazurencu-Marinescu-Pele [11] to include the CVaR. The rest of the paper is organized as follows: in Section 2, we present the details of the methodology; in Section 3, we present our empirical study describing the dataset, the results and the corresponding comments; finally, in Section 4, we conclude the paper.

## 2. Methodology

In this section, we review the methodology to estimate the entropy of the symbolic intraday distribution of logreturns through the STSA, a logistic model connecting the daily VaR and the entropy, and a forecasting model for the daily VaR using the entropy based on a quantile regression published by Pele and Mazurencu-Marinescu-Pele [11]. In addition, we introduce the two main contributions of this paper: a logistic model connecting the daily CVaR and the entropy, and a forecasting model for the daily CVaR using the entropy based on a modified quantile regression model. It is also important to mention that the Bitcoin exchange rate is hereinafter referred to as Bitcoin price.

### 2.1. Entropy of Symbolic Intraday Logreturns

In the intraday context, it is usual to consider a set of days $d \in \{1,\ldots,D\}$ and each day equally partitioned in $M$ time bins. Consequently, for a day $d$ and a time bin $m \in \{1,\ldots,M\}$, we associate a price $P_{d,m}$ and a logprice $p_{d,m} = \ln P_{d,m}$. Then, the intraday logreturn of an asset is defined as follows:

$$r_{d,m} = p_{d,m} - p_{d,m-1}; d = 1,\ldots,D; m = 2,\ldots M. \tag{1}$$

For the empirical study of this paper, it is possible to define $r_{d,1} = p_{d,1} - p_{d-1,M}; d = 2,\ldots,D$ because the Bitcoin is continuously traded. However, it is important to point that for other kind of assets, it would be better to ignore the logreturn $r_{d,1}$. In addition, $r_{1,1}$ is not defined.

The intraday logreturns is usually very noisy. The idea behind the STSA technique [16] to produce low-resolution data from high-resolution data. In particular, STSA is a transformation of a real number sequence to a binary sequence. In our case, the STSA transformation is applied to the intraday logreturn to obtain the symbolic intraday logreturn. The symbolic intraday logreturn is defined as follows:

$$s_{d,m} = \begin{cases} 1, & r_{d,m} \leq 0 \\ 0, & r_{d,m} > 0 \end{cases}. \tag{2}$$

Basically, the symbolic intraday logreturn is a binary sequence of 0s representing increasing prices and 1s representing decreasing prices.

Based on the Shannon entropy definition [17], the entropy of the symbolic intraday logreturns is defined as follows:

$$h_d = -\pi_d \log_2 \pi_d - (1 - \pi_d) \log_2 (1 - \pi_d), \tag{3}$$

where $\pi_d = \Pr(s_{d,m} = 1)$ and $1 - \pi_d = \Pr(s_{d,m} = 0)$. It is possible to notice that the entropy of the symbolic intraday logreturns is a daily entropy. In addition, we estimate $\pi_d, d = 2, \ldots, D$ using the sample frequency $\sum_{m=1}^{M} s_{d,m}/M, d = 2, \ldots, D$ and $\pi_1$ using the sample frequency $\sum_{m=2}^{M} s_{1,m}/(M-1)$.

## 2.2. Entropy and Daily VaR and CVaR

Intuitively, the entropy of the symbolic intraday logreturns is higher at the presence of higher uncertainty in the returns and lower at the presence of lower uncertainty in the returns. Consequently, the likelihood of extreme negative daily logreturns is explained by higher values of entropy. In [11], it was verified that the entropy is positively correlated to the likelihood of extreme negative daily logreturns and the relation between VaR and entropy was modeled using the following logistic regression model:

$$\Pr(y_d = 1) = \frac{e^{b_0 + b_1 h_d}}{1 + e^{b_0 + b_1 h_d}}, \tag{4}$$

where $b_0$ and $b_1$ are constants to be estimated;

$$y_d = \begin{cases} 1, & r_d \leq -\text{VaR}_\alpha \\ 0, & r_d > -\text{VaR}_\alpha \end{cases}, d = 2, \ldots, D \tag{5}$$

are the indicators of the lower tails of the daily logreturns; $r_d = \ln P_d - \ln P_{d-1}, d = 2, \ldots, D$ are the daily logreturns; $P_d$ is the closing price of day $d$; and $\text{VaR}_\alpha$ is the daily value at risk at the significance level $\alpha \in ]0,1[$ defined by

$$\Pr(r_d \leq -\text{VaR}_\alpha) = \alpha \tag{6}$$

or, alternatively,

$$\text{VaR}_\alpha = -\inf\{z|F(z) \geq \alpha\}, \tag{7}$$

where $F(\cdot)$ is the cumulative distribution function of the daily logreturns.

In this paper, the hypothesis is also that the entropy is positively correlated to the likelihood of extreme negative daily logreturns and we model the relation between CVaR and entropy using the following logistic regression model:

$$\Pr(u_d = 1) = \frac{e^{c_0 + c_1 h_d}}{1 + e^{c_0 + c_1 h_d}}, \tag{8}$$

where $c_0$ and $c_1$ are constants to be estimated;

$$u_d = \begin{cases} 1, & r_d \leq -\text{CVaR}_\alpha \\ 0, & r_d > -\text{CVaR}_\alpha \end{cases}, d = 2, \ldots, D \tag{9}$$

are the indicators of the lower tails of the daily logreturns; and $CVaR_\alpha$ is the daily conditional value at risk at the significance level $\alpha \in\, ]0,1[$ defined by

$$CVaR_\alpha = -\frac{1}{\alpha} \int_{-\infty}^{-VaR_\alpha} z f(z)\, dz, \tag{10}$$

where $f(\cdot)$ is the continuous probability density function of the daily logreturns.

### 2.3. Forecasting Model for Daily VaR and CVaR

Pele et al. 2017 and Pele et al. 2019 [9,11] considered a quantile regression model to forecast the daily VaR using the entropy as the explanatory variable. The forecasting model for the daily $VaR_\alpha$ at day $k+w+1$ using the entropy of the day $k+w$ is given by:

$$\widehat{VaR}_{\alpha,k+w+1} = -\hat{b}_0^k - \hat{b}_1^k h_{k+w}, \tag{11}$$

where $\hat{b}_0^k$ and $\hat{b}_1^k$ are estimated using a quantile regression model between the dependent variable $r_d$ and the independent variable $h_{d-1}$ for $d \in \mathcal{W}_w(k)$;

$$\mathcal{W}_w(k) = \begin{cases} \{2,\ldots,w\}, & k = 0 \\ \{k+1,\ldots,k+w\}, & k = 1,2,\ldots \end{cases}. \tag{12}$$

Based on Koenker and Bassett [18], we consider the following optimization problem for the quantile regression estimation:

$$\left\{\hat{b}_0^k, \hat{b}_1^k\right\} = \arg\min \sum_{d \in \mathcal{W}_w(k)} \rho_\alpha \left(r_d - b_0^k - b_1^k h_{d-1}\right), \tag{13}$$

where

$$\rho_\alpha(z) = z\left(\alpha - \mathbb{I}_{\mathbb{R}_{<0}}(z)\right) \tag{14}$$

is the asymmetric absolute loss function and

$$\mathbb{I}_\mathcal{A}(z) = \begin{cases} 1, & z \in \mathcal{A} \\ 0, & z \notin \mathcal{A} \end{cases}. \tag{15}$$

is the indicator function.

Our forecasting model for the daily $CVaR_\alpha$ at day $k+w+1$ using the entropy of the day $k+w$ is given by:

$$\widehat{CVaR}_{\alpha,k+w+1} = -\hat{c}_0^k - \hat{c}_1^k h_{k+w}, \tag{16}$$

where $\hat{c}_0^k$ and $\hat{c}_1^k$ are estimated using a quantile regression model between the dependent variable $r_d$ and the independent variable $h_{d-1}$ for $d \in \mathcal{W}_w(k)$. We consider the following optimization problem for the quantile regression estimation:

$$\left\{\hat{c}_0^k, \hat{c}_1^k\right\} = \arg\min \sum_{d \in \mathcal{W}_k} \rho_{\alpha^*} \left(r_d - c_0^k - c_1^k h_{d-1}\right), \tag{17}$$

where

$$\alpha^* = \hat{F}_{w,k}\left(-\frac{1}{\alpha}\int_{-\infty}^{\inf\{x|\hat{F}_{w,k}(x) \geq \alpha\}} z \hat{f}_{w,k}(z)\, dz\right) \tag{18}$$

is the significance level, $\hat{F}_{w,k}(\cdot)$ is the empirical cumulative distribution function of the logreturns estimated using the time window $\mathcal{W}_w(k)$ and $\hat{f}_{w,k}(\cdot)$ is the empirical density function of the logreturns estimated using the time window $\mathcal{W}_w(k)$.

## 3. Empirical Study

### 3.1. Bitcoin

There are several time series prices for Bitcoin depeding on the digital currency exchange and the currency used in the trading process. In order to compare our results to that obtained by Pele and Mazurencu-Marinescu-Pele [11], we adopt the BTC/USD exchange rate from Gemini Trust Company, LLC (Gemini). Gemini is a digital currency exchange and custodian that allows customers to buy, sell, and store digital assets. In particular, we consider the intraday closing prices of the minute-by-minute time bins and the time period from 8 October 2015 until 29 May 2019. According to Feng et al. [19] apud Pele and Mazurencu-Marinescu-Pele [11], the market capitalization, the daily transaction volume and the liquididy of Bitcoin before 2015 was not good.

For illustration purposes, in Figure 1, we present the Bitcoin's daily closing prices; in Figure 2, we present the Bitcoin's daily close-to-close logreturns; and, finally, in Figure 3, we present the empirical probability density and cumulative distribution functions of the Bitcoin's daily close-to-close logreturns. It is possible to notice the huge increase in the Bitcoin's prices until the end of 2017, the high volatility of the Bitcoin's logreturns and the change over time of the volatility pattern. In addition, it is also possible to notice the existence of extreme values in the distribution of Bitcoin's logreturns. In the following sections, we present the entropy of the symbolic intraday distribution of Bitcoin's logreturns, the logistic model connecting the daily CVaR and the entropy, and a forecasting model for the daily CVaR using the entropy based on a modified quantile regression model.

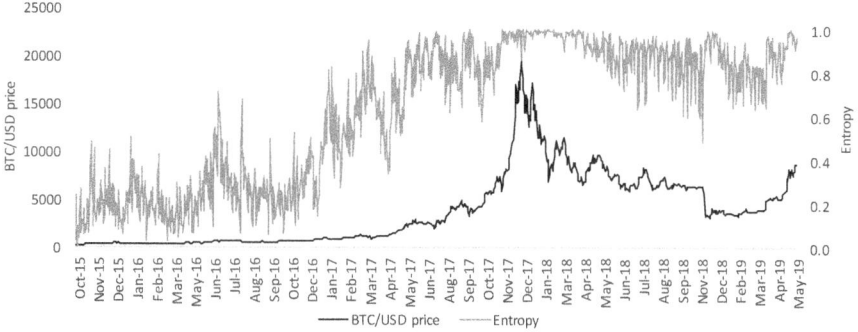

**Figure 1.** Bitcoin's daily closing prices and entropies.

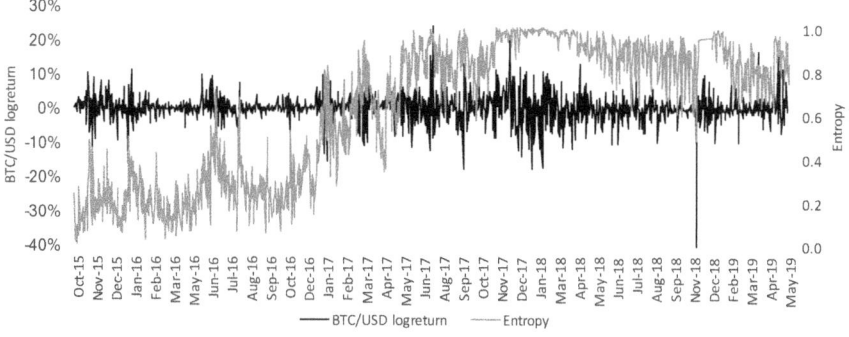

**Figure 2.** Bitcoin's daily close-to-close logreturns and entropies.

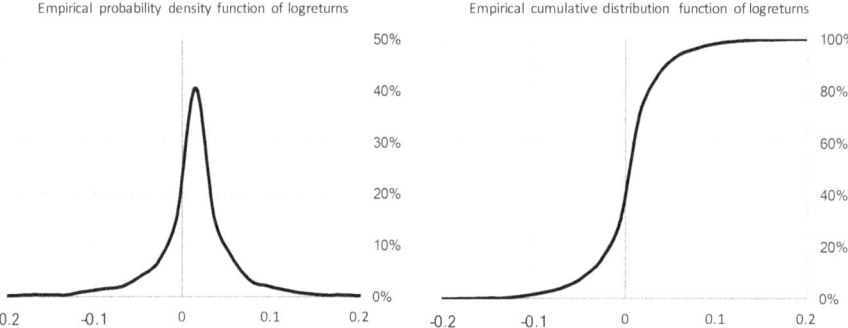

**Figure 3.** The empirical probability density function and the empirical cumulative distribution distribution function of the Bitcoin's daily close-to-close logreturns.

*3.2. Entropy and Daily CVaR*

In Figures 1 and 2, we also present the entropy of the symbolic intraday distribution of Bitcoin's logreturns. As it was mentioned in Section 2, the entropy of the symbolic intraday logreturns is higher at the presence of higher uncertainty in the returns and lower at the presence of lower uncertainty in the returns. In [9,11], they have tested the hypothesis that the daily logprice of Bitcoin is positively correlated to the entropy of the symbolic intraday distribution of Bitcoin's logreturns. However, we state that the logprice time series because of its level, non-stationarity and trend cause possible problems to the hypothesis verification. Consequently, we test the following model:

$$|r_d| = a_0 + a_1 h_d + \varepsilon_d, \quad (19)$$

where $\varepsilon_d$ is the error term. Our hypothesis about Equation (19) is that the absolute value of daily logreturn of Bitcoin is positively correlated to the entropy of the symbolic intraday distribution of Bitcoin's logreturns. The estimation results of Equation (19) are shown in Table 1. It is possible to notice that the estimated coefficient $a_1$ of the entropy is positive and significant supporting our hypothesis.

**Table 1.** Estimation results of Equation (19).

| Parameter | Estimation | p-Value | Standard Error |
|---|---|---|---|
| $a_0$ | 0.006 | 0.000 | 0.005 |
| $a_1$ | 0.032 | 0.000 | 0.008 |
| $R^2$ | 0.132 | | |

In this paper, we propose the study of the relation between entropy of the symbolic intraday distribution of Bitcoin's logreturns and the likelihood of extreme negative daily logreturns represented by the daily CVaR using Equation (8). The estimation results are shown in Tables 2 and 3 for $\alpha = 1\%$ and $\alpha = 5\%$, respectively. It is possible to notice that the estimated coefficients $c_1$ of the entropy for both $\alpha = 1\%$ and $\alpha = 5\%$ are positive and significant supporting the hypothesis that entropy is positively correlated to the likelihood of extreme values of daily logreturns.

**Table 2.** Estimation results of Equation (8) for $\alpha = 1\%$.

| Parameter | Estimation | p-Value | Standard Error |
|---|---|---|---|
| $c_0$ | −9.133 | 0.002 | 1.316 |
| $c_1$ | 8.253 | 0.001 | 3.488 |

Table 3. Estimation results of Equation (8) for $\alpha = 5\%$.

| Parameter | Estimation | p-Value | Standard Error |
|---|---|---|---|
| $c_0$ | −6.961 | 0.001 | 0.592 |
| $c_1$ | 7.800 | 0.001 | 1.605 |

## 3.3. Forecasting Daily CVaR

Let $\hat{\text{CVaR}}_{\alpha,\mathcal{W}_w(k)}$ be the historical daily CVaR at the significance level $\alpha$ calculated in the time window $\mathcal{W}_w(k)$. The forecasting model for the daily CVaR$_\alpha$ at day $k+w+1$ using $\hat{\text{CVaR}}_{\alpha,\mathcal{W}_w(k)}$ is given by:

$$\hat{\text{CVaR}}_{\alpha,k+w+1} = \hat{\text{CVaR}}_{\alpha,\mathcal{W}_w(k)}. \tag{20}$$

In order to study the forecasting performance of the daily CVaR$_\alpha$, we estimate Equations (16) and (20) using a rolling window approach with a window length $w = 250$ trading days. For comparison purposes, in Tables 4 and 5, we present a backtest of models (16) and (20) for significance levels $\alpha = 1\%$ and $5\%$, respectively. The performance of the models is compared with the historical daily CVaR at the significance level $\alpha$ calculated in the time window $\mathcal{W}_w(k+1)$. In particular, we consider the mean absolute error (MAE) and the root mean squared error (RMSE) between $\hat{\text{CVaR}}_{\alpha,k+w+1}$ and $\hat{\text{CVaR}}_{\alpha,\mathcal{W}_w(k+1)}$. As it is possible to notice from our empirical results using Bitcoin, the use of entropy in the forecasting of the daily CVaR of the next day seems to be better than the naive use of the historical CVaR.

Table 4. Backtest results of daily CVaR at the significance level $\alpha = 1\%$.

| Model | MAE | RMSE |
|---|---|---|
| Forecasting using entropy | $5.26 \times 10^{-5}$ | $7.28 \times 10^{-4}$ |
| Forecasting using historical CVaR | $3.56 \times 10^{-4}$ | $4.52 \times 10^{-3}$ |

Table 5. Backtest results of daily CVaR at the significance level $\alpha = 5\%$.

| Model | MAE | RMSE |
|---|---|---|
| Forecasting using entropy | $1.04 \times 10^{-4}$ | $5.42 \times 10^{-4}$ |
| Forecasting using historical CVaR | $3.16 \times 10^{-4}$ | $1.51 \times 10^{-3}$ |

## 4. Conclusions

In this paper, we have two main contributions: a logistic model connecting the daily CVaR and the entropy, and a forecasting model for the daily CVaR using the entropy based on a modified quantile regression model. Basically, we extend the study performed by Pele and Mazurencu-Marinescu-Pele [11] to include the CVaR. In [9,11], they have tested the hypothesis that the daily logprice of Bitcoin is positively correlated to the entropy of the symbolic intraday distribution of Bitcoin's logreturns. However, since the logprice time series is in level and presents a non-stationarity behavior and a trend, the verification of their hypothesis becomes infeasible. Consequently, the hypothesis we verify is that the absolute value of daily logreturn of Bitcoin is positively correlated to the entropy. In addition, we also verify that entropy is positively correlated to the likelihood of extreme values of Bitcoin's daily logreturns using a logistic regression model based on CVaR and the use of entropy to forecast the Bitcoin's daily CVaR of the next day performs better than the naive use of the historical CVaR.

**Author Contributions:** H.H.T. and S.X.A. developed the models, implemented the calculations and wrote the paper; J.M.S. and C.O.R. provided scientific supervision. All authors have read and approved the final manuscript.

**Acknowledgments:** The authors would like to thank the anonymous reviewers for their valuable comments and suggestions to improve the quality of the paper. The authors are grateful for the support of POLI-USP, the Polytechnic School of the University of São Paulo; IME-USP, the Institute of Mathematics and Statistics of the

University of São Paulo; and FAPESP, the State of São Paulo Research Foundation, through the Research Center for Gas Innovation under the grant 2014/50279-4.

**Conflicts of Interest:** The authors declare no conflict of interest.

## Abbreviations

The following abbreviations are used in this manuscript:

| | |
|---|---|
| CVaR | Conditional Value at Risk |
| GARCH | Generalized Autoregressive Conditional Heteroskedasticity |
| MAE | Mean Absolute Error |
| RMSE | Root Mean Squared Error |
| STSA | Symbolic Time Series Analysis |
| VaR | Value at Risk |

## References

1. Rockafellar, R.T.; Uryasev, S.P.O ptimization of conditional value-at-risk. *J. Risk* **2000**, *2*, 21–42.
2. Love, R.F.; Morris, J.G.; Wesolowsky, G.O. *Facilities Location: Models & Methods*; NY: North-Holland, The Netherlands, 1988.
3. Sarykalin, S.; Serraino, G.; Uryasev, S.P. Value-at-risk vs. conditional value-at-risk in risk management and optimization. *Tutorials Oper. Res.* **2008**, 270–294. doi:10.1287/educ.1080.0052.
4. Yamai, Y.; Yoshiba, T. Comparative analyses of expected shortfall and value-at-risk: their validity under market stress. *Monet. Econ. Stud.* **2002**, *20*, 181–237.
5. Rockafellar, R.T.; Uryasev, S.P. Conditional value-at-risk for general loss distributions. *J. Bank. Financ.* **2002**, *26*, 1443–1471.
6. Philippatos, G.C.; Wilson, C. Entropy, market risk and the selection of efficient portfolios. *Appl. Econ.* **1972**, *4*, 209–220.
7. Ebrahimi, N.; Maasoumi, E.; Soofi, E.S. Ordering univariate distributions by entropy and variance. *J. Econ.* **1999**, *90*, 317–336.
8. Dionisio, A.; Menezes, R.; Mendes, D.A. An econophysics approach to analyze uncertainty in financial markets: an application to the Portuguese stock market. *Eur. Phys. J. B* **2006**, *50*, 161–164.
9. Pele, D.T.; Lazar, E.; Dufour, A. Information entropy and measures of market risk. *Entropy* **2017**, *19*, 226.
10. Billio, M.; Casarin, R.; Costola, M.; Pasqualini, A. An entropy-based early warning indicator for systemic risk. *J. Int. Financ. Mark. Inst. Money* **2016**, *45*, 42–59.
11. Pele, D.T.; Mazurencu-Marinescu-Pele, M. Using high-frequency entropy to forecast bitcoins daily value at risk. *Entropy* **2019**, *21*, 102.
12. Zhang, W.; Wang, P.; Li, X.; Shen, D. Some stylized facts of the cryptocurrency market. *Appl. Econ.* **2018**, *50*, 5950–5965.
13. Hu, A.; Parlour, C.A.; Rajan, U. Cryptocurrencies: stylized facts on a new investible instrument. *Work. Pap.* **2018**.
14. Saa, O.T.; Stern, J.M. Auditable Blockchain Randomization Tool. *arXiv* **2019**, arXiv:1904.09500.
15. Colucci, S. On Estimating Bitcoin Value at Risk: A Comparative Analysis. *Work. Pap.* **2018**. doi:10.2139/ssrn.3236813.
16. Daw, C.; Finney, C.; Tracy, E. A review of symbolic analysis of experimental data. *Rev. Sci. Instrum.* **2003**, *74*, 915–930.
17. Shannon, C.E. A mathematical theory of communication. *Bell Syst. Tech. J.* **1948**, *27*, 379–423.
18. Koenker, R.; Bassett, G. Regression quantiles. *Econometrica* **1978**, *46*, 33–50.
19. Feng, W.; Wang, Y.; Zhang, Z. Can cryptocurrencies be a safe haven: A tail risk perspective analysis. *Appl. Econ.* **2018**, *50*, 4745–4762.

© 2019 by the authors. Licensee MDPI, Basel, Switzerland. This article is an open access article distributed under the terms and conditions of the Creative Commons Attribution (CC BY) license (http://creativecommons.org/licenses/by/4.0/).

*Proceedings*

# The Nested_fit Data Analysis Program [†]

**Martino Trassinelli**

Institut des NanoSciences de Paris, CNRS, Sorbonne Université, 4 Place Jussieu, 75005 Paris, France; martino.trassinelli@insp.jussieu.fr

† Presented at the 39th International Workshop on Bayesian Inference and Maximum Entropy Methods in Science and Engineering, Garching, Germany, 30 June–5 July 2019.

Published: 28 November 2019

**Abstract:** We present here `Nested_fit`, a Bayesian data analysis code developed for investigations of atomic spectra and other physical data. It is based on the nested sampling algorithm with the implementation of an upgraded lawn mower robot method for finding new live points. For a given data set and a chosen model, the program provides the Bayesian evidence, for the comparison of different hypotheses/models, and the different parameter probability distributions. A large database of spectral profiles is already available (Gaussian, Lorentz, Voigt, Log-normal, etc.) and additional ones can easily added. It is written in Fortran, for an optimized parallel computation, and it is accompanied by a Python library for the results visualization.

**Keywords:** nested sampling; bayesian evidence; model comparison; atomic spectra

---

## 1. Introduction

`Nested_fit` is a general purpose parallelized data analysis code for the evaluation of *Bayesian evidence* and parameter probability distributions for given data sets and modeling function. The computation of the Bayesian evidence is based on the nested sampling algorithm [1–3], for the integration of the likelihood function over the parameter space. This integration is obtained reducing the $J$-dimensional volume (where $J$ is the number of parameters) in a one-dimensional integral by a clever exploration of the parameter space. In `Nested_fit`, this exploration is obtained with a search algorithm for new parameter values called *lawn mower robot*, which has been initially developed by L. Simons [4] and modified here for a better exploration of multimodal problems.

`Nested_fit` has been developed over the past years to analyze several sets of experimental data from, mainly, atomic physics experiments. For this reason, it has some special feature well adapted to the analysis of atomic spectra as specific line profiles, possibility to study correlated spectra at the same time, eg. background and signal-plus-background spectra, and with a likelihood function built considering a Poisson statistics per each channel, well adapted to low-statistics data.

In the next section we will describe the general structure and feautres of `Nested_fit`. In Section 3 we shortly introduce the basic concepts of Bayesian model comparison and the nested sampling method. The specific algorithm for the parameter space exploration for the nested sampling is presented in details in Section 4. An example of application of `Nested_fit` is presented in Section 5 for the analysis of single two-body electron capture ion decay. A conclusive section will end the article, where recent application of `Nested_fit` to different atomic physics analysis are mentioned.

## 2. General Structure of the Program

The general structure of the program is represented in Figure 1. The main input files are two: the file `nf_input.dat`, where all computation input parameters are included, and the data file, which name is indicated in the parameter input file. The function name in the input file indicates the model to be used for the calculation of the likelihood function. Several functions are already defined in the function library for modelling spectral lines: Gaussian, Log-normal, Lorentzian, Voigt (Gaussian and Lorentzian convolution), Gaussian convoluted with an exponential (for asymmetric peaks), etc. Additional functions can be easily defined by the users in the dedicated routine (USERFNC). Differently from the version presented in Ref. [5] (V. 0.7), in the new version discussed here (V. 2.2) non-analytical or simulated profile models can be implemented. In this case, one or more additional files have to be provided by the users. These external data, which can have some noise like the case of simulated data, are interpolated by B-splines using FITPACK routines [6]. The B-spline parameters are stored and used as profile/model with the total amplitude and a possible offset as free parameters. An additional feature of this new program version, is the possibility to analyze data with error bars. This option has to be indicated in the input file.

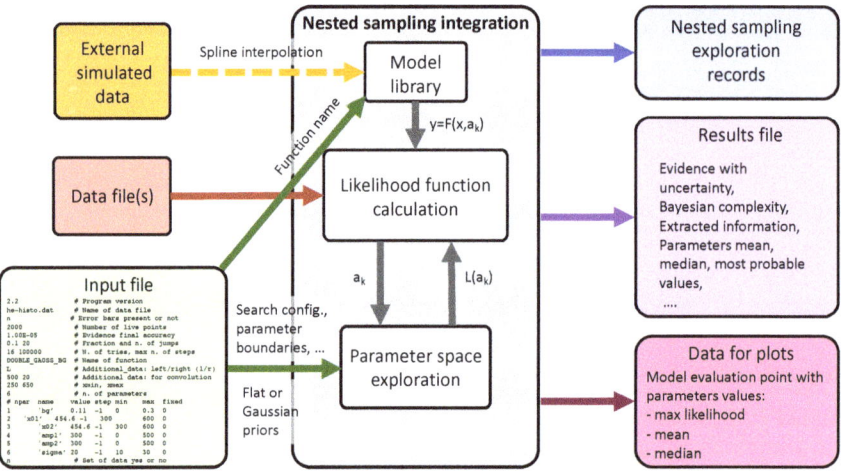

**Figure 1.** Scheme of the Nested_fit program.

Several data sets can be analyzed at the same time by selecting the option "set of data: YES". This is particularly important for the correct study of physically correlated spectra at the same time, e.g., background and signal-plus-background spectra. This is done using a global user-defined function with common parameters of specific models for each spectrum. In the case of multiple data files, the program read an additional input parameter file nf_input_set.dat for the additional datafile names to analyze and data ranges to consider.

The exploration of the parameter space and the corresponding evaluation of the likelihood function is done implementing the nested sampling algorithm [1–3]. If the data are in the format $(channel, counts)$, a Poisson distribution for each channel is assumed for the likelihood function. If the data has error bars $(channel, y, \delta y)$, a Gaussian distribution is assumed (new feature in V. 2.2).

The main analysis results are summarized in the output file nf_output_res.dat . Here the details of the computation (n. of live points, n. of trials, n. of total iteration) can be found as well as the final evidence value and its uncertainty $E \pm \delta E$, the parameter values $\hat{a}$ corresponding to the maximum of the likelihood function, the mean, the median, the standard deviation and the confidence intervals (68%, 95% and 99%) of the posterior probability distribution of each parameter. The information gain $\mathcal{H}$ and the Bayesian complexity $\mathcal{C}$ are also provided in the output.

Data for plots and for further analyses are provided in the files nf_output_data_*.dat. These files contain the original input data together with the model function values corresponding to the parameters with the highest likelihood function value (nf_output_data_max.dat) or the parameter mean value (`nf_output_data_mean.dat`) or median value (`nf_output_data_median.dat`) with the corresponding residuals and error bars. Additional `nf_output_fit_*.dat` files contain a model evaluation with higher density than the original data for graphical presentation purpose.

The step-by-step details of the nested sampling exploration are provided in the file `nf_output_points.dat` that contains the live points used during the parameter space exploration, their associated likelihood values and posterior probabilities. From this file, the different parameter probability distributions and joint probabilities can be built from the marginalization of the unretained parameters. For this purpose, a special dedicated Phython library `Nested_res` has been developed. Additional informations can be found in Ref. [5].

## 3. Implementation of the Nested Sampling for the Evidence Calculation

For a given data set(s) $\{x_i, y_i\}$ and model(s) $\mathcal{M}$, the Bayesian evidence $P(\{x_i, y_i\}|\mathcal{M}, I)$ is extracted for the evaluation of the probability to the different models them-selves:

$$P(\mathcal{M}|\{x_i, y_i\}, I) \propto P(\{x_i, y_i\}|\mathcal{M}, I) \times P(\mathcal{M}|I), \quad (1)$$

where $P(\mathcal{M}|I)$ is the prior probability of each model (assumed constant if not specific preferences for the model is present) and $I$ indicates the background information. The Bayesian evidence is the integral value of the likelihood function over the entire parameter space defined by the priors $P(a|\mathcal{M}, I)$:

$$E(\mathcal{M}) \equiv P(\{x_i, y_i\}|\mathcal{M}, I) = \int P(\{x_i, y_i\}|a, \mathcal{M}, I) P(a|\mathcal{M}, I) d^J a = \int L^{\mathcal{M}}(a) P(a|\mathcal{M}, I) d^J a, \quad (2)$$

where $J$ is the number of the parameters of the considered model, and where we explicitly show the dependency of likelihood function $L^{\mathcal{M}}(a)$ on the model $\mathcal{M}$.

The calculation of the Bayesian evidence is made with the nested sampling, similarly to other available codes [2,7–10]. Nested sampling allows for reducing the above integral in the one-dimensional integral

$$E(\mathcal{M}) = \int_0^1 \mathcal{L}(X) dX, \quad (3)$$

where $X$ is defined by the relation

$$X(\mathcal{L}) = \int_{L(a) > \mathcal{L}} P(a|I) d^J a. \quad (4)$$

Equation (3) can be numerically calculated using the rectangle integration method subdividing the $[0,1]$ interval in $M+1$ segments with an ensemble $\{X_m\}$ of $M$ ordered points $0 < X_M < ... < X_2 < X_1 < X_0 = 1$. We have then

$$E(\mathcal{M}) \approx \sum_m \mathcal{L}_m \Delta X_m, \quad (5)$$

where $\mathcal{L}_m = \mathcal{L}(X_m)$ and $\Delta X_m = X_m - X_{m+1}$. The evaluation of $\mathcal{L}_m$ is obtained by the exploration of the likelihood function with a Monte Carlo sampling via a subsequence of steps. For this, we use a collection of $K$ parameter values $\{a_k\}$ that we call *live points*. More details on the nested sampling algorithm and its implementation can be found in Refs. [1–3,7–10]. The specific implementation of nested sampling in `Nested_fit` is presented in details in Ref. [5].

The bottleneck of the nested sampling algorithm is the search of new points within the $J$-dimensional volume defined by $L > \mathcal{L}_m$. Different methods are commonly employed to accomplish this difficult task. One efficient method is the ellipsoidal nested sampling [7]. It is based on the approximation of the iso-likelihood contour defined by $L = \mathcal{L}_m$ by a $J$-dimensional ellipsoid calculated from the covariance matrix of the live points. The new point is then selected within the ellipsoidal volume (with an enlargement factor selected by the user). This method, well adapted for unimodal posterior distribution has also been extended to multimodal problems [8,9], i.e., with the presence of distinguished regions of the parameter space with high values of the likelihood function. Other search algorithms are based on Markov chain Monte Carlo (MCMC) methods [10] and the recent *Galilean Monte Carlo* [11,12], particularly adapted to explore the regions close to the boundary of $V_{L>\mathcal{L}_m}$ volumes. Nested_fit program is based on an improved version of the *lawn mower robot* method, originally developed by L. Simons [4] and presented in details in the next section.

## 4. The Lawn Mower Robot Search Algorithm

A schematic view of the improved *lawn mower robot* algorithm is represented in Figure 2. To cancel the correlation between the starting point and the final point, a series of $N$ jumps are made in this volume. The different stages of the algorithm are

1. Choose randomly a starting point $a_{n=0} = a_0$ from the available live points $\{a_{m,k}\}$ as starting point of the Markov chain where $n$ is the number of the jump. The number of tries $n_t$ (see below) is set to zero.
2. From the values $a_{n-1}$, find a new parameter sets $a_n$ where each $j$th parameter is calculated by $(a_n)_j = (a_{n-1})_j + f\, r_j \sigma_j$, where $\sigma_j$ is the standard deviation of the live points of the nested sampling computation step relative to the $j$th parameter, $r_j \in [-1,1]$ is a sorted random number and $f$ is a factor defined by the user.

    (a) If $L(a_n) > \mathcal{L}_m$ and $n < N$, go to the beginning of step 2 with an increment of the jump number $n = n + 1$.
    
    (b) If $L(a_n) > \mathcal{L}_m$ and $n = N$, $a_{n=N}$ is new *live point* to be included in the new set $\{a_{m+1,k}\}$.
    
    (c) If $L(a_n) < \mathcal{L}_m$ and $n < N$ and the number of tries $n_t$ is less than the maximum allowed number $N_t$, go back to beginning of step 2 with an increment of the number of tries $n_t = n_t + 1$.
    
    (d) If $L(a_n) < \mathcal{L}_m$ and $n < N$ and $n_t = N_t$ a new parameter set $a_0$ has to be selected. Instead than choosing one of the existing live points, $a_0$ is built from distinct $j$th components from different live points: $(a_0)_j = (a_{m,k})_j$ where $k$ is randomly chosen between 1 and $K$ for each $j$. Then $a_{n=0} = a_0$ and go to the beginning of step 2.

Step 2c, the main improvement of the original lawn mower robot algorithm, makes the algorithm well adapted to problems with multimodal parameter distributions allowing easy jump between high-likelihood regions. The value of $N_t$ is fixed in the code ($N_t$ = 10,000 in the present version). The other parameters can be provided by the input file.

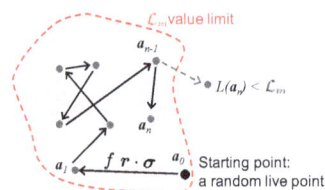

**Figure 2.** Scheme of lawn mower robot algorithm.

## 5. An Application to Low-Statistics Data

To show the capabilities of Nested_fit, we present in this section its implementation on a particular critical case corresponding to a debated experiment. In 2008 it was observed an unexpected modulation in the two-body electron capture decay of single H-like $^{142}_{61}$Pm ions to the stable $^{142}_{60}$Nd bare nucleus, with a monochromatic electron-neutrino emission [13]. The same modulation frequency, but with much smaller amplitude, was found in 2010 data [14] but not in the latest campaign in 2014 [15] where much more events have been recorded.

The unstable ions are produced by collision with a solid target and then injected in a storage ring where they are cooled down. In the storage ring, the decay time of single ions is measured from changes of the Schottky noise frequency induced by the ion revolution. The H-like $^{142}_{61}$Pm ion and $^{142}_{60}$Nd bare nucleus masses correspond in fact to different revolution frequencies. From the accumulated data of single decay events, the decay probability per unit of time can be measured. An example of the data collected in 2010 is presented in Figure 3 (up).

The observed modulation of the expected exponential decay has not yet a clear explanation. A possible connection with neutrino masses differences is speculated in the literature. The determination of the presence or not of a modulation is a perfect case for implementing Bayesian model comparison with Nested_fit.

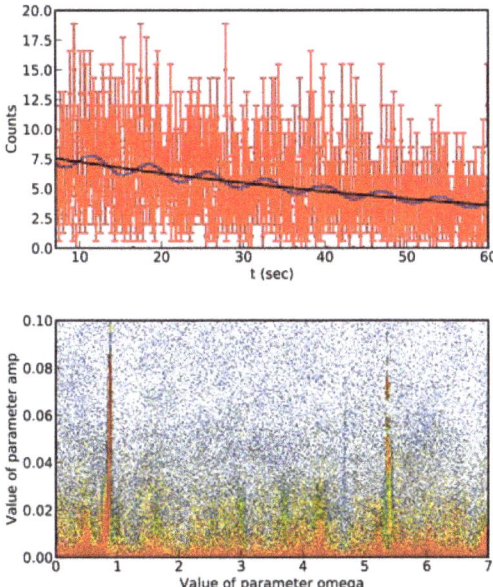

**Figure 3.** Top: Data relative to the single decay of H-like $^{142}_{61}$Pm to $^{142}_{60}$Nd bare nucleus obtained with a binning of 0.08 s. The profile curves relative to pure exponential and exponential with modulation models are also represented. Bottom: 2D histogram of the joint probability of the amplitude $a$ and pulsation $\omega$ of the model with modulation. Red, yellow and green colors represent approximatively the regions corresponding to 68%, 95% and 66% confidence intervals. Both figures are obtained by Python nested_res.py library that accompany Nested_fit program.

When a possible modulation of the exponential decay is assumed, the likelihood function corresponding to 2010 data presents several maxima. This reflect the periodicity nature of the considered function, which can manifest itself via different harmonics, and the low number of available counts per channel. The difficulties to deal with these multiple likelihood maxima pushed in fact the creation of the improved *lawn mower robot* algorithm.

In Figure 3 (top) we present the collected data together with the exponential and modulated exponential functions corresponding to the most probable parameter set. The output result from Nested_fit are presented in Table 1 where model 1 and 2 represent the absence of presence of modulation. For each model, values of the evidence, Bayesian complexity and extracted information are provided, as well as model probabilities. The uncertainty of the probabilities is related to the uncertainty of the evidence. As example of probability distribution, we present in Figure 3 (bottom) the joint probability of the amplitude $a$ and pulsation $\omega$ of the modulation in model 2. The 2D histogram (obtained with Python nested_res.py library that accompany Nested_fit program) is constructed by marginalization on the other parameters. As it can be seen, different maxima are visible, which make difficult the convergence of the nested sampling method. The improved *lawn mower robot* algorithm can deal with this kind of situation, even if the computation time is sometime long (several days in a single CPU).

Table 1. Summary of the results provided from Nested_fit for the two considered models. The parameter values are given in terms of most probable value and 95% confidential interval (CI).

| | Model 1 | Model 2 |
|---|---|---|
| Function | $y = N_0 e^{-t/\tau}$ | $y = N_0 e^{-t/\tau}[1 + a\sin(\omega t + \phi)]$ |
| $\log_e$(Evidence) | $-1594.11 \pm 0.30$ | $-1594.60 \pm 0.36$ |
| Probability | 34.2–41.9% | 58.1–68.8% |
| Complexity | 2.05 | 15.19 |
| Extracted information [nat] | 4.76 | 6.32 |
| $\omega$ (CI 95%) [rad s$^{-1}$] | – | $0.89(0.17 - 6.86)$ |
| $a$ (CI 95%) | – | $9.2 \times 10^{-2}(2.2 \times 10^{-4} - 7.2 \times 10^{-2})$ |
| $\phi$ (CI 95%) [rad] | – | $3.84(0.18 - 6.14)$ |

As it can be observed, the assigned probability to each model are similar and the confidential intervals for the parameter relative to the modulation model are very large. These two aspects reflect the difficulty to treat this problem where the acquired data are not sufficient to provide a marked preference for one model with or without modulation (see Ref. [15] for a more extended discussion). Even if apparently unsatisfying, this result avoid however possible over-interpretation of the data commonly encountered when classical methods are employed, as recently discussed in Ref. [16] in the context of nuclear physics.

## 6. Conclusions

We presented here the program Nested_fit, a general purpose parallelized data analysis code for the evaluation of Bayesian evidence and other statistically relevant outputs. It uses the nested sampling method with the implementation of the improved lawn mower robot algorithm for the evaluation of the Bayesian evidence. Nested_fit has been developed over the past years for the analysis of several sets of atomic experimental data that strongly contribute to the code evolution. We would like to mention in particular the analysis of low-statistics X-ray spectra of He-like uranium [5,17], X-ray spectra of pionic atoms [18,19], electron photoemission spectra from nano-particles [20,21], single-ion decay spectra [15] and response function of crystal X-ray spectrometers (in progress).

Compared to the version reported in Ref. [5], the presented version (V. 2.2) shows additional important features: i) the possibility to interpolate and use computed or simulated external profiles and ii) the implementation of Gaussian likelihood function for data with error bars.

Future developments of Nested_fit will be focussed on the implementation of new exploration methods for the live point evolution of the nested sampling [8,9,11,12]. More precisely, the main goal is the improvement the efficiency for the exploration of the parameter space where the likelihood function presents several local maxima.

**Funding:** This research received no external funding.

**Acknowledgments:** The author thanks the Alexander von Humboldt Foundation that provide the financing to attend the MaxEnt2019 conference. Moreover, the author would like to express once again his deep gratitude to Leopold M. Simons who introduced the author to the Bayesian data analysis and without whom this work could not have been started. The development of this program would not be possible without the close interactions and discussions with many collaborators that the author would like to thank as well: N. Winckler, R. Grisenti, A. Lévy, D. Gotta, Y. Litvinov, J. Machado, N. Paul and all members of the Pionic Hydrogen, FOCAL and GSI Oscillation collaborations and the ASUR group at INSP.

**Conflicts of Interest:** The authors declare no conflict of interest.

## References

1. Skilling, J. Nested Sampling. *AIP Conf. Proc.* **2004**, *735*, 395–405.
2. Sivia, D.S.; Skilling, J. *Data Analysis: A Bayesian Tutorial*, 2nd ed.; Oxford University Press: Oxford, UK, 2006.
3. Skilling, J. Nested sampling for general Bayesian computation. *Bayesian Anal.* **2006**, *1*, 833–859.
4. Theisen, M. Analyse der Linienform von Röntgenübergängen nach der Bayesmethode. Master Thesis, Faculty of Mathematics, Computer Science and Natural Sciences, RWTH Aachen University, Aachen, Germany, 2013.
5. Trassinelli, M. Bayesian data analysis tools for atomic physics. *Nucl. Instrum. Methods B* **2017**, *408*, 301–312.
6. Dierckx, P. *Curve and Surface Fitting with Splines*; Oxford University Press: Oxford, UK, 1995.
7. Mukherjee, P.; Parkinson, D.; Liddle, A.R. A Nested Sampling Algorithm for Cosmological Model Selection. *Astrophys. J. Lett.* **2006**, *638*, L51.
8. Feroz, F.; Hobson, M.P. Multimodal nested sampling: An efficient and robust alternative to Markov Chain Monte Carlo methods for astronomical data analyses. *Mon. Not. R. Astron. Soc.* **2008**, *384*, 449–463.
9. Feroz, F.; Hobson, M.P.; Bridges, M. MultiNest: An efficient and robust Bayesian inference tool for cosmology and particle physics. *Mon. Not. R. Astron. Soc.* **2009**, *398*, 1601–1614. doi:10.1111/j.1365-2966.2009.14548.x.
10. Veitch, J.; Vecchio, A. Bayesian coherent analysis of in-spiral gravitational wave signals with a detector network. *Phys. Rev. D* **2010**, *81*, 062003.
11. Skilling, J. Bayesian computation in big spaces-nested sampling and Galilean Monte Carlo. *AIP Conf. Proc.* **2012**, *1443*, 145–156.
12. Feroz, F.; Skilling, J. Exploring multi-modal distributions with nested sampling. *AIP Conf. Proc.* **2013**, *1553*, 106–113.
13. Litvinov, Y.A.; Bosch, F.; Winckler, N.; Boutin, D.; Essel, H.G.; Faestermann, T.; Geissel, H.; Hess, S.; Kienle, P.; Knöbel, R.; et al. Observation of non-exponential orbital electron capture decays of hydrogen-like $^{140}$Pr and $^{142}$Pm ions. *Phys. Lett. B* **2008**, *664*, 162–168.
14. Kienle, P.; Bosch, F.; Bühler, P.; Faestermann, T.; Litvinov, Y.A.; Winckler, N.; Sanjari, M.; Shubina, D.; Atanasov, D.; Geissel, H.; et al. High-resolution measurement of the time-modulated orbital electron capture and of the decay of hydrogen-like $^{142}$Pm$^{60+}$ ions. *Phys. Lett. B* **2013**, *726*, 638–645.
15. Ozturk, F.C.; Akkus, B.; Atanasov, D.; Beyer, H.; Bosch, F.; Boutin, D.; Brandau, C.; Bühler, P.; Cakirli, R.B.; Chen, R.J.; et al. Recision Test of Purely Exponential Electron Capture Decay of Hydrogen-Like $^{142}$ Pm Ions. *arXiv* **2019**, preprint arXiv:1907.06920.
16. King, G.; Lovell, A.; Neufcourt, L.; Nunes, F. Direct Comparison between Bayesian and Frequentist Uncertainty Quantification for Nuclear Reactions. *Phys. Rev. Lett.* **2019**, *122*, 232502.
17. Trassinelli, M.; Kumar, A.; Beyer, H.F.; Indelicato, P.; Märtin, R.; Reuschl, R.; Stöhlker, T. Doppler-tuned Bragg spectroscopy of excited levels in He-like uranium: A discussion of the uncertainty contributions. *J. Phys. CS* **2009**, *163*, 012026.
18. Trassinelli, M.; Anagnostopoulos, D.F.; Borchert, G.; Dax, A.; Egger, J.P.; Gotta, D.; Hennebach, M.; Indelicato, P.; Liu, Y.W.; Manil, B.; et al. Measurement of the charged pion mass using X-ray spectroscopy of exotic atoms. *Phys. Lett. B* **2016**, *759*, 583–588. doi:10.1016/j.physletb.2016.06.025.
19. Trassinelli, M.; Anagnostopoulos, D.; Borchert, G.; Dax, A.; Egger, J.P.; Gotta, D.; Hennebach, M.; Indelicato, P.; Liu, Y.W.; Manil, B.; et al. Measurement of the charged pion mass using a low-density target of light atoms. *EPJ Web Conf.* **2016**, *130*, 01022.

20. Papagiannouli, I.; Patanen, M.; Blanchet, V.; Bozek, J.D.; de Anda Villa, M.; Huttula, M.; Kokkonen, E.; Lamour, E.; Mevel, E.; Pelimanni, E.; et al. Depth Profiling of the Chemical Composition of Free-Standing Carbon Dots Using X-ray Photoelectron Spectroscopy. *J. Phys. Chem. A* **2018**, *122*, 14889–14897.
21. Villa, M.D.A.; Gaudin, J.; Amans, D.; Boudjada, F.; Bozek, J.; Grisenti, R.E.; Lamour, E.; Laurens, G.; Macé, S.; Nicolas, C.; et al. Assessing the surface oxidation state of free-standing gold nanoparticles produced by laser ablation. *Langmuir* **2019**, submitted.

© 2019 by the author. Licensee MDPI, Basel, Switzerland. This article is an open access article distributed under the terms and conditions of the Creative Commons Attribution (CC BY) license (http://creativecommons.org/licenses/by/4.0/).

*Proceedings*

# A New Approach to the Formant Measuring Problem [†]

**Marnix Van Soom *** and **Bart de Boer ***

Artificial Intelligence Laboratory, Vrije Universiteit Brussel, Pleinlaan 2, 1050 Brussels, Belgium
* Correspondence: marnix@ai.vub.ac.be, bart@ai.vub.ac.be
† Presented at the 39th International Workshop on Bayesian Inference and Maximum Entropy Methods in Science and Engineering, Garching, Germany, 30 June–5 July 2019.

Published: 25 December 2019

**Abstract:** Formants are characteristic frequency components in human speech that are caused by resonances in the vocal tract during speech production. They are of primary concern in acoustic phonetics and speech recognition. Despite this, making accurate measurements of the formants, which we dub "the formant measurement problem" for convenience, is as yet not considered to be fully resolved. One particular shortcoming is the lack of error bars on the formant frequencies' estimates. As a first step towards remedying this, we propose a new approach for the formant measuring problem in the particular case of steady-state vowels—a case which occurs quite abundantly in natural speech. The approach is to look at the formant measuring problem from the viewpoint of Bayesian spectrum analysis. We develop a pitch-synchronous linear model for steady-state vowels and apply it to the open-mid front unrounded vowel [ɛ] observed in a real speech utterance.

**Keywords:** Bayesian inference; general linear model; steady-state; vowel; formant; acoustic phonetics

---

## 1. Introduction

Formants are characteristic frequency components in human speech that are caused by resonances in the vocal tract (VT) during speech production and occur both in vowels and consonants. Fant (1960) [1] systematized the then relatively young science of acoustic phonetics with his acoustic theory of speech production, often called the source-filter model, which has since become the dominant paradigm. At that time, the source-filter model, which is formulated in the language of linear time-invariant system theory, justified the practice of deriving formants from power spectra as prominent local maxima in the power spectral envelope of appropriately windowed and processed speech signals. From this point of view each formant is characterized by three parameters describing the local maximum associated with it: the maximum's center frequency (called the formant frequency), its bandwidth and its peak amplitude.

The concept of a formant is fundamental to phonetics and automated speech processing. For example, formants are considered to be primary features for distinguishing vowel classes, speech perception and for inferring speaker identity, sex and age. Despite this fundamental status—and despite a long history of work on vowel formants starting out with [2]—the issue of making accurate measurements of the formant parameters, which we dub "the formant measurement problem" for convenience, is as yet not considered to be fully resolved (e.g., [3–5]). Accordingly, a large amount of formant measurement methods exist in the literature, of which most rely on linear predictive (LPC) analysis. The fundamental cause underlying the formant measurement problem is that most of these methods yield formant frequency estimates (the main quantity of interest) that are sensitive to various user-made choices, such as the form and length of the tapering window form or the number of poles in LPC analysis (e.g., [6,7]). In other words, measuring formants currently requires rather careful fine-tuning while speech is notorious for its variability [8]. In addition, there currently seems to be no way to put error bars on the formant frequency, bandwidth and amplitude measurements.

In this paper an attempt is made to tackle the formant measuring problem for an important special case, i.e., steady-state vowels (SSVs). The usual definition of the SSV is the steady-state portion of a vowel, i.e., the time interval in which the VT configuration can be taken to be approximately fixed on the time scale of the pitch period, which is on the order of 5 ms. The fixed VT implies that the SSV is characterized by formants with unchanging frequency and bandwidth through time. In contrast, the SSV model in this paper allows the formant amplitudes and pitch periods to change over time, but this change is expected to be small due to the steady state. As such, SSVs can be recognized in natural speech as a semi-periodic string of typically about 3 to 5 pitch periods [9].

By modeling SSVs in the time domain with a pitch-synchronous linear model, it becomes possible to apply the machinery of Bayesian spectrum analysis [10,11] to the formant measuring problem. Our approach shows several promises:

- Ability to derive error bars on the formant frequencies, bandwidths and peak amplitudes.
- Elimination of windowing and averaging procedures. A typical method to measure formants in a SSV is to slide over the signal with a tapering window, estimate the formant frequency, bandwidth and peak amplitude in each window, and then to average these estimates over the windows [9]. In our approach, the pitch-synchronous nature of the model eliminates any windowing procedure (and thus various user-made choices) by making use of the pitch period as a natural time scale [12]. In addition, the formant frequencies and bandwidths are estimated simultaneously in each period, which can be understood as a generalized averaging operation over pitch periods ([11] Section 7.5).
- "Automatic" model order determination. This is done by inferring the most probable model order given the SSV (and the model). This can be contrasted with traditional LPC analysis, where the number of poles must be decided by the user on the basis of several well-established guidelines, but where the final judgment ultimately remains qualitative. However, in the current approach, the proposed model (including the prior pdfs) is still too simple to guarantee satisfactory model order determination in all cases.

Compared to standard LPC analysis, there are three main disadvantages of our approach:

- Limited applicability: we only model SSVs, though possible extensions are discussed in the conclusion of the paper.
- For our approach it is necessary to determine the pitch periods in advance. There are several algorithms available (e.g., waveform matching [13] or ML estimation [14]) for this task, but ideally this should be a part of the SSV model itself.
- Though the inference algorithm described below is efficient and relatively fast compared to typical problems in numerical Bayesian inference, it is still much slower than LPC analysis. For example, all calculations for the SSV [ɛ] discussed below took about half a minute.

## 2. SSV Model

The model of a SSV proposed here is inspired by Ladefoged [15]'s picture of speech; that is, the pulses in the speech waveforms coincide with glottal closing instants (GCIs). The GCIs causing the pulses in the speech waveforms is illustrated by the electroglottograph (EGG)—see Figure 1.

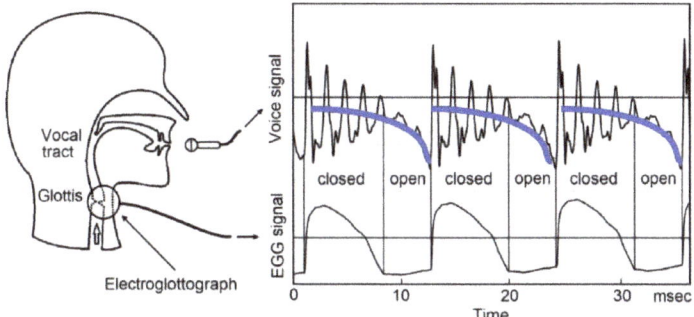

**Figure 1.** (After [16]; reproduced with kind permission.) The EGG signal is the electrical conductance between two electrodes placed on the neck. When the glottis is closed, the measured conductance is high, and vice versa. The EGG signal is displayed in the bottom panel. In the top panel the synchronized speech waveform is plotted and three individual pitch periods are shown. From comparing the top and bottom panel, it is evident that the pulses in the speech waveform occur when the EGG signal rises sharply; i.e., when the glottis closes. Additionally, there are two phases (closed-open) within each pitch period in which the damping of the speech waveform seemingly changes; this is primarily due to the glottal waveform which becomes prominent at the end of the pitch periods as the glottis is reaching its maximum aperture. The blue baseline drawn in the top panel by simple visual inspection is related to the time derivative of the glottal flow (dGF). A second and smaller effect causing the pitch periods to exhibit two phases is the extra coupling to the subglottal cavities (such as the lungs) of the acoustic waves in the VT when the glottis opens; this causes these waves to decay faster.

## 2.1. Individual Pitch Periods

According to Ladefoged, in each pitch period of length $T$ these pulses excite the VT such that the resulting speech waveform consists of a superposition of $Q$ decaying resonances

$$\sum_{j=1}^{Q}\{B_j\cos(\omega_j t)+C_j\sin(\omega_j t)\}\exp(-\alpha_j t). \quad (0\le t<T) \qquad (1)$$

This is a model of the formants in the time domain where $Q$ is the number of formants (i.e., damped sinusoids), $B_j$ and $C_j$ are the amplitudes, and $\omega_j$ and $\alpha_j$ are the frequency [rad Hz] and the decay constant [Hz] of formant $j$, respectively. Thus if we take $Q=2$, we model the first two formants of a pitch period which are denoted as F1 and F2. The frequency of F1 is $\omega_1$ and its decay constant is $\alpha_1$, and similarly for F2.

Ladefoged's model has considerable merit [12,15], but it does not incorporate the effects of the glottal flow derivative (dGF) during the open phase of the pitch period. During this phase, the glottal flow weakly excites the VT which can be roughly approximated with a simple differentiation ([17] p. 3). For this reason, we consider the dGF instead of the glottal flow itself. In Figure 1 the baselines marked in blue roughly indicate the dGF trends, which are clearly not negligible in the open phase of the pitch period as they become the dominant effect in that phase. There are indications that this a general effect [12,17]. As the dGF can be taken to vary relatively slowly, we propose to model it by a polynomial of order $P-1$. Thus the model function for one pitch period of length $T$ is

$$f(t;P,Q,A_k,B_j,C_j,\Omega)=\sum_{k=1}^{P}A_k L_{k-1}(t)+\sum_{j=1}^{Q}\{B_j\cos(\omega_j t)+C_j\sin(\omega_j t)\}\exp(-\alpha_j t), \quad (0\le t<T) \qquad (2)$$

where the $L_{k-1}(t)$ are the numerically convenient Legendre polynomials and the $A_k$ are their amplitudes and $\Omega\equiv(\omega_1\cdots\omega_Q,\alpha_1\cdots\alpha_Q)$.

With the above model for individual pitch periods we can use Bayesian spectrum analysis to obtain the posterior distribution of the $\Omega$, which are the parameters of interest as they describe the $Q$ formants. Suppose the SSV consists of $n$ predetermined pitch periods such that the SSV is defined by the string $\{D_1 D_2 \cdots D_i \cdots D_n\}$, i.e.,

$$D_i = \{d_i[t]\} \quad (t = 0, 1, 2, \cdots, N_i - 1) \tag{3}$$

is the waveform of the $i$th pitch period which consists of $N_i$ samples. Then we assume that

$$d_i[t] = f(t; P, Q, A_k, B_j, C_j, \Omega) + e_i[t] \quad (t = 0, 1, 2, \cdots, N_i - 1), \tag{4}$$

where $e_i[t] \sim N(0, \sigma^2)$, i.e., the pdf for the errors is white noise with constant power. An example of a fit of the model function $f(t)$ in Equation (4) to one pitch period $i$ is shown in Figure 2.

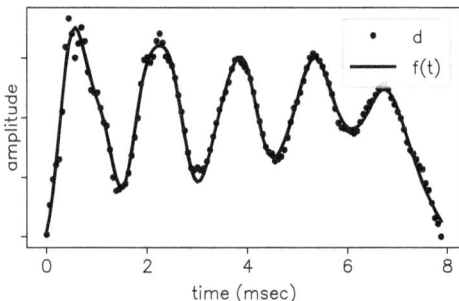

**Figure 2.** One pitch period extracted from a SSV /ɛ/ plotted as samples (the dots) together with the model function fit $f(t; P, Q, \hat{A}_k, \hat{B}_j, \hat{C}_j, \hat{\Omega})$ (smooth line).

Since the model function $f(t; P, Q, \Omega_i)$ is a linear combination of basis functions (which depend on $\Omega$), this model is an instance of the well-known general linear model [18]. It is possible to marginalize over the amplitudes $A_k, B_j, C_j$ together with the noise level $\sigma$: assuming uniform priors, we arrive at the typical Student-t distribution (written in non-standard form) ([11] p. 35)

$$p(\Omega|P, Q, D_i, I) \propto \left[1 - \frac{m\overline{h_i^2}(\Omega)}{N_i \overline{d_i^2}}\right]^{\frac{m-N_i}{2}}. \quad \text{(pitch period)} \tag{5}$$

Here $m = P + 2Q$ is the number of basis functions, $\overline{d_i^2} = (1/N_i) \sum_{t=0}^{N_i-1} d_i[t]^2$ and $\overline{h_i^2}(\Omega)$ is the sufficient statistic for the problem, obtained in the standard way [19] from the projections of the data $d_i[t]$ on the orthogonalized basis functions which are linear combinations of the basisfunctions $\{L_{k-1}(t), \cos(\omega_j t) \exp(-\alpha_j t), \sin(\omega_j t) \exp(-\alpha_j t)\}$ appearing in Equation (2). The sufficient statistic $\overline{h_i^2}(\Omega)$ can be seen as a generalization of the Schuster periodogram [10], to which it reduces in the case that $P = 0$ and $\alpha_j = 0$ ($1 \le j \le Q$) and the frequencies $\omega_j$ are well separated (i.e., $|\omega_j - \omega_k| \gg 2\pi/N_i$ for $1 \le j < k \le Q$).

2.2. Multiple Pitch Periods: SSV

We model an SSV as a string of independent pitch periods that share the same formants (see Figure 3); that is, the formant frequencies and decay constants $\Omega = (\omega_1 \cdots \omega_Q, \alpha_1 \cdots \alpha_Q)$, as well as the noise power $\sigma^2$ are kept fixed across the pitch periods, as well as the order parameters $P$ and $Q$.

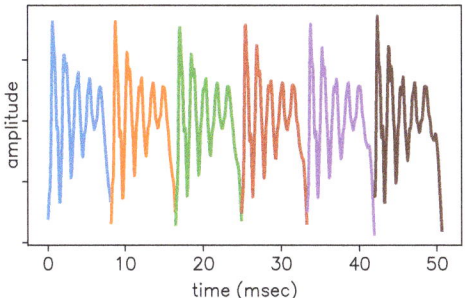

**Figure 3.** A model of the SSV (in this case an /ɛ/, which is also used in the next section) consists of several model functions $f(t; P, Q, A_{i,k}, B_{i,j}, C_{i,j}, \Omega)$ strung together, as shown in this plot by giving each such model function its own color. The model functions are fitted to the individual pitch periods but are "frustrated" by their having to share the formant frequencies and decay constants $\Omega$.

The other parameters are again marginalized over, once again assuming uniform priors, and we arrive at the simplest generalization of Equation (5):

$$p(\Omega|P, Q, D_1 \cdots D_n, I) \propto \prod_{i=1}^{n} p(\Omega|P, Q, D_i, I). \quad \textbf{(SSV)} \qquad (6)$$

The assumption of independent pitch periods—while knowing that these are in fact perfectly correlated, being almost periodical—has an important effect on the error bars of the final estimates $\hat{\Omega}$ which is comparable to using uninformative priors; the effect is to make the error bars conservative, because the uncertainties in the amplitudes etc. transfer to the posterior marginal error bars on the frequencies and decay constants in $\hat{\Omega}$. Any correlation put into the model can only decrease this uncertainty, and thus decrease the magnitude of the error bars [11]. Bretthorst has shown this explicitly for the case of correlated errors $e_i[t]$ which in our story would apply to adjacent pitch periods [20]. Since the frequency estimates are already sharp (see the next section for an example) from the viewpoint of what one is used to in acoustic phonetics [3], this issue (i.e., overly conservative error bars for the frequency estimates) seems perfectly acceptable as it buys us very convenient analytical expressions for the posterior $p(\Omega|P, Q, D_1 \cdots D_n, I)$. However, the error bars on the decay rates and the amplitudes are quite broad, and improving the accuracy of their estimates is desirable in certain applications.

### 2.3. Estimation

We perform a Gaussian approximation at the maximum a posteriori point $\hat{\Omega}$, which is found by optimizing Equation (6) with the Levenberg-Marquardt algorithm [21]. This approximation lets us estimate the posterior covariances (of which the diagonal gives the desired error bars) and lets us crudely estimate the posterior probability of the polynomial order and number of formants $(P, Q)$. "Crudely" because our priors are uninformative, so we can hardly expect model comparison as guided by $p(P, Q|D_1 \cdots D_n, I)$ to give satisfactory results (relative to all the things we know about the data—correlations, acceptable physical forms, etc.—but did not tell probability theory).

## 3. Application on a Steady-State Portion of [ɛ]

We apply the model to a steady-state portion of the vowel [ɛ] (the second 'e' in "etcetera") consisting of $n = 6$ pitch periods shown in Figure 4.

Before applying the model, the order parameters $P$ and $Q$ must be chosen. For simplicity we set $Q = 3$, i.e., we are interested in the first three formants, which is a typical case. Figure 5 shows the posterior $p(P|Q = 3, D_1 \cdots D_6, I)$, where now the former parameters of interest $\Omega$ have been integrated out using the Gaussian approximation. The preferred value of $P$ is clearly $P = 4$. Unfortunately, the choice $(P = 4, Q = 3)$ yields unphysical results as the inferred sinusoids are not damped but actually grow considerably during the pitch periods. The same goes for $P = 3$. These unphysical results can arise because we did not restrict $\alpha_j > 0$ in the parameter space. The choices $P = 5, 6, 7, 8$ do yield physical results, and the estimates for the $Q = 3$ formants depend only slightly on the actual value of $P$ in this range (remember that $P - 1$ is the order of the Legendre polynomial describing the slowly varying baseline).

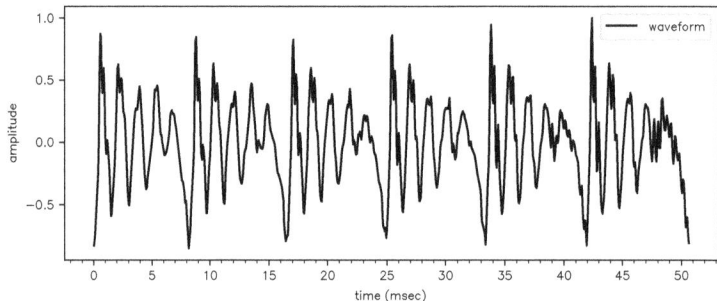

**Figure 4.** Steady-state portion of [ɛ] consisting of six pitch periods. Extracted from the CMU ARCTIC database [22], speaker BDL, sentence a0001.wav, from 2.847 to 2.898 sec at 16 kHz (no downsampling was done).

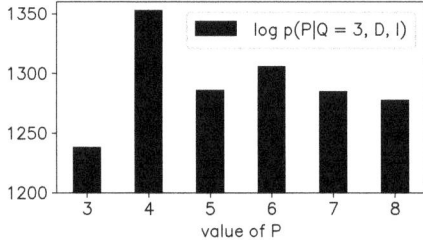

**Figure 5.** The posterior probability $p(P|Q = 3, D_1 \cdots D_6, I)$ for the SSV /ɛ/.

Finally, Figure 6 shows the results for $(P = 6, Q = 3)$. Despite the fact that we can estimate the formant frequencies, bandwidths and peak amplitudes, we only compare the formant frequencies to a standard LPC analysis. The reason is that the comparison of the bandwidths and peak amplitudes between our model and the LPC model is not well defined, as on the one hand the LPC bandwidths depend on the size of the LPC window used; and on the other hand the formant amplitudes are allowed to vary between pitch periods in our model. The actual formant frequency estimates are (at two standard deviations):

- $(F1)_{est} = 658 \pm 2$ Hz at $-2.0 \pm 0.1$ dB/ms
- $(F2)_{est} = 1463 \pm 10$ Hz at $-2.9 \pm 0.5$ dB/ms
- $(F3)_{est} = 2660 \pm 10$ Hz at $-3.0 \pm 0.7$ dB/ms

The formant frequency estimates calculated with LPC on the entire steady-state portion is:

- $(F1)_{LPC} = 670$ Hz
- $(F2)_{LPC} = 1491$ Hz
- $(F3)_{LPC} = 2771$ Hz

We calculated these with Praat [23], a popular tool in acoustic phonetics. The discrepancies might seem large (on the order of 50 Hz) but this is actually quite acceptable in acoustic phonetics. While much more work is needed to understand the discrepancy, the basic reason is quite clear: LPC analysis can be interpreted as an all-pole expansion of the spectrum of a (windowed) segment of speech to estimate the formants in that window. But Jaynes [10] and Bretthorst [11] showed clearly that the spectrum is only an optimal estimator for frequency content (more precisely: spectral lines) if six conditions are met ([11] p. 20), of which two are very clearly violated in the estimation of formant frequencies in SVVs. The two conditions state that the data must not contain a constant component and that there is no evidence of a low frequency. But these conditions do not hold because of the slowly varying dGF components, perhaps to the point that the spectrum is not just a suboptimal estimator, but a potentially misleading one. This misleading nature (i.e., local maxima in spectra do not designate actual harmonic content) has been shown for economical data [24], where data often need to be "detrended", which is exactly what we do here.

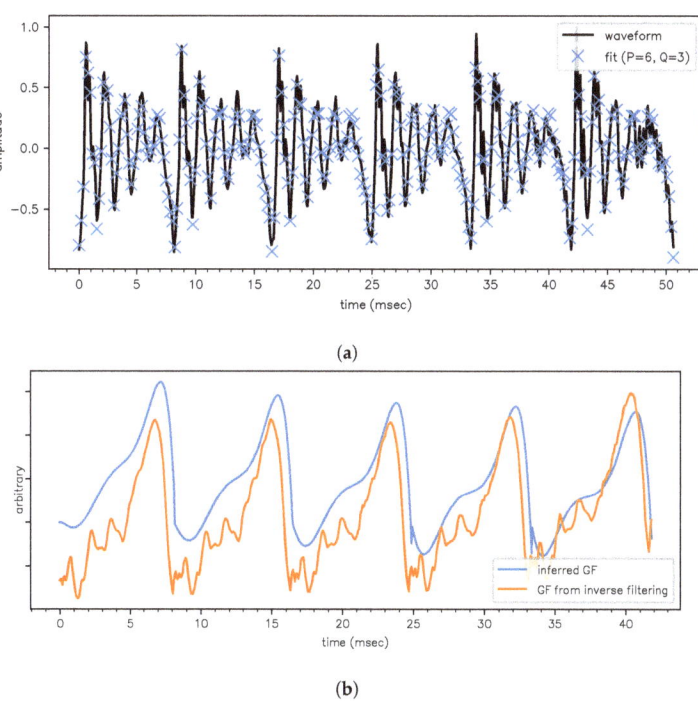

**Figure 6.** *Cont.*

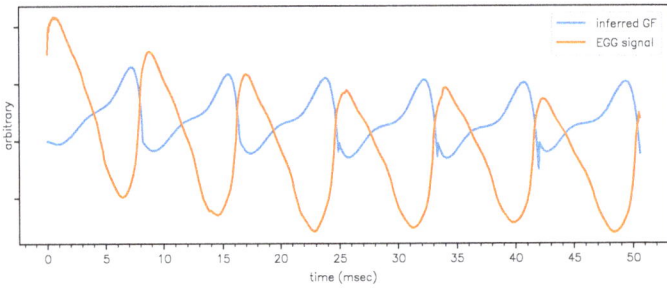

(c)

**Figure 6.** Application of the model ($P = 6, Q = 3$) to a SSV /ɛ/ in order to estimate the frequency and decay constants of F1, F2, F3. (**a**) Fit to the data shown in Figure 4. (**b**) We can estimate the dGF waveform via the amplitudes $\hat{A}_k$, and integrating this estimate yields the glottal flow (GF) waveform. As an independent qualitative check, in this subplot the GF is compared to a GF obtained via inverse filtering [25]. (**c**) The estimated GF waveform (the same as in plot (c)) is now compared to the EGG signal which was simultaneously recorded with the /ɛ/ waveform [22]. A high value of the EGG means that the glottis is closed, so the GF must be small. A low value of the EGG means that the glottis is open, and the GF should be big (as air from the lungs can escape outward). This anticorrelation is perfectly observed.

## 4. Conclusions

Though we have shown that this approach holds some promises, a lot more work is needed before anything definitive can be asserted; in particular work is underway in which the method will be systematically compared to a "ground truth" to test its validity using a speech production simulation with known glottal flow and formants. In addition, the model would benefit greatly from making the estimation of the pitch periods part of the inference, rather than requiring it as a given; using more informative priors is also expected to improve the model order inference (e.g., it is known that there are typically about three resonances of significance, for a human vocal tract, below about 3500 Hz ([26] p. 20)). Finally, for shorter pitch periods (i.e., higher fundamental frequency of speech) the model is expected to deteriorate, as we do not allow the damped sines to "leak" into the next pitch period. Thus a more elaborate model would have to include nearest-neighbor interactions between adjacent pitch periods.

Applications of high-accuracy formant measurements equipped with error bars should be plentiful, but two examples worthwhile are forensic speaker identification (e.g., [27]) and medical diagnosis (e.g., [28]). Remarkably, the field of forensic phonetics seems to have adopted a Bayesian methodology [29,30], so perhaps in the future this model could deliver quantities directly for use in their likelihood ratios [6].

### 4.1. Possible Extensions

This approach can be extended to non-stationary vowels (i.e., non-SSVs) by explicitly modeling the formant tracks (e.g., by parametrization or by a free-form model [31]); one would then get error-bars on the tracks. The pitch period model (Equation (5)) might be extendable to a restricted class of consonants called plosives (such as [k]), insofar the plosive mechanism is qualitatively similar to the glottal closure [32].

**Author Contributions:** Conceptualization and writing: M.V.S. and B.d.B.; methodology and analysis: M.V.S.

**Funding:** This research was supported by the Research Foundation Flanders (FWO) under grant number G015617N.

**Conflicts of Interest:** The authors declare no conflict of interest.

## References

1. Fant, G. *Acoustic Theory of Speech Production*; Mouton: Den Haag, The Netherlands, 1960.
2. Hermann, L. Phonophotographische Untersuchungen. *Pflügers Arch. Eur. J. Physiol.* **1889**, *45*, 582–592.
3. Fulop, S.A. *Speech Spectrum Analysis*; Signals and Communication Technology; Springer: Berlin, Germany, 2011; OCLC: 746243279.
4. Kent, R.D.; Vorperian, H.K. Static Measurements of Vowel Formant Frequencies and Bandwidths: A Review. *J. Commun. Disord.* **2018**, *74*, 74–97, doi:10.1016/j.jcomdis.2018.05.004.
5. Maurer, D. *Acoustics of the Vowel*; Peter Lang: Bern, Switzerland, 2016.
6. Harrison, P. Making Accurate Formant Measurements: An Empirical Investigation of the Influence of the Measurement Tool, Analysis Settings and Speaker on Formant Measurements. Ph.D. Thesis, University of York, York, UK, 2013.
7. Vallabha, G.K.; Tuller, B. Systematic Errors in the Formant Analysis of Steady-State Vowels. *Speech Commun.* **2002**, *38*, 141–160, doi:10.1016/S0167-6393(01)00049-8.
8. Peterson, G.E.; Barney, H.L. Control Methods Used in a Study of the Vowels. *J. Acoust. Soc. Am.* **1952**, *24*, 175–184.
9. Rabiner, L.R.; Schafer, R.W. Introduction to Digital Speech Processing. *Found. Trends Signal Process.* **2007**, *1*, 1–194, doi:10.1561/2000000001.
10. Jaynes, E.T. *Bayesian Spectrum and Chirp Analysis*; Springer: Dordrecht, The Netherlands, 1987; pp. 1–29.
11. Bretthorst, G.L. *Bayesian Spectrum Analysis and Parameter Estimation*; Springer Science & Business Media: Berlin, Germany, 1988.
12. Chen, C.J.; Miller, D.A. Pitch-Synchronous Analysis of Human Voice. *J. Voice* **2019**, doi:10.1016/j.jvoice.2019.01.009.
13. Boersma, P. Should Jitter Be Measured by Peak Picking or by Waveform Matching. *Folia Phoniatr. Et Logop.* **2009**, *61*, 305–308.
14. Wise, J.; Caprio, J.; Parks, T. Maximum likelihood pitch estimation. *IEEE Trans. Acoust. Speech Signal Process.* **1976**, *24*, 418–423.
15. Ladefoged, P. *Elements of Acoustic Phonetics*; University of Chicago Press: Chicago, IL, USA, 1996.
16. Miller, D.G.; Schutte, H.K. Characteristic Patterns of Sub-and Supraglottal Pressure Variations within the Glottal Cycle. In Proceedings of the Transcripts of the Thirteenth Symposium on Care of the Professional Voice, New York, NY, USA, 1984; pp. 70–75.
17. Doval, B.; D'Alessandro, C.; Henrich, N. The Spectrum of Glottal Flow Models. *Acta Acust. United Acust.* **2006**, *92*, 1026–1046.
18. Ó Ruanaidh, J.J.K.; Fitzgerald, W.J. *Numerical Bayesian Methods Applied to Signal Processing*; Statistics and Computing; Springer New York: New York, NY, USA, 1996, doi:10.1007/978-1-4612-0717-7.
19. Fitzgerald, P.; Godsill, S.J.; Kokaram, A.C. Bayesian Methods in Signal and Image. *Bayesian Stat.* **1999**, *6*, 239–254.
20. Bretthorst, G.L. Bayesian Spectrum Analysis on Quadrature NMR Data with Noise Correlations. In *Maximum Entropy and Bayesian Methods*; Springer: Berlin, Germany, 1989; pp. 261–273.
21. Press, W.H.; Teukolsky, S.A.; Vetterling, W.T.; Flannery, B.P. *Numerical Recipes in C: The Art of Scientific Computing*, 2nd Ed.; Cambridge University Press: New York, NY, USA, 1992.
22. Kominek, J.; Black, A.W. The CMU Arctic speech databases. In Proceedings of the Fifth ISCA Workshop on Speech Synthesis, Pittsburgh, PA, USA, June 14-16, 2004.
23. Boersma, P.; Weenink, D.J.M. Praat, a system for doing phonetics by computer. *Glot Int.* **2002**, *5*, 341–345.
24. Sanchez, J. Application of Classical, Bayesian and Maximum Entropy Spectrum Analysis to Nonstationary Time Series Data. In *Maximum Entropy and Bayesian Methods*; Springer: Berlin, Germany, 1989; pp. 309–319.
25. Alku, P. Glottal Inverse Filtering Analysis of Human Voice Production—A Review of Estimation and Parameterization Methods of the Glottal Excitation and Their Applications. *Sadhana* **2011**, *36*, 623–650, doi:10.1007/s12046-011-0041-5.
26. Rabiner, L.R.; Juang, B.H.; Rutledge, J.C. *Fundamentals of Speech Recognition*; PTR Prentice Hall: Englewood Cliffs, NJ, USA, 1993; Volume 14.

27. Becker, T.; Jessen, M.; Grigoras, C. Forensic Speaker Verification Using Formant Features and Gaussian Mixture Models. In Proceedings of the Ninth Annual Conference of the International Speech Communication Association, Brisbane, Australia, September 22-26, 2008.
28. Ng, A.K.; Koh, T.S.; Baey, E.; Lee, T.H.; Abeyratne, U.R.; Puvanendran, K. Could Formant Frequencies of Snore Signals Be an Alternative Means for the Diagnosis of Obstructive Sleep Apnea? *Sleep Med.* **2008**, *9*, 894–898, doi10.1016/j.sleep.2007.07.010.
29. Bonastre, J.F.; Kahn, J.; Rossato, S.; Ajili, M. Forensic Speaker Recognition: Mirages and Reality. *S. Fuchs/D* **2015**, 255, doi10.3726/978-3-653-05777-5.
30. Nolan, F. Speaker Identification Evidence: Its Forms, Limitations, and Roles. In Proceedings of the Conference "Law and Language: Prospect and Retrospect", Levi, Finnish, Lapland, 2001; pp. 1–19.
31. Sivia, D.; Skilling, J. *Data Analysis: A Bayesian Tutorial*; OUP: Oxford, UK, 2006.
32. Chen, C.J. *Elements of Human Voice*; World Scientific: Singapore, 2016.

 © 2019 by the authors. Licensee MDPI, Basel, Switzerland. This article is an open access article distributed under the terms and conditions of the Creative Commons Attribution (CC BY) license (http://creativecommons.org/licenses/by/4.0/).

*Proceedings*

# The Spin Echo, Entropy, and Experimental Design [†]

**Keith Earle** [*,‡] **and Oleks Kazakov** [‡]

Department of Physics, University at Albany, Albany, NY 12222, USA; okazakov@albany.edu
* Correspondence: kearle@albany.edu; Tel.: +1-518-442-4503
† Presented at the 39th International Workshop on Bayesian Inference and Maximum Entropy Methods in Science and Engineering, Garching, Germany, 30 June–5 July 2019.
‡ These authors contributed equally to this work.

Published: 19 February 2020

**Abstract:** The spin echo experiment is an important tool in magnetic resonance for exploring the coupling of spin systems to their local environment. The strong couplings in a typical Electron Spin Resonance (ESR) experiment lead to rapid relaxation effects that puts significant technical constraints on the kinds of time domain experiments that one can perform in ESR. Recent developments in high frequency ESR hardware have opened up new possibilities for utilizing phase-modulated or composite phase slice (CPS) pulses at 95 GHz and higher. In particular, we report preliminary results at 95 GHz on experiments performed with CPS pulses in studies of rapidly relaxing fluid state systems. In contemporary ESR, this has important consequences for the design of pulse sequences where, due to finite excitation bandwidths, contributions from the Hamiltonian dynamics and relaxation processes must be considered together in order to achieve a quantitative treatment of the effects of selective, finite bandwidth pulses on the spin system under study. The approach reported on here is generic and may be expected to be of use for solid state and fluid systems. In particular we indicate how our approach may be extended to higher frequencies, e.g., 240 GHz.

**Keywords:** Magnetic Resonance; entropy; optimization; nested sampling

## 1. Introduction

ESR is a powerful analytical tool for studying structure and dynamics in a broad range of solid, and fluid systems. ESR studies are sensitive to processes on nanosecond to microsecond time scales and complement the range of time scales that are accessible to other magnetic resonance techniques, such as Nuclear Magnetic Resonance (NMR). In the hands of NMR spectroscopists, time domain, or pulse techniques, have proven to be highly useful and flexible for the study of specific interactions that are not readily resolved by frequency domain, or continuous wave, techniques. The development of CPS pulse techniques in NMR, which address the limitations of real, non-ideal pulses, benefitted from the ready availability of suitable radio frequency amplifiers, phase modulators, and receiver technology. Equivalents suitable for ESR rely on the more complex microwave or even millimeter wave technology, due to the faster time scales and excitation frequencies relevant for ESR vis-à-vis NMR. In the ESR case, the bandwidth of a typical ESR spectrum and the limited power output, relatively speaking, has limited the adoption of CPS techniques. Recent advances in millimeter wave technology have called for a reassessment of this conventional wisdom, and we report on intial successes in applying CPS pulse technology at 95 GHz. The behavior of spin systems and their response to coherent perturbations may be quantified by the von Neumann entropy. Space limitations prevent anything more than a cursory discussion of this important concept. In order to develop CPS pulse sequences, extensive computations are necessary to model CPS pulse parameters, and early work used drastic approximations in order to explore a reasonable range of CPS pulse parameter space. With the advent of cheap, fast computing, these limitations are not so stringent as before, and our initial success provides the impetus for developing more computationally intensive parameter optimization methods for higher frequency ESR pulse work, where spectral extent and relaxation effects are expected to play an even more significant role. This contribution presents theoretical and experimental results that point the way to significant future applications. In section 2 we discuss the necessary theoretical background for understanding the spin evolution equation of motion and its solution in practical cases. We have tried to indicate where approximations are made and their expected domain of validity. Preliminary results demonstrating the utility of the theoretical formalism and its application to CPS pulse specification is given in Section 3. We put these results in broader context and indicate directions for future work in Section 4.

## 2. Spin Evolution

The Stochastic Liouville Equation (SLE) is an important tool for studying the response of a spin system to external magnetic perturbations. The SLE achieves this by quantifying the properties of the spin system via the density matrix formalism and modeling external perturbations by incorporating their effects in an equation of motion for the density matrix. The resulting equation of motion consists of two parts which may be interpreted in terms of classical analogues. The conservative, Hamiltonian dynamics may be understood in terms of a classical dipole model, where the response is governed by a torque equation. Couplings of the spin system to its surroundings are governed by dissipative processes that may be thought of as frictional torques. These considerations may be quantified as follows

$$\frac{\partial}{\partial t}\rho(t) = i\frac{i}{\hbar}[\mathcal{H}(t), \rho(t)] - \Gamma(\rho(t) - \rho_0). \tag{1}$$

In Equation (1), the density matrix $\rho(t)$ is perturbed from its equilibrium value $\rho_0$ by a time-dependent Hamiltonian $\mathcal{H}(t)$ which describes the conservative dynamics. The dissipative contributions to the dynamics modeling the recovery of the spin system to equilibrium are quantified by the relaxation operator $\Gamma$. As it stands, Equation (1) is very general. For the purposes of this contribution, we will apply it to a system of weakly interacting spin 1/2 magnetic dipoles perturbed by a time-dependent Zeeman interaction. The expectation value of the spin angular momentum is computed via the defining relation

$$\langle \mathbf{S}(t) \rangle = \text{Tr}\{\rho(t)\mathbf{S}\} \tag{2}$$

The expectation value of the spin computed from Equation (2) may be related to the observed magnetic dipole by identifying

$$\mu(t) = \gamma \langle \mathbf{S}(t) \rangle. \tag{3}$$

The constant of proportionality is the gyromagnetic ratio which is proportional to $1/m$ where $m$ is the mass of the relevant spin bearing charge. In ESR this is usually an unpaired electron which may arise from a crystalline defect, or be introduced via chemical means into a molecular system of interest. From a practical perspective, electron magnetic moments are typically on the order of 1000 times larger than nuclear magnetic moments due to the $1/m$ factor, and this has important consequences for the technology used for perturbing the ESR spin system and monitoring its recovery to equilibrium.

The Zeeman interaction of a magnetic dipole $\mu$ in an applied time-dependent magnetic field $\mathbf{H}$ is quantified by the following Hamiltonian

$$\mathcal{H}(t) = -\gamma \mathbf{S} \cdot \mathbf{H}(t). \tag{4}$$

Equation (4) is the quantum mechanical generalization of the potential energy of a classical dipole in an applied magnetic field. Note that the density matrix for a system of spin 1/2 dipoles may be expanded in a set of spin 1/2 angular momentum operators. The commutator term in Equation (1) will then depend on commutators of the form

$$[S_q, S_r] = i\epsilon_{qrs} S_s. \tag{5}$$

In Equation (5), $\epsilon_{qrs}$ is antisymmetric in all of its indices. Classically this corresponds to a cross-product and motivates the identification of the commutator term as a torque term in Equation (1). Note that we are using a Cartesian representation of the spin operators $\{S_x, S_y, S_z\}$ here.

For the purposes of this report, we will assume a very simple form for the relaxation operator and introduce two phenomenological relaxation times. The recovery of the spin system to equilibrium will be quantified by the spin-lattice relaxation time $T_1$. The $T_1$-dependent terms govern the dissipative response of the longitudinal component of the magnetization, or expectation value of $S_z$. In general one neither expects nor observes that the transverse components of the magnetization, or expectation values of $S_x$ and $S_y$, respond on the $T_1$ time scale. In fact, the transverse relaxation time $T_2$ generally satisfies the inequality $T_2 \ll T_1$. It is useful to note that one may construct more complicated relaxation operators in order to model dissipative processes, such as rotational diffusion, but that is beyond the scope of the work reported on here.

Upon computing the expectation value of the magnetization in a time-dependent field, and ignoring relaxation for the moment, the SLE becomes a torque equation for the magnetization of the spin system which can be written in the following form

$$\frac{d}{dt}\boldsymbol{\mu}(t) = \boldsymbol{\mu}(t) \times \gamma \mathbf{H}(t). \tag{6}$$

Equation (6) describes the conservative Hamiltonian dynamics. When the dissipative terms are included, we also need to compute the expectation value of the equilibrium magnetization $\gamma \text{Tr}\{\rho_0 \mathbf{S}\} \equiv \mu_0 \hat{z}$. When the relaxation terms are included, the equation of motion for the magnetization assumes the following matrix form

$$\frac{\partial}{\partial t} \begin{bmatrix} \mu_x(t) \\ \mu_y(t) \\ \mu_z(t) \end{bmatrix} = \begin{bmatrix} -1/T_2 & \gamma H_z(t) & -\gamma H_y(t) \\ -\gamma H_z(t) & -1/T_2 & \gamma H_x(t) \\ \gamma H_y(t) & \gamma H_x(t) & -1/T_1 \end{bmatrix} \begin{bmatrix} \mu_x(t) \\ \mu_y(t) \\ \mu_z(t) \end{bmatrix} + \begin{bmatrix} 0 \\ 0 \\ \mu_0/T_1 \end{bmatrix}. \tag{7}$$

These are the Bloch equations, derived from the SLE, Equation (1). Note that Equation (7) is a special case of Equation (1). Equation (1) is more general than Equation (7), but Equation (7) has sufficient generality for the work reported on here. Note that Equation (7) is a set of coupled, linear, inhomogeneous first order differential equations. In order to cast this set of equations into a more tractable form, it is useful to specialize the form of the time-varying magnetic field and introduce some commonly applied approximations valid when the z component of the applied magnetic field is taken to be static and much larger in magnitude than any of the other components.

## 2.1. Rotating Wave Approximation

A common experimental situation is to suppose that a static field is applied along the laboratory z axis and a linearly polarized magnetic field is applied along the laboratory x axis. It is convenient to resolve the linearly polarized perturbing field into counter-rotating circularly polarized components. When this is done, the applied field is

$$\mathbf{H} = (H_1 \cos \omega t \hat{x} + H_1 \sin \omega t \hat{y}, H_1 \cos \omega t \hat{x} - H_1 \sin \omega t \hat{y}, H_0 \hat{z}) \tag{8}$$

When $H_0$ is large compared to $H_1$, it is standard practice to drop the counterrotating component of the applied magnetic field and perform a coordinate transformation that renders both the circularly polarized applied field and the static applied field constant. This is the content of the rotating wave approximation (RWA) and the transformation to the rotating frame. A good discussion of these points may be found in Allen and Eberly [1]. When these transformations are made, Equation (7) takes the following matrix form

$$\frac{d}{dt}\begin{bmatrix} \mu_x \\ \mu_y \\ \mu_z \end{bmatrix} = \begin{bmatrix} -1/T_2 & -\Delta & 0 \\ \Delta & -1/T_2 & \omega_1 \\ 0 & -\omega_1 & -1/T_1 \end{bmatrix} \begin{bmatrix} \mu_x \\ \mu_y \\ \mu_z \end{bmatrix} + \begin{bmatrix} 0 \\ 0 \\ \mu_0/T_1 \end{bmatrix}. \qquad (9)$$

In Equation (9), $\Delta$ is the detuning from resonance defined by the frequency of the perturbing field and the static field $\Delta = \omega - \gamma H_0$. In addition $\omega_1 = \gamma H_1$. Equation (9) may be solved for constant fields, which we have achieved by exploiting the RWA and transforming to the rotating frame, by Laplace transform techniques[2]. The solution of this coupled set of equations when relaxation effects can be ignored is known as the Rabi Solution [3].

## 2.2. Quantum Mechanics of Two Level Systems

In this work, we will exploit the close relationship between the SLE, Equation (1) and the Bloch Equations, Equation (7). For the present purpose, note that for a two-level system the quantum mechanical spin operators, as well as the density matrix $\rho$ may be represented as $2 \times 2$ matrices. The density matrix is a Hermitian operator and satisfies the constraint that $\text{Tr}\{\rho\} = 1$, so that not all elements of $\rho$ are independent. For a $2 \times 2$ density matrix, there are in fact three independent parameters, which may be identified with the expectation values of the components of the magnetic dipole moments in the magnetic resonance problem. For more details on the density matrix formalism, Slichter [4] is a good reference. As the components of the magnetization vector are sufficient to specify completely the density matrix for a two-level, or spin 1/2, system, one may think of the transformation that diagonalizes the density matrix as a series of rotations. A useful procedure in practice is to perform a rotation about the $\hat{z}$ direction that rotates the expectation values of $S_x$ and $S_y$ into the $x - z$ plane, and then perform a rotation about the $\hat{y}$ axis to transform the expectation value of $\mathbf{S}$ to the $\hat{z}$ axis. This diagonalization process is equivalent to finding the eigenvalues of the density matrix. Once the eigenvalues of the density matrix are known, one may compute matrix-valued functions in a straight-forward fashion, using, e.g., Sylvester's Theorem. A good discussion of this point is given by Merzbacher [5]. In particular, one quantity of interest is the von Neumann entropy

$$S = -\text{Tr}\{\rho \ln(\rho)\} \qquad (10)$$

Equation (10) is the extension of the Shannon entropy to a more general quantum mechanical setting. It is a useful tool for examining the response of spin systems to various manipulations by pulsed fields. For the present purpose, it is useful to note that the trace in Equation (10) is invariant under the matrix diagonalization process, and so the eigenvalues of the density matrix are particuarly useful here. For a two-level system, in fact,

$$S = -\frac{(1+\lambda)}{2} \ln \frac{(1+\lambda)}{2} - \frac{(1-\lambda)}{2} \ln \frac{(1-\lambda)}{2} \qquad (11)$$

Here $\lambda$ is twice the expectation value of $\mathbf{S}$ for a spin 1/2 system. Note that the von Neumann entropy is symmetric under the transformation $\lambda \to -\lambda$. This symmetry gives rise to an apparent paradox in so-called inversion recovery experiments, but space constraints preclude further discussion.

## 2.3. Pulses and Magnetic Resonance

Many contemporary ESR experiments are performed using pulse techniques for which the density operator formalism developed above is highly useful. In general, one applies a strong time-varying pulse in the lab frame to perturb the magnetization from equilibrium and then monitors the subsequent time development of the magnetization to study details of structure and dynamics in the relevant spin system. Notice that the formalism that we developed above is most readily used when the applied fields are static, but recall that we have already made such a transformation by using the RWA and changing coordinate systems to the rotating frame. In addition, the foramlism developed above is also applicable to applied fields that are piece-wise continuous, that is, the fields in the rotating frame may be treated as constant including step discontinuities. This is an example of the flexibility of the formalism in practical applications. Consider a situation where there is a slight detuning from exact resonance, $\Delta \neq 0$ in Equation (9). Suppose further, that at time $t = 0$ the magnetic dipole is aligned with the applied static field and has equilibrium magnetization $\mu = (0, 0, \mu_0)$. Now suppose that the co-rotating component of the perturbing field acts for a time $t_{\pi/2}$ so that $\omega_1 t_{\pi/2} = \pi/2$. As we shall see, this corresponds to rotating the equilibrium magnetization by $\pi/2$ degrees in the $y - z$ plane. At time $t_{\pi/2}$, the perturbing field is turned off, and the spins are allowed to precess freely. In order to quantify this behavior, it is useful to have an explicit expression for the Rabi solution for piecewise continuous fields.

The Rabi solution is an analytic solution that accounts for detunings from resonance and strong pulses in the rotating frame but does not account for relaxation effects during the pulse. For sufficiently strong pulses this would not be a serious limitation, but ESR applications typically require a more complete treatment. With these limitations in mind, the Rabi solution for a finite detuning and strong constant field along the rotating frame x axis may be written in the following form

$$\begin{bmatrix} \mu_x(t) \\ \mu_y(t) \\ \mu_z(t) \end{bmatrix} = \begin{bmatrix} \cos^2 \chi + \sin^2 \chi \cos \Omega t & -\sin \chi \sin \Omega t & -\cos \chi \sin \chi (1 - \cos \Omega t) \\ \sin \chi \sin \Omega t & \cos \Omega t & \cos \chi \sin \Omega t \\ -\cos \chi \sin \chi (1 - \cos \Omega t) & -\cos \chi \sin \Omega t & \sin^2 \chi + \cos^2 \chi \cos \Omega t \end{bmatrix} \begin{bmatrix} \mu_x(0) \\ \mu_y(0) \\ \mu_z(0) \end{bmatrix} \quad (12)$$

In Equation (12) the following quantities are defined: $\tan \chi = \Delta/\omega_1$, $\Omega^2 = \Delta^2 + \omega_1^2$. For strong pulses and detunings that are not too large, one may make the useful approximations: $\Omega \approx \omega_1$, $\cos \chi \approx 1$, $\sin \chi \approx \Delta/\omega_1 \ll 1$. These approximations allow us to develop a useful approximate matrix equation valid to order $\Delta/\omega_1 \ll 1$. Under the condition that $\omega_1 t_{\pi/2} = \pi/2$, Equation (12) becomes

$$\begin{bmatrix} \mu_x(t_{\pi/2}) \\ \mu_y(t_{\pi/2}) \\ \mu_z(t_{\pi/2}) \end{bmatrix} = \begin{bmatrix} 1 & -\Delta/\omega_1 & -\Delta/\omega_1 \\ \Delta/\omega_1 & 0 & 1 \\ -\Delta/\omega_1 & -1 & 0 \end{bmatrix} \begin{bmatrix} \mu_x(0) \\ \mu_y(0) \\ \mu_z(0) \end{bmatrix} \quad (13)$$

For the case of exact resonance $\Delta = 0$, it is easy to see that if $\mu(0) = (0, 0, \mu_0)$ then $\mu(t_{\pi/2}) = (0, \mu_0, 0)$, corresponding to a rotation of the magnetization by $\pi/2$ around the $\hat{x}$ axis.

In the absence of relaxation terms, or for times short enough that no significant relaxation has occured, it is also useful to develop the Rabi solution for free precession, that is, for the case $\omega_1 = 0$. A straightforward computation shows that for the case $\Delta \neq 0$,

$$\begin{bmatrix} \mu_x(t) \\ \mu_y(t) \\ \mu_z(t) \end{bmatrix} = \begin{bmatrix} \cos \Delta t & -\sin \Delta t & 0 \\ \sin \Delta t & \cos \Delta t & 0 \\ 0 & 0 & 1 \end{bmatrix} \begin{bmatrix} \mu_x(0) \\ \mu_y(0) \\ \mu_z(0) \end{bmatrix} \quad (14)$$

Note that in the absence of equilibrium inducing relaxation terms, $\mu_z(t)$ is a constant of the motion under free precession. Thus, Equation (14) describes free precession in the plane transverse to $\hat{z}$. We may compute the precession of the transverse magnetization for a time $\tau$ after a $\pi/2$ pulse by inserting the solution vector, the left hand side of Equation (13), into the right hand side of Equation (14). A straightforward calculation yields the following result $\mu(\tau) = \mu_0(-\sin(\Delta(\tau + 1/\omega_1)), \cos(\Delta(\tau + 1/\omega_1)), 0)$. One may also compute the effect of a subsequent $\pi$ pulse followed by a period of free precession of duration $\tau'$. The result of the calculation to order $\Delta/\omega_1$ is as follows

$$\begin{bmatrix} \mu_x(\tau') \\ \mu_y(\tau') \\ \mu_z(\tau') \end{bmatrix} = \mu_0 \begin{bmatrix} \sin(\Delta(\tau + 1/\omega_1 - \tau')) \\ -\cos(\Delta(\tau + 1/\omega_1 - \tau')) \\ 2(\Delta/\omega_1) \sin(\Delta(\tau + 1/\omega_1)) \end{bmatrix} \quad (15)$$

Note that if $\tau' = \tau + 1/\omega_1$ then the transverse spin response is independent of the detuning, and we say that the magnetization has been refocused along the $-\hat{y}$ axis. The significance of this result for practical magnetic resonance spectroscopy is hard to overstate. In general, the detuning can arise from a homogeneous process which affects all spins equally, or an inhomogeneous process which has different effects on different spin cohorts. For many practical applications in magnetic resonance one may tune a spectrometer so that the spectral response is centered on a symmetric, inhomogeneous distribution of detunings caused by magnetic inhomogeneities. The total response of the system must be integrated over this inhomogeneous distribution which acts as a source of broadening that reduces the observable magnetization. When the detunings are all refocused in the $x - y$ planse as in Equation (15), however, the transverse spin response becomes independent of the detuning and one observes the total transverse magnetization of the resonance-active spins. Note further that during a period of free precession in the absence of relaxation effects the magnetization is a vector of constant magnitude and thus the eigenvalues of the density matrix are also constant. Thus during free precession in the absence of relaxation, the dynamics is conservative, and no change in entropy occurs.

## 2.4. Practical Considerations in ESR

In ESR, the relevant timescales for relaxation processes can be on the order of nanoseconds, and the spectral extent in frequency units can be much larger than $\omega_1$. When this is so, numerical solution of Equation (7) or its density matrix equivalent Equation (1) are needed to make quantitative predictions of the response of the spin system. With a view towards establishing a formalism that may be extended to more complex systems, we return to the density matrix formalism and develop a formalism based on irreducible spherical tensor operators and treat the matrix elements of $\rho$ as elements of a column vector in the space of transitions, the so-called Liouville space. We may make the RWA as before and achieve an equivalent of the rotating coordinate frame by performing a Floquet analysis of the time-dependent matrix elements of the density matrix. In practice, it is found that the deviation of the density matrix from its equilibrium value is the relevant quantity. When these transformations are carried out, it is found that the matrix representation of Equation (1) takes the following form,

$$\frac{\partial}{\partial t}\begin{bmatrix} C_1(t)/\sqrt{2} \\ C_0(t)/2 \\ C_{-1}(t)/\sqrt{2} \end{bmatrix} = \begin{bmatrix} -i\Delta - 1/T_2 & -i\sqrt{2}de^{i\phi} & 0 \\ -i\sqrt{2}de^{-i\phi} & -1/T_1 & i\sqrt{2}de^{-i\phi} \\ 0 & i\sqrt{2}de^{-i\phi} & i\Delta - 1/T_2 \end{bmatrix} \begin{bmatrix} C_1(t)/\sqrt{2} \\ C_0(t)/2 \\ C_{-1}(t)/\sqrt{2} \end{bmatrix} + i\frac{qd}{\sqrt{2}}\begin{bmatrix} e^{i\phi} \\ 0 \\ -e^{-\phi} \end{bmatrix}. \quad (16)$$

Equation (16) has been specialized to the case of an ensemble of spin 1/2 dipoles, but it is in a form that may be generalized along the lines discussed in Stillman and Schwartz [2]. The new notational features include $d \equiv \omega_1/2$ the transition moment, and $q$ the normalization of the equilibrium density matrix. In addition, the phase factors $\phi \in \{0, \pi/2, \pi, 3\pi/2\}$ allow for more general perturbing fields corresponding to excitations along the $\pm x$ or $\pm y$ axes in the rotating frame. As Equation (16) incorporates detunings and relaxation times of arbitrary magnitude, it is the appropriate generalization of the approximate results discussed in Section 2.3. In order to make the connection to quantities observed in the laboratory, we note that $C_1(t)$ is the expectation value of the operator $S_x + iS_y \equiv S_+$. Note that if $\phi \in \{0, \pi\}$ the matrix appearing in Equation (16) is complex symmetric. In practical applications, it is found that these values of $\phi$ are the relevant ones [6]. When the matrix representation of Equation (16) is complex symmetric, efficient methods of solution exist, particularly when the spin system is more complex than the simple case treated here [7]. For the purposes of this work, we retain the general form of Equation (16) in order to treat arbitrary pulse phases in the rotating frame. When the perturbing field characterized by $d$ is piecewise continuous, one may solve Equation (16) in each domain where $d$ is constant (in the rotating frame) and use continuity arguments to propagate the solution to the next region of continuous $d$. As noted by Stillman and Schwartz [2], the matrix representation of Equation (16) takes the following form

$$\dot{\mathbf{c}} = \mathbf{A}\mathbf{c} + \mathbf{q}. \tag{17}$$

The Laplace transform solution of Equation (17) in each region of constant $d$ may be shown to be [2]

$$\mathbf{c}(t) = \exp(\mathbf{A}t)\mathbf{c}(0) - \mathbf{A}^{-1}[\mathbf{1} - \exp(\mathbf{A}t)]\mathbf{q} \tag{18}$$

In those regions where $d = 0$ the spins evolve under free precession and the solution of Equation (18) reduces to the case covered in Section 2.3 if relaxation effects can be ignored. Note that the solution of Equation (18) requires a matrix inversion and the computation of a matrix exponential in general. This is typically a computational resource intensive calculation. For the simple case treated here, closed form expressions may be computed for $\mathbf{A}^{-1}$ and $\exp(\mathbf{A}t)$. An explicit form for the matrix inverse may be given as follows

$$\mathbf{A}^{-1} = \frac{1}{\Delta} \begin{bmatrix} (i\Delta - 1/T_2)/T_1 - 2d^2 & -i\sqrt{2}de^{i\phi}(i\Delta - 1/T_2) & -2d^2 e^{i\phi} \\ -i\sqrt{2}de^{-i\phi}(i\Delta - 1/T_2) & -(1/T_1)^2 - (\Delta)^2 & i\sqrt{2}de^{i\phi}(-i\Delta - 1/T_2) \\ -2d^2 e^{-2i\phi} & i\sqrt{2}de^{-i\phi}(-i\Delta - 1/T_2) & (-i\Delta - 1/T_2)/T_1 - 2d^2 \end{bmatrix}. \tag{19}$$

Note that if $\phi \in \{0, \pi\}$ then $\mathbf{A}^{-1}$ is complex symmetric, as is $\mathbf{A}$. The quantity $\Delta = ((\Delta)^2 + (1/T_2)^2)/T_1 + 4d^2/T_2$. This determinant is also relevant for finding the eigenvalues of $\mathbf{A}$. Once the eigenvalues of $\mathbf{A}$ are available, the methods given in Apostol [8] or Merzbacher [5] may be used to compute $\exp(\mathbf{A}t)$. For this simple problem, the eigenvalues of $\mathbf{A}$ may be found using the methods of Nickalls [9]. Python scripts based on these considerations were developed in the Earle laboratory and are available on request.

## 3. Results

In order to accommodate the significant detunings that can occur for ESR experiments, it is possible to use CPS pulses first developed in NMR [10] to achieve an optimum response of the spin system. The first application of CPS pulses to ESR was reported by Crepeau, et al. [6]. They developed a multiparameter search algorithm over the microwave phase $\phi$ and pulse slice duration to determine the optimum pulse parameters with desired spin system response. Their numerical simulations accounted for the distorting effects of detuning and a resonant structure with a finite excitation bandwidth but did not account for relaxation effects. Under these conditions, they were able to determine CPS pulse parameters that: minimized the amount of z magnetization following a $\pi/2$ pulse; optimized the amount of y magnetization after a $\pi/2$ pulse; and optimized the amount of z magnetization after a $\pi$ pulse. This work was carried out at X-band, corresponding to a microwave frequency of 9.25 GHz.

Advances in millimeter wave technology have allowed these experiments to be revisited at higher frequencies. An ESR pulse spectrometer at 95 GHz [11] was recently retrofitted with a phase-agile front end and improved sensitivity. These upgrades motivated a reexamination of CPS pulses for high field work. As a starting point, the parameters determined empirically by Crepeau, et al. [6] were scaled to the larger $\omega_1$ available at 95 GHz.

Initial studies of CPS pulses following hardware and software upgrades to the ACERT 95 GHz high-power/cw spectrometer Figure 1 (right panel) were conducted culminating in the succesful demonstration of a CPS $\pi/2$ pulse at 95 GHz as shown in Figure 1 (left panel). Additional studies were performed to evaluate the performance of an arbitrary waveform generator (AWG) (WavePond DAx22000), in conjunction with the rebuilt transceiver (ELVA-1 MKII) analog/vector modulator and receiver sections. For these studies, a high power Extended Interaction Klystron (EIK) with a nominal output of 1 kW was deployed in the spectrometer TX arm as required for high power evaluation. In order to monitor the power returning to the receiver in the RX arm, a corner reflector was used at reference plane 'W' as a proxy for the Fabry-P??rot resonator assembly typically used in a 95 GHz ESR experiment. The tests verify the performance of a CPS sequence of successive phases and timings from an early CPS paper [6] that demonstrates the enhancement potential of the spectrometer's AWG pulse capability.

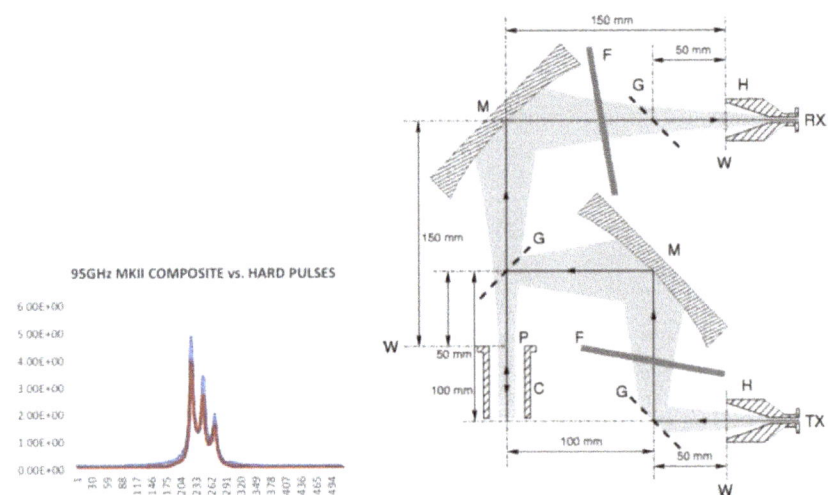

**Figure 1. Left** Panel: Comparison of a CPS $\pi/2$ pulse response (red trace) to a standard full excitation hard pulse (blue trace). The three line spectrum is of a nitroxide spin label in the motional narrowing regime where the Zeeman interaction is split by a further hyperfine interaction with an $I = 1$ Nitrogen nucleus. In this range of motion, the three hyperfine lines are only weakly coupled and evolve approximately independently. Thus their contributions to the spectrum may be simply summed. **Right** Panel: A schematic of the spectrometer used for these experiments. A Fabry-Pérot resonant structure (not shown) is coupled to reference plane 'W' to perform ESR experiments.

## 4. Discussion

Figure 1 shows the results of a scaled emulation of an effective $\pi/2$ composite-phase sequence described by Crepeau, et al. [6]. The sequence reported on here consists of the sequence 9 ns/$\phi = 0$, 5 ns/$\phi = \pi$, 4 ns/$\phi = 0$, 3 ns/$\phi = \pi$, for a total composite pulse length of 21 ns. Reproduction of this composite pulse is of reasonably high fidelity, with the exception of some observed interference due to multiple reflections that contaminate the final (3 ns) slice. With the $\omega_1$ enhancing Fabry-Pérot resonator assembly coupled to reference plane 'W' shown in the right panel of Figure 1, it was determined that the CPS pulse at an output power level approximately 50% of $P_{o,max}$ produced spectral coverage (red trace) and an effective $\omega_1$ value virtually indistinguishable from a 5 ns EIK $P_{o,max}$ hard pulse (blue trace). This result clearly demonstrates that a CPS pulse can successfully emulate a hard pulse of significantly shorter duration and higher peak power. Work is currently under way to extend this preliminary result to CPS $\pi$ pulses and thus spin echo pulse sequences. Further applications to pulse spectroscopy at higher frequencies, e.g., 240 GHz are also under way. In order to determine suitable CPS pulse characteristics at 240 GHz, we are revisiting the parameter optimization problem discussed in Section 3. This is especially timely as significant advances in computational hardware since the original studies of Crepeau, et al. [6] have allowed more computationally intensive search algorithms to be usefully employed. Work is under way to extend the Laplace transform techniques reported on in this work to the CPS pulse optimization problem. For this purpose, we are developing a nested sampling approach to determine the CPS parameters that optimize the relevant cost functions for $\pi/2$ and $\pi$ pulses. Extensions of the methods covered here to longer pulse sequences involving three optimally phased pulses for exploring multidimensional pulse ESR spectroscopy are also under way.

**Acknowledgments:** The resources of the ACERT facility at Cornell University were used during the course of preparation of this manuscript.

**Author Contributions:** K.E. performed the theoretical analysis and O.K. developed the analysis software. K.E. wrote the paper.

**Conflicts of Interest:** The authors declare no conflict of interest.

## References

1. Allen, L.; Eberly, J.H. *Optical Resonance and Two-Level Atoms*; Dover: New York, NY, USA, 1987; pp. 41–51.
2. Stillman, A.E.; Schwartz, R.N. Theory of Electron Spin Echoes in Nonviscous and Viscous Liquids. In *Time Domain Electron Spin Resonance*; Kevan, L., Schwartz, R.N., Eds.; Wiley: New York, NY, USA, 1979; c. 5.
3. Allen, L.; Eberly, J.H. *Optical Resonance and Two-Level Atoms*; Dover: New York, NY, USA, 1987; pp. 56–61.
4. Slichter, C.P. *Principles of Magnetic Resonance*; Springer: New York, NY, USA, 1996; pp. 157–189.
5. Merzbacher, E. Matrix Methods in Quantum Mechanics. *Am. J. Phys.* **1968**, *36*, 814–821.
6. Crepeau, R.H.; Dulčić, A.; Gorcester, J.; Saarinen, T.R.; Freed, J.H. Composite Pulses in Time-Domain ESR. *J. Magn. Reson.* **1989**, *84*, 184–190.
7. Schneider, D.J.; Freed, J.H. Spin Relaxation and Motional Dynamics. In *Lasers, Molecules, and Methods*; Hirschfelder, J.G., Wyatt, R.E., Coalson, R.D., Eds.; Wiley: Hoboken, NJ, USA, 1989; c. 10.
8. Apostol, T.M. Some Explicit Formulas for the Exponential Matrix exp(A$t$). *Am. Math. Mon.* **1969**, *76*, 289–292.
9. Nickalls, R.W.D. A New Approach to Solving the Cubic: Cardano's Solution Revealed. *Math. Gaz.* **1993**, *77*, 354–359.
10. Levitt, M.H. Compensation for Pulse Imperfections in NMR Spin Echo Experiments. *J. Magn. Reson.* **1982**, *43*, 65–80.
11. Hofbauer, W.K.; Earle, A.; Dunnam, C.R.; Moscicki, J.K.; Freed, J.H. High-power 95GHz Pulsed Electron Spin Resonance Spectrometer. *Rev. Sci. Inst.* **2004**, *75*, 1194–1208.

© 2020 by the authors. Licensee MDPI, Basel, Switzerland. This article is an open access article distributed under the terms and conditions of the Creative Commons Attribution (CC BY) license (http://creativecommons.org/licenses/by/4.0/).

MDPI  
St. Alban-Anlage 66  
4052 Basel  
Switzerland  
Tel. +41 61 683 77 34  
Fax +41 61 302 89 18  
www.mdpi.com

*Proceedings* Editorial Office  
E-mail: proceedings@mdpi.com  
www.mdpi.com/journal/proceedings

www.ingramcontent.com/pod-product-compliance
Lightning Source LLC
LaVergne TN
LVHW071938080526
838202LV00064B/6629